VOLUME ONE HUNDRED AND NINETY

PROGRESS IN MOLECULAR BIOLOGY AND TRANSLATIONAL SCIENCE

Precision Medicine

VOLUME ONE HUNDRED AND NINETY

PROGRESS IN MOLECULAR BIOLOGY AND TRANSLATIONAL SCIENCE

Precision Medicine

Edited by

DAVID B. TEPLOW
David Geffen Sch of Medicine-UCLA
Los Angeles, CA, United States

ELSEVIER

ACADEMIC PRESS
An imprint of Elsevier

Academic Press is an imprint of Elsevier
50 Hampshire Street, 5th Floor, Cambridge, MA 02139, United States
525 B Street, Suite 1650, San Diego, CA 92101, United States
The Boulevard, Langford Lane, Kidlington, Oxford OX5 1GB, United Kingdom
125 London Wall, London EC2Y 5AS, United Kingdom

First edition 2022

ISBN: 978-0-323-99784-3
ISSN: 1877-1173

For information on all Academic Press publications
visit our website at https://www.elsevier.com/books-and-journals

Publisher: Zoe Kruze
Acquisitions Editor: Ashlie M. Jackman
Developmental Editor: Jhon Michael Peñano
Production Project Manager: James Selvam
Cover Designer: Matthew Limbert

Typeset by STRAIVE, India

Contents

Contributors *ix*
Preface *xiii*

1. Introduction of medical genomics and clinical informatics integration for p-Health care **1**

Li Tong, Hang Wu, May D. Wang, and Geoffrey Wang

1. Introduction 2
2. State-of-the-art advanced data integration 16
3. Case studies 25
4. Privacy and security 27
5. Conclusion 29

References 30

2. Precision diagnostics in cancer: Predict, prevent, and personalize **39**

Judy S. Crabtree and Lucio Miele

1. Introduction 40
2. NGS-based technologies 43
3. Therapy selection 49
4. Clinical trial design 51
5. Conclusions 52

References 53

3. Artificial intelligence and machine learning in precision medicine: A paradigm shift in big data analysis **57**

Mehar Sahu, Rohan Gupta, Rashmi K. Ambasta, and Pravir Kumar

1. Introduction 58
2. Artificial intelligence and big data: Changes in the landscape of precision healthcare and medicine 61
3. The emergence of AI in precision medicine 63
4. Synergies between artificial intelligence and precision medicine 64
5. Promises of implementing AI/ML in healthcare 69
6. The integration of nanomaterials in AI-based precision medicine 73
7. The perspective of big data analytics in precision medicine 75
8. AI and personalized medicine: A case of neurological disease 76
9. Obstacles in translating big data into healthcare 79
10. Opportunities for translating big data into healthcare 81

11. Artificial intelligence and machine learning in precision medicine: An industry perspective 82
12. Future challenges and perspectives of artificial intelligence in precision medicine 87
Conflict of interest 89
Acknowledgment 89
Author's contribution 89
References 89

4. Precision medicine with multi-omics strategies, deep phenotyping, and predictive analysis **101**
Zeeshan Ahmed

1. Introduction 102
2. Multi-omics strategies & precision medicine 104
3. Deep phenotyping & precision medicine 107
4. Discussion & future recommendations 111
Acknowledgments 115
Competing interests 115
Author contribution 115
Biographical note 115
Funding information 116
References 116

5. Progress in molecular biology and translational science addressing the needs of nano-rare patients **127**
Stanley T. Crooke

1. Introduction 128
2. Parsing patient populations 129
3. Mutation focused drug discovery 131
4. Prerequisites to respond to the needs of nano-rare patients 131
5. Creating an industrialized approach to the non-profit treatment of nano-rare patients 135
6. The future 141
7. Remaining challenges 144
References 145

6. Precision medicine in epilepsy **147**
Ryan James McGinn, Erica Leah Von Stein,
Jacqueline Elizabeth Summers Stromberg, and Yi Li

1. Introduction 148
2. Precision therapy in genetic epilepsy syndromes 148
3. Precision therapy with personalized pharmacogenomics 173

4. Precision therapy with neuromodulation 174
5. Conclusion 177
Funding 177
References 177

7. EGFR, NF-κB and noncoding RNAs in precision medicine 189
Yanli Li, Xiaomin Liu, and Zhongliang Ma

1. Part 1: Epidermal growth factor receptor (EGFR) 191
2. EGFR as a target in lung cancer treatment 192
3. Acquired resistance of EGFR TKIs 192
4. Precision medicine of EGFR in lung cancer 194
5. Part 2: Nuclear factor-κappa B signaling 195
6. NF-κB signaling pathway and its activation 195
7. Part 3: Noncoding RNAs and EGFR in precision medicine 197
8. Conclusion and future perspectives 207
References 209

8. Evolution of biomarker research in autoimmunity conditions for health professionals and clinical practice 219
Arthur Silverstein, Anton Dudaev, Maria Studneva, John Aitken, Sofya Blokh, Andrew David Miller, Sofia Tanasova, Noel Rose†, John Ryals, Christoph Borchers, Anders Nordstrom, Marina Moiseyakh, Arturo Solís Herrera, Nikita Skomorohov, Trevor Marshall, Alan Wu, R. Holland Cheng, Ksenia Syzko, Philip D. Cotter, Marianna Podzyuban, William Thilly, Paul David Smith, Paul Barach, Khaled Bouri, Yehuda Schoenfeld, Eiji Matsuura, Veronika Medvedeva, Ilya Shmulevich, Liang Cheng, Paul Seegers, Yekaterina Khotskaya, Keith Flaherty, Steven Dooley, Eric J. Sorenson, Michael Ross, and Sergey Suchkov

1. Introductory 221
2. Abs with catalytic activity (catAbs or abzymes) 226
3. Progressing toward personalized and precision healthcare services and challenges in applying biomarkers for clinical autoimmunity conditions 231
4. Specific autoimmunity conditions subareas 243
5. Biomarkers in clinical practice: Biomarkers as applicable to personalized dietary protocols in autoimmune conditions 270
6. Conclusion 273
References 274

Index *277*

† Deceased

Contributors

Zeeshan Ahmed
Institute for Health, Health Care Policy and Aging Research, Rutgers, The State University of New Jersey; Department of Medicine, Rutgers Robert Wood Johnson Medical School, Rutgers Biomedical and Health Sciences, New Brunswick, NJ, United States

John Aitken
Otakaro Pathways, Christchurch, New Zealand

Rashmi K. Ambasta
Molecular Neuroscience and Functional Genomics Laboratory, Delhi Technological University (Formerly Delhi College of Engineering), Shahbad Daulatpur, Delhi, India

Paul Barach
Wayne State University, Detroit, IL, United States

Sofya Blokh
Filatov's Moscow Clinical Pediatric Hospital, Moscow, Russia

Christoph Borchers
Lady Davis Institute, Segal Cancer Proteomics Centre, McGill University, Montreal, QC, Canada

Khaled Bouri
Georgetown University, Washington, DC, United States

Liang Cheng
Genitourinary Pathology Service, The Urologic Pathology Fellowship, and Molecular Diagnostics and Molecular Pathology Laboratories, Indiana University School of Medicine, Indianapolis, IN, United States

R. Holland Cheng
College of Biological Sciences, UC Davis, Davis, CA, United States

Philip D. Cotter
UCSF, San Francisco; ResearchDx and PacificDx, Irvine, CA, United States

Judy S. Crabtree
Department of Genetics, Louisiana State University Health Sciences Center, New Orleans, LA, United States

Stanley T. Crooke
n-Lorem Foundation, Carlsbad, CA, United States

Steven Dooley
Medical Faculty Mannheim, Heidelberg University, Heidelberg, Germany

Anton Dudaev
MGUPP, Moscow, Russia

Keith Flaherty
Harvard Medical School; Mass General Cancer Center, Massachusetts General Hospital, Boston, MA, United States

Rohan Gupta
Molecular Neuroscience and Functional Genomics Laboratory, Delhi Technological University (Formerly Delhi College of Engineering), Shahbad Daulatpur, Delhi, India

Arturo Solís Herrera
University of Florida, Gainesville, FL, United States

Yekaterina Khotskaya
Clinical Decision Support Management, The Curematch, Inc, San Diego, CA, United States

Pravir Kumar
Molecular Neuroscience and Functional Genomics Laboratory, Delhi Technological University (Formerly Delhi College of Engineering), Shahbad Daulatpur, Delhi, India

Yanli Li
Lab for Noncoding RNA & Cancer, School of Life Sciences, Shanghai University, Shanghai, China

Yi Li
Department of Department of Neurology & Neurological Sciences, Stanford University School of Medicine, Stanford, CA, United States

Xiaomin Liu
Lab for Noncoding RNA & Cancer, School of Life Sciences, Shanghai University, Shanghai, China

Zhongliang Ma
Lab for Noncoding RNA & Cancer, School of Life Sciences, Shanghai University, Shanghai, China

Trevor Marshall
University of Adelaide, Adelaide, SA; Murdoch University, Murdoch, WA, Australia; The Autoimmunity Research Foundation, Thousand Oaks, CA, United States

Eiji Matsuura
Neutron Therapy Research, Center Collaborative Research Center for Okayama Medical Innovation Center & Department of Cell Chemistry, Okayama University, Okayama, Japan

Ryan James McGinn
Department of Department of Neurology & Neurological Sciences, Stanford University School of Medicine, Stanford, CA, United States

Veronika Medvedeva
I.M. Sechenov First Moscow State Medical University (Sechenov University), Moscow, Russia

Lucio Miele
Department of Genetics, Louisiana State University Health Sciences Center, New Orleans, LA, United States

Andrew David Miller
Mendel University and Veterinary Research Institute, Brno, Czech Republic

Marina Moiseyakh
MGUPP, Moscow, Russia

Anders Nordstrom
Center for Proteomics, Faculty of Medicine, Department of Molecular Biology, Umea University, Umea, Sweden

Marianna Podzyuban
MGUPP, Moscow, Russia

Noel Rose†
Johns Hopkins Bloomberg School of Public Health, Baltimore, MD, United States

Michael Ross
UCLA Medical Center Obstetrics & Gynecology, Los Angeles, CA, United States

John Ryals
Mathematics Department at Central Georgia Technical College, Warner Robins, GA, United States

Mehar Sahu
Molecular Neuroscience and Functional Genomics Laboratory, Delhi Technological University (Formerly Delhi College of Engineering), Shahbad Daulatpur, Delhi, India

Yehuda Schoenfeld
Sheba Medical Center in Tel Hashomer, Sackler Faculty of Medicine, Tel-Aviv University, Tel-Aviv, Israel

Paul Seegers
Stichting PALGA (Pathologisch Anatomisch Landelijk Geautomatiseerd Archief), Houten, The Netherlands

Ilya Shmulevich
Institute of Systems Biology (ISB), Seattle, WA, United States

Arthur Silverstein
Department of the History of Medicine, Johns Hopkins University, Baltimore, MD, United States

Nikita Skomorohov
MGUPP, Moscow, Russia

Paul David Smith
Newberry College, Newberry, SC, United States

Eric J. Sorenson
Mayo Clinic, Rochester, MN, United States

Maria Studneva
MGUPP, Moscow, Russia; EPMA (European Association for Predictive, Preventive and Personalized Medicine), Brussels, Belgium

† Deceased

Sergey Suchkov
MGUPP; A.I. Evdokimov Moscow State University of Medicine & Dentistry, Moscow, Russia; EPMA (European Association for Predictive, Preventive and Personalized Medicine), Brussels, Belgium; ISPM (International Society for Personalized Medicine), Tokyo, Japan; PMC (Personalized Medicine Coalition), Washington, DC; New York Academy of Sciences, New York, NY, United States; Association for Medical Education in Europe (AMEE), Dundee, United Kingdom

Jacqueline Elizabeth Summers Stromberg
Department of Department of Neurology & Neurological Sciences, Stanford University School of Medicine, Stanford, CA, United States

Ksenia Syzko
MGUPP, Moscow, Russia

Sofia Tanasova
A.I. Evdokimov Moscow State University of Medicine & Dentistry, Moscow, Russia

William Thilly
Massachusetts Institute of Technology, MIT, Cambridge, MA, United States

Li Tong
The Research Center for Biomedical Information Technology, SIAT, Shenzhen, China

Erica Leah Von Stein
Department of Department of Neurology & Neurological Sciences, Stanford University School of Medicine, Stanford, CA, United States

Geoffrey Wang
The Research Center for Biomedical Information Technology, SIAT, Shenzhen, China

May D. Wang
The Research Center for Biomedical Information Technology, SIAT, Shenzhen, China

Alan Wu
San Francisco General Hospital; UCSF, San Francisco, CA, United States

Hang Wu
The Research Center for Biomedical Information Technology, SIAT, Shenzhen, China

Preface

William Osler (1849–1919), the Canadian physician, humanitarian, and teacher, one of the four founders of the Johns Hopkins School of Medicine and one of its first professors of medicine, once said to his students, "Listen to your patient. He is telling you the diagnosis," to which he also added, "And she just might tell you the best management, too." Other than the so-called Hippocratic Oath, "diagnose and treat" remains the most fundamental tenet of clinical medicine. New methods of diagnosis and treatment have greatly improved the delivery of medical care, but each physician still starts by treating a patient. How has this been done in the past? Not by creating novel treatment plans for each individual, but by considering the patient a statistical ensemble of a human and then implementing the standard treatment approach indicated for this ensemble. No more. A revolution in medicine is occurring thanks to advances in molecular biology and computer science. The "one-size-fits-all" approach is now being supplanted by precision medicine, a broad concept involving individualized tailoring of therapies guided by genetics.

An example is the sequencing of the DNA of tumor cells, which allows scientists and physicians to identify causative mutations. Knowing this provides several treatment options. One of the most promising is DNA editing. Technologies such as CRISPR–Cas9 may replace single mutant nucleotides with their normal counterparts, excise and replace short altered DNA sequences, or turn genes on and off. These methods expand the treatment armamentarium of physicians because they enable therapies that may target not only tumorigenic mutations within a single gene but also in genes that may be parts of an entire metabolic pathway. One nullifies the effect of a cancer-causing mutation by altering upstream or downstream genes and gene products within a pathway.

A patient presenting with lung cancer, for example, will not require surgery, radiation, chemotherapy, or other painful and debilitating treatments. Instead, gene defects already present in the patient's DNA, or in their tumor tissue, will be identified, as will the sequence information used by new "clinical molecular biology" laboratories to create individualized gene-editing treatments. The fear and morbidity associated with chemotherapy will no longer be issues. The patient will come to the clinic and be seen by an RN or other healthcare provider during a 5-min appointment, at which time they will receive a single IM injection or pill and then continue with their normal day. Any tumors will disappear after the treatment is done. That's it.

The logical extension of this strategy is genetic analysis and modification of an individual's complete genome, either at birth or when diseases of genetic origin occur. This could allow the prevention of diseases geneticists know will occur during a particular person's life. Are we there yet? Almost. The authors of this volume tell us how close for a range of conditions, including epilepsy, autoimmune disease, and cancer. "Big data" contributions to precision medicine's efforts for disease prevention, diagnosis, and treatment are also reviewed. Precision medicine is developing so rapidly that one might heed the following metaphorical advice: "Don't change the channel or you may miss it!"

David B. Teplow

CHAPTER ONE

Introduction of medical genomics and clinical informatics integration for p-Health care

Li Tong, Hang Wu, May D. Wang, and Geoffrey Wang*
The Research Center for Biomedical Information Technology, SIAT, Shenzhen, China
*Corresponding author: e-mail address: jeff.wang@gatech.edu

Contents

1. Introduction 2
   1.1 Clinical informatics 6
   1.2 Genomic and genomic medicine 9
2. State-of-the-art advanced data integration 16
   2.1 Raw data level integration 20
   2.2 Feature level integration 21
   2.3 Decision level integration 23
3. Case studies 25
   3.1 Predicting length of stay after infant cardiac surgery by integrating proteomic data with genomic variations using majority voting 25
4. Privacy and security 27
5. Conclusion 29
References 30

Abstract

Achieving predictive, precise, participatory, preventive, and personalized health (abbreviated as p-Health) requires comprehensive evaluations of an individual's conditions captured by various measurement technologies. Since the 1950s, analysis of care providers' and physicians' notes and measurement data by computers to improve healthcare delivery has been termed clinical informatics. Since the 2010s, wide adoptions of Electronic Health Records (EHRs) have greatly improved clinical informatics development with fast growing pervasive wearable technologies that continuously capture the human physiological profile in-clinic (EHRs) and out-of-clinic (PHRs or Personal Health Records) to bolster mobile health (mHealth). In addition, after the Human Genome Project in the 1990s, medical genomics has emerged to capture the high-throughput molecular profile of a person. As a result, integrated data analytics is becoming one of the fast-growing areas under Biomedical Big Data to improve human healthcare outcomes. In this chapter, we first introduce the scope of data integration and review applications, data sources, and tools for clinical informatics and

Progress in Molecular Biology and Translational Science, Volume 190
ISSN 1877-1173
https://doi.org/10.1016/bs.pmbts.2022.05.002

medical genomics. We then describe the data integration analytics at the raw data level, feature level, and decision level with case studies, and the opportunity for research and translation using advanced artificial intelligence (AI), such as deep learning. Lastly, we summarize the opportunities in biomedical big data integration that can reshape healthcare toward p-health.

1. Introduction

Delivering predictive, precise, participatory, preventive, and personalized health, abbreviated as p-Health, is the primary goal of future healthcare systems that can significantly improve care quality while reducing cost. To accomplish this goal, in-clinic Electronic Health Records (EHRs), out-of-clinic Personal Health Records (PHRs),[1] and high throughput genomic data could be assessed and validated concertedly through translational data analytics pipelines consisting of six steps: data collection, data quality control, feature extraction, knowledge modeling, decision making, and action-taking. As shown in Fig. 1, this chapter will examine the challenges and opportunities in data integration analytics, from clinical informatics to genomics medicine for p-Health, at three levels—raw data, feature, and decision.

Clinical informatics was first introduced in the 1950s when the U.S. National Bureau of Standards and the U.S. Air Force used digital computers to develop expert systems (such as MYCIN and Internist-I[2]) and computerized medical records management systems. In 1959, Ledley and colleagues discussed the idea of using computational reasoning to aid medical diagnostic processes for the first time and brought out the EHR idea.[3] EHR systems improve the communication among physicians, providers, and patients, the quality of clinical decisions, and the delivery of care significantly in hospital routine practices.[4] However, clinical informatics progress was slow due to the low EHR adoption rate. In 2008, the American Hospital Association (AHA) Annual Survey reported that only 13.4% of the non-Federal acute care hospitals in the U.S. had adopted basic or comprehensive EHR systems, and only 1.6% have an EHR system with clinical decision support.[5] With the new policy, "the Health Information Technology for Economic and Clinical Health (HITECH) Act 2009",[6] in 2015, the EHR system adoption rate had increased to about 80% in the US, with 34.4% of these hospitals having EHR systems with clinical decision support[5] and, by CDC's estimate, this number reached approximately 90% in 2019.[7]

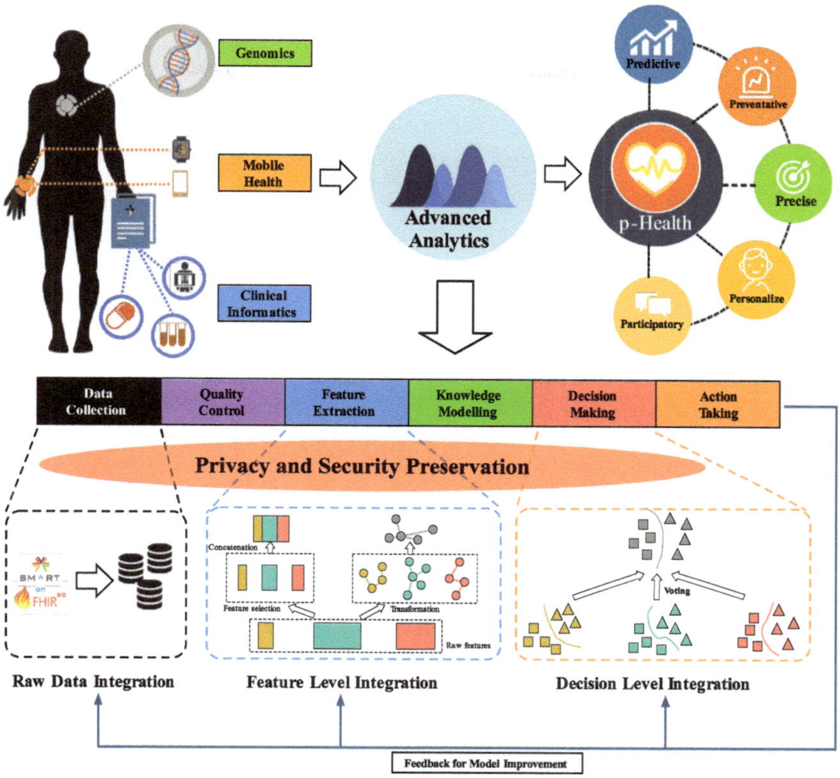

Fig. 1 Enabling p-Health by data integration. Integrating clinical informatics with genomics and mobile health enables p-Health through advanced data analytics. The typical data analysis pipeline consists of data collection, quality control, feature extraction, knowledge modeling, decision making, and action taking. For translation to clinical applications, privacy and security preservation needs to be implemted through each step of the pipeline. Along this pipeline, data integration can be achieved at raw data level, feature level, and decision level, respectively.

The development of low-cost sensors embedded in either mobile phones or wearable devices has recently enabled out-of-clinic care data to be captured in PHRs, in addition to in-clinic EHRs. When the individual data of anthropometric measurement and patient response were collected and analyzed through the wristband-type activity sensors, the out-of-clinic PHR were conveniently analyzed and concluded in addition to the EHRs. For example, the integration of low cost sensor (PHR) data have given clinicians and patients an option to actively share lifestyle and behavior data within the EHR workflow to support clinical interventions,[8] even though the obstructive sleep apnea related outcomes were insignificant in this case.

In addition, low-cost microwave sensors with high sensitivity were also used for monitoring blood glucose levels of diabetics. The glucose levels measured are as sensitive as 70–120 mg/dL in blood with a 0.94 MHz/(mg/dL) resolution at several transmission resonances.[9] The prototype has shown the shift in AI-powered wearable signal processing microwave sensors for PHRs toward their commercialization. The continuous monitoring of a person's daily physiology and communication has enabled personalized, preventive, predictive health for early disease warning and participatory patient education, bridging the gap between in-clinic and out-of-clinic care. Thus, in this chapter, the term "EHRs" is used for both traditional in-clinic EHR and out-clinic PHR.

Since the completion of Human Genome Project in 2001,[10] applying genomics to customize clinical care (or medicine) has become possible. Genomics focuses on discovering the structure, function, and editing of genomes.[11] Based on the definition of the National Human Genome Research Institute (NHGRI), "Genomic Medicine" is an emerging medical discipline that uses the genomic information of an individual for their clinical care and health outcomes,[12] and NHGRI "Genomics" covers the study of direct information about DNA or RNA, excluding the study of downstream derived products (e.g., proteomics, metabolomics).[12] DNA sequences capture genomic variations at the single nucleotide level, e.g., single nucleotide polymorphisms (SNPs),[13] and at the chromosome level, as with structure variations (SVs).[14] RNA sequences contain genomic variations from gene expression or alternative splicing events.[15] Human diseases result from the complex interactions between genotypes and environment,[16] therefore incorporating molecular level information such as genetic variations is essential for the accomplishment of precision medicine in p-Health. The unique genetic information of individuals can reveal disease status and responses to treatment (e.g., more than 80% of rare disease are caused by genetic mutations, and genomics can play an important role in the diagnosis[17]). The genomic variations detected in DNA or RNA sequences are genotypes that complements phenotypes in genomic medicine. With the development of high-throughput next-generation sequencing (NGS) technologies, genomic data such as the whole genome of an individual, can now be sequenced for as little as $1000 in a few days.[18] As of February 2022, the NIH Genetic Testing Registry contained over 22,023 conditions with 18,741 genes genetic tests.

As shown in Fig. 1, integrating "clinical informatics" and "genomic medicine" presents challenges, including data harmonization, data quality control, and advanced data analytics to build an integrated intelligent clinical decision support system for p-Health. To overcome these challenges, data integration is performed with a six-step data analytic pipeline at three levels—raw data, feature, and decision. We will first define and then review the clinical informatics and genomic medicine in Sections 1.1 and 1.2, respectively. Then in Section 2, we discuss state-of-the-art data integration methods. In Section 3, we will address the Privacy and Security issues of the p-Health. At the end in Section 4, we conclude with the future of advanced data integration from clinical informatics to genomics for p-Health (Table 1).

Table 1 Concept glossary for clinical informatics and genomics.

Term	Definition	Ref.
Clinical Informatics	The application of informatics and information technology to deliver healthcare services.	19
Electronic Health/ Medical Records (EHRs/EMRs)	An EHR/EMR is an electronic version of a patient's medical history, that is maintained by the provider over time, and may include all of the key administrative clinical data relevant to that persons care under a particular provider, including demographics, progress notes, problems, medications, vital signs, past medical history, immunizations, laboratory data and radiology reports.	20
Health Informatics	The interdisciplinary study of the design, development, adoption, and application of IT-based innovations in healthcare services delivery, management and planning.	21
Personal Health Records (PHRs)	PHR contains health information that is managed by the individual and can be accessed and managed by authorized users in a private, secure, and confidential environment.	1
Artificial intelligence (AI)	The study of "intelligent agents": any device that perceives its environment and takes actions that maximize its chance of success at some goal.	22
Machine Learning	The subfield of computer science that gives "computers the ability to learn without being explicitly programmed."	23
Deep Learning	Deep learning allows computational models that are composed of multiple processing layers to learn representations of data with multiple levels of abstraction.	24

Continued

Table 1 Concept glossary for clinical informatics and genomics.—cont'd

Term	Definition	Ref.
NHGRI "Genomic Medicine"	An emerging medical discipline that uses genomic information about an individual as part of their clinical care and the health outcomes.	12
NHGRI "Genomics"	the study of direct information about DNA or RNA, excluding the study of downstream derived products (e.g., proteomics, metabolomics).	12
Genome	The complete set of DNA in an organism.	11
Transcriptome	A collection of all the gene readouts (RNA) present in a cell.	11
-Omics	The study of the –ome. For example, genomics and transcriptomics are the studies of the genome and transcriptome respectively.	25
Single Nucleotide Polymorphism (SNP)	A variation at a single position in a DNA sequence among individuals.	13
Structural variation (SV)	Genomic alterations that involve segments of DNA that are larger than 1 kb, and can be microscopic or submicroscopic.	14
Copy number variation (CNV)	Copy number variation of DNA segments ranging from kb to Mb. CNV is one type of SVs.	26
Alternative splicing	During gene expression, different combinations of splice sites can be joined to each other, resulting one gene coding for multiple proteins.	15
Next generation sequencing (NGS)	A new generation of non-Sanger-based sequencing technologies. These high throughput sequencing technologies can sequence DNA and RNA at much faster speed and lower cost compared to Sanger sequencing.	27

1.1 Clinical informatics

Clinical informatics uses data analytics to gain new insight from individual and population health records and to improve clinical decision-making by combining data-derived knowledge and domain expert knowledge (Fig. 2). Typical data sources include the following.

Conventional individual health records, which is the first challenge in clinical informatics, contain various types of data, from structured billing codes to unstructured clinical notes. Throughout the hospital stay of patients, administrative information (e.g., demographics, gender information, and diagnostic codes for billing and public health reporting purposes), auxiliary clinical

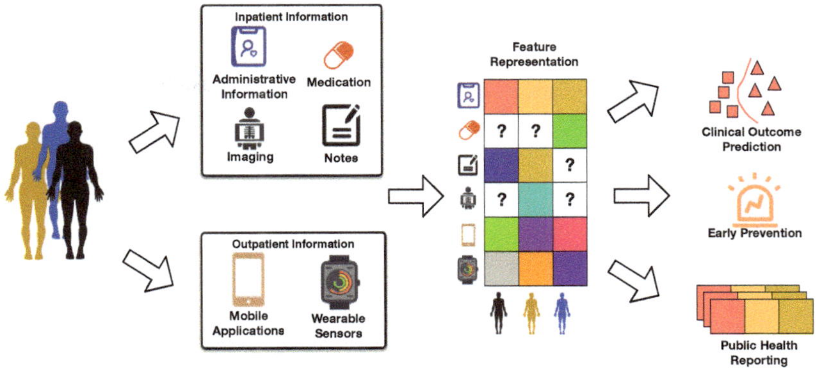

Fig. 2 An overview of clinical informatics. In-clinic and out-of-clinic health information contribute to the primary data sources for clinical informatics. Data from different sources are combined to form the feature representation of each patient, for later clinical endpoints such as clinical outcome prediction, early prevention, and public health reporting.

data (e.g., lab tests, medical and radiological imaging, medications, genetics), continuous physiological data (e.g., bedside monitoring data in intensive care units), and unstructured clinical notes record comprehensive information about patients' clinical status and outpatient care information. The *public health data* are more abstract and often contain aggregated statistics of diseases in geographical regions and times (e.g., the National Center for Health statistics (NCHS) collects 2.6 million death certificates each year that record demographics and causes of death within the US population. Recently, billions of mobile phones and wearable sensors (e.g., FitBit) and behavior imaging have enabled daily health monitoring in *outpatient care*. Especially during the COVID-19 pandemic, the United States Centers for Medicare & Medicaid Services (CMS) and health insurance companies and mobile health service providers, have all stepped up their support for mobile health care.[28] PHRs collect data from these applications and provide a more personalized evaluation of patients. Within the context of this chapter, EHRs are considered to comprise data sources from traditional EHRs, physiological and imaging data, and out-of-clinic as "extended" EHRs to represent heterogeneous types of data existing in clinical informatics. Other data sources that contribute to health include *ontology*, such as Unified Medical Language System (UMLS) and SNOMED CT.[29] Physicians and clinical researchers use nodes to represent medical concepts and edges between nodes to encode the relationships between concepts, such as what diseases can lead to certain symptoms and what medications can help alleviate specific symptoms.

Analyzing these data has benefits to both population and individual care for p-Health. Recent progress is shown in Table 2. At the population care

Table 2 Highlights of progress made in clinical informatics.

Task	Data collection and quality control	Feature extraction	Knowledge modeling and decision making
Re-admission Prediction	Models using administrative data[44] Models using real-time administrative data,[45] Models adding survey or chart review data[46]	Mainly raw values with missing values imputed	Logistic regression,[47] graphical models,[48] random forests,[49] and neural networks[50]
Mortality Prediction/ Survival analysis	Models using EHR data,[51] Models using clinical trials data[52]	Raw values regression[53]	Cox regression[53] and logistic regression, risk of mortality score[51]
Diagnosis Prediction	Models using EHR data[54]	Raw values and transformed values,[55] Deep learning features[56,57]	Recurrent neural networks,[56] multiple layer perceptron, decision trees, nearest neighbors, generalized linear model[58]
Epidemics Prediction	Models using public health data[59] Models incorporating social media data[60]	Mainly use raw reported values	Bayesian data analysis, text mining on social media,[60] time series mining[59]
Phenotyping and Risk Factors Identification	Models using EHR data,[61] Models using claims data,[62] Models using clinical trials data[63]	Raw values, Deep learning features[57,64]	Matrix/tensor factorization,[65] Bayesian networks, unsupervised clustering[55]
Causal Inference	Models using simulated data,[66] Models using observational data[67]	Raw data, Deep learning features[68]	Nearest neighbor matching,[68] decision trees[66] and random forests,[66] panel data, propensity scores, causal networks[69]
Mobile Health Software Applications	Models using mobile data[70]	N/A	Apple watch, Samsung gear, Jawbone, FitBit (devices); WebMD (referencing app); AmWell, PatientIO (Telemedicine); SugarSense (Diabetes)
Mobile health data analytics	Wearable sensors and smartphones	Raw values[71] Deep learning features[72]	Time series[73] Visual analytics[74]

level, health policymakers are interested in understanding epidemics from a statistical point of view across spatial and temporal dimensions.[30] A great amount of evidence has shown that EHRs have enabled multi-institute collaborations on data sharing, disease understanding, and, most importantly, disease dealing. The efficiency of EHRs has enabled p-Health on a personal level and has supported international collaboration in terms of developing effective vaccines and creating herd immunity. It has been one of the most efficient anti-pandemic practices in human history.[31–38] Aggregated EHRs help explore the evolution and spreading of diseases for resource allocation, which may be used to predict and prevent the outbreak of epidemics. Death certificates that contain causes of death can assist in understanding the diseases spread in the nation. At the individual care level, diagnosing patients' conditions and deciding prognosis are crucial steps. Clinical informatics can provide decision support for in-clinic physicians and care providers,[39] where clinical outcome (hospital re-admission, mortality) prediction helps prevent adverse in-clinic clinical events. Data prediction enabled through mobile sensors give early warning of medical conditions and provides physicians with out-of-clinic participatory health. EHR-associated scores have been used to predict the patients' clinical courses and outcomes and support clinical decisions. It also has been reported that respiratory support information from EHRs have been consistent with mortality prediction.[37,40] Besides prediction, grouping similar patients' records[41] and simulation[42] can suggest reference treatment options to assist clinical action-taking.

To gain more insight in clinical informatics, public datasets and tools are summarized in Table 3. Due to privacy and other health regulatory issues, only a limited number of datasets are publicly available, and comprehensive data sets may be obtained through research collaboration with specific clinical institutions. As the data available in EHRs mainly reflect phenotypes, to truly achieve personalized and precise health,[42] it is critical to expand from clinical informatics to genome medicine.[43]

1.2 Genomic and genomic medicine

Genomics can analyzes all genes and their interactions in each person at any point in time. Genomic medicine uses such genomic information to prevent, diagnose, and treat diseases clinically, i.e., for p-Health (Fig. 3). Human diseases result from complex interactions of genotypes and environmental factors.[16] Personalized information embedded in genomics can help physicians predict disease risk factors and patients' responses to treatments.

Table 3 Selected databases and tools for clinical informatics.

Purpose	Data source	Dataset description
Selected databases or projects for clinical informatics		
Individual Health	Medical Information Mart for Intensive Care III (MIMIC III)[75]	Clinical, notes, waveform information for about 58,000 admissions in intensive care units
Individual Health	The Alzheimer's Disease Neuroimaging Initiative (ADNI)[76]	Imaging, clinical, genetics for a cohort about 2000 in an Alzheimer's disease study
Individual Health Records	Parkinson's Progression Markers Initiative[77]	Imaging, clinical, biomarkers for a cohort of about 800 in a Parkinson's disease study
Public Health	CMS 2008–2010 Data Entrepreneurs' Synthetic Public Use File (DE-SynPUF)	About 40 million synthetic claims data
Public Health	Mortality data[78]	Death certificates containing demographics and causes of death for about 2 million each year in the U.S.
Purpose	**Description**	**Open source tools**
Selected tools for clinical informatics		
Data Storage and Management	Storing health data for hospital management or personal access	FHIRBase,[79] EHRServer[80]
General Predictive Modeling	Building predictive pipelines using patients' health records	PARAMO[81]
Visual and Analytics	Visualizing health data for easy interpretation and data understanding	EHDViz[82]
Domain Knowledge Modeling	Representing domain knowledge as ontology for knowledge-guided analysis	UMLS[29]
Mobile Health	Using data collected from mobile devices for assisted clinical decision making	Open mHealth[83]

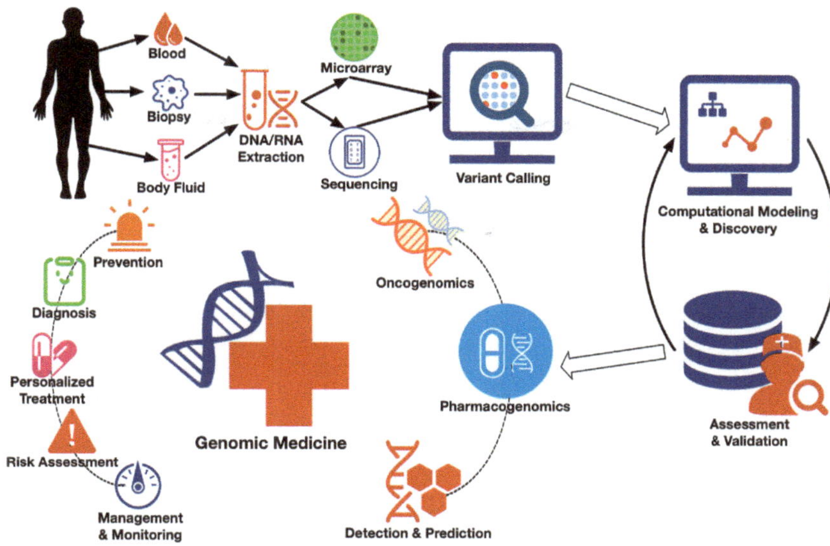

Fig. 3 An overview of genomic medicine. DNAs/RNAs are firstly extracted from the patient's blood, biopsy, or body fluid samples, and then genotyped by either microarray or sequencing. The raw genomic data are further analyzed with bioinformatics approaches to generate a list of genomic variants. AI-based computational modeling is applied for disease related genomic variants discovery with the guidance of current knowledge bases. The identified genomic biomarkers will be evaluated and validated by domain experts and will in turn enrich the knowledge bases. Ongeogenomics, pharmacogenomics, and disease risk predication are three major components of genomic medicine, which enables disease prevention, diagnosis, personalized treatment, risk assessment, and health management and monitoring.

For example, familial mutations in tumor suppressor genes such as BRCA1 and BRCA2 increase the risk of female breast and ovarian cancers.[84] Test for these traits can prevent, diagnose, and treat breast and ovarian cancers early.

As shown in Table 4, genomic medicine has mainly been applied to three areas: *oncogenomics, pharmacogenomics, and disease risk prediction*. Oncogenomics and pharmacogenomics are the most studied examples for applying personalized genomics.[85] Disease risk prediction is especially beneficial for those diseases with genetic markers of high clinical impacts. Various cancer known to be caused by gene mutations[86] have been studied extensively with personalized genomics to identify cancer-related genetic variations for personalized diagnosis and treatment.[87] For example, Nguyen et al. have applied personal genomics to relate the outcomes for breast different cancer subtypes after breast-conserving therapy.[88] Zheng et al. have associated genetic variations with prostate cancer.[89] *Pharmacogenomics* is the study of how a person's

Table 4 Highlights of progress made in genomic medicine.

Task	Applications	Selected examples	Genetic variants	Computational methods and knowledge modeling
Oncogenomics	Identify oncogenes (understanding genetic mechanisms for cancer generation and progression)	Colorectal cancer[98] Prostate cancer[89] Ovarian cancer[99]	SNPs[89,99] CNVs[98] DNA Methylation (e.g., altered methylation of CpG islands)[98] Gene expression[88]	Genome-wide association studies (GWAS)[99] Cox proportional-hazards model[98] Logistic-regression[89] Statistic tests (Chi-square test, Student's *t*-test, log-rank test, Fisher's exact test)[88,98]
	Personalized treatment	Breast cancer[88] Colorectal cancer[98]		
Pharmacogenomics	Optimize treatments (personalized treatments, drug safety, optimal dosing)	Cystic Fibrosis[100] Malignant melanoma with BRAF[90]	SNPs[90,101] Gene expression[102] Gene fusion[103]	GWAS[101] Statistical tests[90]
	Drug discovery (targeted, efficient trials)	Rheumatoid arthritis[101] Breast cancer with HER2[102] Lung cancer with EML4–ALK fusion[103]		
Disease diagnosis or risk prediction	Infectious diseases	Ebola[104]	SNPs[104,105]	GWAS[105]
	Rare diseases	Severe intellectual disability[106]		
	Other endpoints	Alzheimer's disease[105]		

genome affects their response to drugs (i.e., the relationship between genetic variations and drug responses). It can guide physicians in choosing optimal drugs for patients based on their genetic variations, which minimizes adverse drug reactions and maximize effectiveness[85] for personalized care. For example, the chemotherapeutic agent Vemurafenib specifically targets melanomas with the BRAF V600E mutation, with efficacy and adverse events evaluated.[90] *Disease risks* vary with genetic variations. For example, 80% of rare diseases have genetic origins, making personalized genomics essential for early screening, diagnosis, and individualized treatment.[17] Multiple rare diseases have genetic biomarkers with sufficiently high clinical impact, and more are discovered each year. In COVID-19 drug discovery, some pharmacogenetic data were collected to repurpose old drugs for COVID-19.[37,40] In addition, in silico modeling and pharmacogenomic markers have also been utilized for these COVID-19 drug repurposing efforts. By combining genetic biomarkers using genome-wide association studies (GWAS) with environmental factors, risk factors of some diseases can be thoroughly assessed, allowing for earlier intervention and better treatment.

Table 5 lists databases and projects for large-scale genomics studies for personalized genomics. These databases or projects focus on: (1) general genetic variants discovery; (2) specific diseases; or (3) specific populations, respectively. *The first group of projects* aims to discover genetic variants and understand their functions. To identify common genetic variations and to discover novel ones, the 1000 Genomes Project,[91] followed by the 10,000 human genomes project,[92] sequenced a large number of human genomes. These projects provide a base for the clinical use of genetic variations. The encyclopedia of DNA elements project (ENCODE)[93] aims to discover the functional elements in the human genome to make sense of genomic data. UK Biobank[94] connects genetic variants with a wide range of diseases and outcomes by providing EHR (with imaging) and genomic data, which is a perfect resource for integrated data analytics. *The second group of projects* aims to better understand and treat specific diseases. These projects include the cancer genome atlas project (TCGA),[95] Alzheimer's Disease neuroimaging initiative (ADNI),[96] and Parkinson's progression markers initiative (PPMI).[77] The TCGA project focuses on utilizing cancer genomics to improve the understanding, prevention, diagnosis, and treatment of various cancers. Similarly, ADNI and PPMI projects are developed for two of the most prevalent neurodegenerative diseases, Alzheimer's disease, and Parkinson's disease, respectively. These disease-oriented databases collect multi-omics data, clinical information (EHR), and medical imaging data

Table 5 Selected databases and tools for genomics.

Database or project	Description	Data elements	Ref.
1000 Genomes Project	The first project to sequence the genomes of a large number of people, which aims at finding most genetic variants with frequencies of at least 1% in the population studied.	Variant calls (VCF format). Alignments (BAM or CRAM), Raw sequence files.	91
Encyclopedia of DNA Elements Project (ENCODE)	A group of international projects aims at building a comprehensive list of functional elements in the human genome.	DNA-seq, RNA-seq, and various DNA-RNA, DNA-protein interaction assays like ChIP-seq	93
UK Biobank	A national and international health resources open for all health researchers. The database is established for improving the prevention, diagnosis, and treatment of diseases.	EMR, Questionnaires for health and lifestyle, Genetic information (whole genome sequencing, exome sequencing, genotyping), Imaging (MRI scan for 100,000 participants)	94
The Cancer Genome Atlas Project (TCGA)	A publicly available dataset with more than 2 TB of genomic data. The project aims at improving the prevention, diagnosis, and treatment of cancer by discovering the key genomic changes in 33 types of cancers.	EMR (clinical information of participants), histopathology slide images, molecular information (mRNA/mi RNA expression, protein expression, copy number, etc.)	95
Alzheimer's Disease Neuroimaging Initiative (ADNI)	An ongoing, longitudinal, multicenter study designed to develop clinical, imaging, genetic, and biochemical biomarkers for the early detection and tracking of Alzheimer's disease (AD).	EMR (demographics, clinical assessment), Genetics data (SNP), MRI image data, PET image data, Image analysis results, chemical biomarkers.	96
Parkinson's Progression Markers Initiative (PPMI)	An observational clinical study to verify progression markers in Parkinson's disease with a comprehensive set of clinical, imaging and bio-sample data.	EMR, Imaging (MRI, PET, SPECT), whole exome sequencing data, biological specimens.	77

Table 5 Selected databases and tools for genomics.—cont'd

Database or project	Description	Data elements	Ref.
Minority Health Genomic and Translational Research Bio-Repository Database (MH-GRID)	A study aims at finding the causes of severe high blood pressure and other cardio-metabolic problems in people of African ancestry	Genes, EMR (diet, sleep, body mass index, stress, access to healthy food and parks).	97

Selected tools and applications for genomics

Task	Description	Selected tools with ref.
Quality control	Identify artifacts in Microarray data or NGS data including low-quality and contaminated reads.	caCORRECT,[107] FastQC[108]
Read alignment	Alignment (mapping) of the short reads generated by NGS to reference sequences (genome or transcriptome)	BWA[109]
SNP detection	Identify SNPs from the aligned reads	GATK HaplotypeCaller[110]
SV detection	Identify SVs from the aligned reads	Hydra-sv[111]
CNV detection	Identify CNVs from the aligned reads	GATK gCNV[110]
Transcript quantification	Estimate gene expression levels	Cufflinks[112]
Normalization	Normalize systematic variations between samples	Cuffnorm[112]
Differential expression	Identify the differentially expressed genes in two or more conditions.	Cuffdiff[112]
Alternative splicing	Identify differentially spliced transcripts	Cufflinks[112]
Prediction and Modeling	Prediction and modeling using Omics data	OmniClassifier,[113] omniBiomarker[114]
Data Management	Data sharing and data integrity maintenance	ArrayWiki[115]

(magnetic resonance imaging (MRI), positron emission tomography (PET), and pathological images) to enable data integration and a better understanding of each disease. *The third group of projects* aims to stratify medicine for specific populations. One example is the minority health genomic and translational research bio-repository database (MH-GRID).[97] Motivated by the fact that high blood pressure affects African Americans more than other racial groups, MH-GRID contains data collected from over a thousand African Americans across the US. Besides genetic data, the MH-GRID collects health-related information such as diet, sleep, body mass index, stress, and access to healthy food and parks.

Table 5 lists selected methods and tools for DNA and RNA analysis for personalized genomic medicine. DNA and RNA can first be sequenced by next-generation sequencing (NGS) platforms and then analyzed by bioinformatics approaches. Developed from the Sanger chain-termination method, NGS techniques parallelize the sequencing process and produce billions of sequences concurrently,[27] by using sequencing-by-synthesis, and have significantly lowered the cost of DNA sequencing compared to the Sanger method itself.[27] With sequences as input, the first step of genomic data analysis is sequencing data quality control to remove potential artifacts and low-quality reads. The sequenced short reads passing quality control are then aligned to the reference genome. The aligned reads are further analyzed to identify various genetic variations, e.g., SNPs, CNVs, and SVs. Based on NHGRI genomic definition presented in Section 1, genetic variation at the transcription level can be characterized by RNA-seq to obtain gene expression, alternative splicing, and gene fusion information. RNA sequencing (RNA-seq) reads are first aligned to the genome (spliced alignment) or transcriptome (un-spliced alignment) and are then quantified and normalized to yield gene expression levels. RNA-seq can capture major transcription-level regulation events like alternative splicing.[15]

The following two sections will focus on data integration, aiming at utilizing both clinical informatics and genomic medicine for improved p-Health.

2. State-of-the-art advanced data integration

Data integration uses multiple sources of information to provide a better understanding of a system (Fig. 4). Wong reviewed requirements and solutions to data integration and warehousing at the raw data access level in biomedicine in EnsEMBL, GenoMax, and SRS.[116] Goble et al. introduced

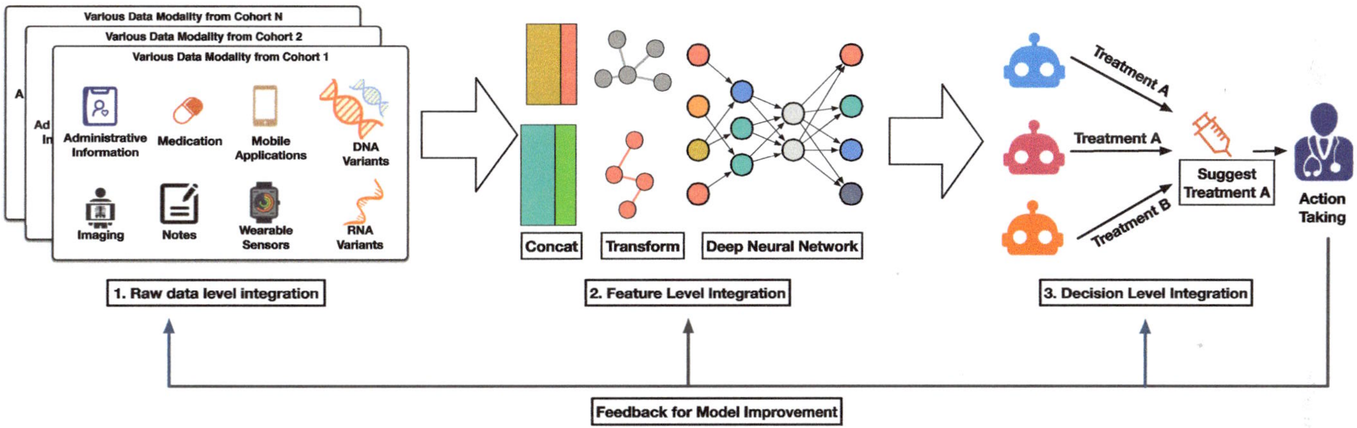

Fig. 4 Summary of advanced data integration analytics. The multimodality health data can be integrated at raw data level, feature level, and decision level. At raw data level, health data from different modalities and cohorts are harmonized. At the feature level, features from different data can be integrated by concatenation, transformation, or deep learning methods. At the decision level, the outputs of multiple models are combined for decision making and further action taking by the doctors.

"a loose federation of bio-nations" to handle the heterogeneity of biomedical data sources and the ontology in data integration.[117] More recently, to obtain new insights, Gomez-Cabrero et al. described how to deal with the diversity of existing omics data types and formats of large datasets.[118] They used dynamic Bayesian networks, self-organizing maps, and network inference methods over popular data sources such as 1000 Genomes Project, ENCODE, TCGA, and ImmGen. Ritchie et al. reviewed -omic data integration by comparing meta-dimensional and multi-stage analysis.[119] Based on current work in biomedical data integration, we identify the following *four challenges* for integrating EHRs and genomic data:

The first challenge is data collection and harmonization. New technologies, such as mobile sensors and DNA sequencing, all require modality-specific techniques for data storage, visualization, and analysis by computing devices. New genomic technologies are emerging every year, ranging from microarray to third-generation sequencing, which makes genomic data heterogeneous with various formats and standards. The NGS has different platforms, such as Illumina, 454 sequencing, or SOLiD sequencing, that lead to more variation. The development of third-generation sequencing by Pacific Biosciences,[120] with a much longer read length will add further variations in genomic data. Validating the data generated by these various technologies is challenging.

The second challenge is data quality. In clinical informatics, EHR data can be unstructured and noisy, with issues such as high percentages of missing values, errors, invalid data, and outliers. Thus, data quality control processes, such as missing data imputation, data conflict resolution, and data transformation, are needed.[121] In genomic medicine, DNA sequencing quality is affected by sample contamination and sequencing errors. Thus, numerous new bioinformatics approaches (e.g., FastQC[108]) are being developed to improve the sequencing quality before downstream analysis. In addition, to establish unified standards for genomic data processing for reliable clinical applications, the US Food and Drug Administration (FDA) coordinates the microarray quality control (MAQC) project[122] and the sequencing quality control (SEQC) project[123] for a comprehensive assessment of microarray and RNA-seq, respectively.

The third challenge is to develop advanced data analytics to extract knowledge from EHR and genomic data. In clinical informatics, feature engineering constructs a meaningful set of representations for patients, including medications, lab tests, and procedures, whereas the past analytics relied on hand-crafted features or input from domain experts, which are very limited in representation and predictive power. The new opportunities are on deep learning to find representations from large data. In genomic medicine, the "curse of dimensionality" (i.e., the feature dimension is significantly larger than the patient sample size) is a well-known challenge. Genome sequencing generates millions of genomic variations such as SNPs, CNVs, gene expression levels, and alternative splicing events for each sample, while the sample size of the study may be several hundred. Directly applying traditional data analysis will result in ill-conditioned feature matrixes, and the solution is to filter out irrelevant variants use feature selection in either a supervised or unsupervised fashion. Lack of interpretability is another major challenge for EHR and genomic data analytics. Predictive models built on EHR data have shown potential in predicting the hospital length of stay and re-admission probability. However, deploying them in daily hospital practice is challenging because physicians do not fully trust the model prediction because the bases for such predications are unknown. Making sense of genomic data is also a bottleneck for translating genomic discoveries into clinical practice.[124] Current data-driven approaches are mainly based on mathematical models, statistical tests, and computational methods. Lack of biological and clinical interpretation weakens the clinical impact of the novel genetic biomarkers discovered. Experimental validation is important for translating discoveries from data mining into biological knowledge and further applications to clinical care.

The fourth challenge concerns converting knowledge into actionable decision-making. Artificial intelligence (AI)[22] started in the 1950s and has undergone a paradigm shift from symbolic logical reasoning to machine learning,[23] including the recent success of deep learning[24] in automatic representation learning. Deep learning has the potential to provide highly accurate predictive modeling, and deep reinforcement learning shows significant breakthroughs in playing video games, dialog systems, and other tasks. However, a major challenge for computer-aided decision-making in clinical informatics is that it is unethical to test different options. Moreover, it is almost impossible in practice to test two treatment options in one individual. Historical data may be used to simulate the effect of each action applied to an individual, which is termed "counterfactual inference." However, to ultimately address the problem, we need to understand the causal relationship between medical events and between treatment options and diseases, for causality beyond correlation studies.

In this section, we review state-of-the-art data integration analytics at the raw data, feature, and decision levels to enable p-Health.

2.1 Raw data level integration

Raw data level integration happens in the data collection and storage stage, when different modalities of data are collected from different institutions and times for the same disease. The multi-modality data from more patients can enable machine learning algorithms to achieve better predictive models.[125] How to integrate different types of data and databases is also challenging. In clinical informatics, there are 186 commercial EHR system vendors in the U.S. Market,[106] with Epic, Cerner, and MEDITECH having the highest market shares, and each of them has its specific data standard and format. As a result, multiple clinical institutions with different EHR vendors cannot exchange information directly.[107] Existing standards such as Health Level Seven International (HL7) and Clinical Document Architecture (CDA)[126] cannot support the interexchange of data from different EHRs automatically for decision support systems, while manual data conversion case-by-case is both time-consuming and prone to errors. Besides, ensuring privacy and security in the raw data integration is also crucial.[127] To integrate data, the international standard body, Health Level Seven International (HL7),[108] has made a collaborative effort to develop a new EHR standard FHIR (Fast Healthcare Interoperability Resources), and its extensions SMART-on-FHIR, to make data from different EHR interoperable. FHIR is a new emerging standard that uses a resource-centric approach (as opposed to document-centric) to specify data elements in two parts: (i) standardized healthcare data models; and (ii) a set of application programming interfaces (APIs) for interacting and modifying these data models.[128] Thus, data resources can exist as online services rather than static files so that applications can read and write these resources in real-time. SMART-on-FHIR apps are based on FHIR but work outside of the constraints of many stakeholders' existing technical and security infrastructures. Thus, opportunities in raw data level integration include extending FHIR Resources, implementing RESTful APIs, and SMART-on-FHIR apps to facilitate the raw data integration process. Examples include death reporting, health record vendor translator, and mobile health apps[129]; and SMART on FHIR Genomics.[130] In genomic medicine, currently, FHIR enables the incorporation of genomic variations into EHRs. One example for establishing standards and APIs for genomic data storing, sharing, and clinical applications is SMART on FHIR Genomics.[109] FHIR app development presents challenges and opportunities for establishing standards and APIs for genomic data storing, sharing, and clinical applications. Such uses have been reported in the medical fields of pathology, ophthalmology, NGS and meta-studies.

2.2 Feature level integration

The next step in the data analytics pipeline following data collection and quality control is feature extraction. The feature level integration includes both concatenation-based and model-based methods.

Directly concatenating raw features to integrate different data modalities can introduce thousands of features, but the challenges for integrating EHRs and genomics features at the feature level are that the information represented by feature vectors may have varying representation power and noise, and the computational challenges introduced by adding features together may lead to model under-fitting or non-convergence. Thus feature selection, such as L1 norm feature selection[131] and minimum-redundancy maximum-relevancy (mRMR) feature selection,[132] is needed.

Model-based feature integration often adopts an encoder-decoder framework to combine features from different modalities. For each modality, modality-specific encoders can map features of every data modality to a joint feature embedding space first. Features combined in the joint space will go through a single decoder for final reporting. Various machine learning algorithms have been researched to accomplish model-based integration. Multiple kernel learning (MKL)[133] learns a kernel transformation for each modality and then combines all kernel transformations with a weighted linear average. Canonical correlation analysis (CCA)[134] finds a new linear feature space, in which features from all modalities have maximum correlation when projected onto the new space. Probabilistic graphical model (PGM)[135] treats each feature as a random variable and performs statistical inference to obtain an integrated feature, which is also represented in the form of latent variables. These techniques are effective in several applications (Table 6). However, the disappearance of individual modality's feature learning and feature integration may result in a meaningless feature representation.

An emerging advanced AI method in feature-level integration is deep learning that can combine feature extraction and prediction, which learns a meaningful representation for high accuracy in the given task.[24] Multimodal deep learning[136] is an early work using the restricted Boltzmann machine, while CCA is also extended with a deep feedforward network in[137] for feature-level integration. In clinical research, the multi-modal analysis using deep learning improves model accuracy in medical diagnosis and imaging analysis.[138–141] More recently, EHRs, SNPs, and MRIs have been integrated with deep autoencoders for improved prediction of Alzheimer's diseases,[142] in which features from multi-modalities are first transformed

Table 6 Selected methods and tools for biomedical data integration.

Categories	Methods	Software or applications with ref.	Data modalities
Raw data level: Ontology Semantic web	Resource description framework	AlzPharm[155]	Neuron properties, pathological changes, drug responses, disease stages.
	Heterogeneous database system	Heterogeneous Database Integration in Biomedicine[156]	Distributed databases
Feature level integration: Concatenation-based integration Transformation-based integration	Grammatical evolution neural network	ATHENA[157]	SNPs, microarray, proteomics, sequence data, biomarkers, clinical data
	Bayesian network	WinBUGS[158]	SNP, mRNA (gene expression), phenotypes
	Multivariate Cox LASSO model	Glmpath[159]	mRNA, DNA methylation, CNV, and microRNA
	Feature extraction, selection, and concatenation	Anduril[160]	Gene expression, splice variant, SNP, CNV, DNA methylation, miRNA, siRNA, proliferation array
	Feature extraction and rule integration	Combining rule and machine learning classifiers[161]	Temporal events, rules, clinical notes
	Kernel-based integration	SKMsmo[162]	Gene expression, protein sequences
		Clinical decision support[163]	Clinical, micro array
	Graph-based semi-supervised learning	Cancer clinical outcome prediction[164]	CNV, DNA methylation, Gene expression, miRNA
		Protein functions prediction[165]	Co-participation in a protein complex, physical interactions, genetic interactions, Pfam domain structure, and gene expression

Table 6 Selected methods and tools for biomedical data integration.—cont'd

Categories	Methods	Software or applications with ref.	Data modalities
Decision level Integration: Voting-based integration Ensemble-based integration	Majority voting/ Weighted majority voting	Predicting protein fold recognition[148]	Pseudo-amino acid composition, predicted secondary structure, hydrophobicity, normalized van der Waals volume, polarity, polarizability
	Bayesian networks	CONEXIC[151]	CNV, and gene expression
		Breast cancer prognosis prediction[152]	Clinical and micro array
	Heterogeneous network	Drug-drug interactions[166]	Drug phenotypic, therapeutic, chemical, and genomic properties

to lower dimensions with deep autoencoders and then concatenated at the intermediate feature level.[143,144]

The opportunities in advanced AI-based feature integration include constructing differentiable encoders and the interpretation of integrated features and predictive models, especially deep learning models. Differentiable encoders can extract features from both EHRs and genomic data so that the whole pipeline can be trained with backpropagation. On feature interpretation, despite the progress in building highly accurate predictive models, physicians cannot trust black-box algorithms if they cannot understand what features contribute to the final prediction and how to link features to their original physiological meanings. Perturbation analysis,[145] influence functions,[146] and visualization of neural network outputs are some pioneering works for model interpretation. Identifying important raw features that contribute to the final prediction or similar patients from historical data are two directions that may help us understand the behavior of current models.

2.3 Decision level integration

Decision level integration first generates multiple base models using each data modality independently as the training sets and then generates a final model by combining individual models trained in the first step.

The challenges of decision level integration are the construction of an accurate base decision model and the combination of base models that can prevent model overfitting while adding more parameters.

The conventional decision-level combination uses simple majority, the weighted majority[147] (to predict protein fold recognition[148]), and ensemble learning. Ensemble learning designs special weighting methods to address the overfitting by adjusting base models with the resampled training set (bagging[149]), sequentially increasing weights of misclassified training data (boosting[150]). For example, the random forest classifier makes decisions based on the multiple decision trees it constructed from the resampled training set. Graphical-model-based approaches, such as Bayesian networks, combine multi-omics data to better understand glioblastoma and breast cancer.[151,152]

In advanced AI methods for decision level integration, there exist three opportunities: construction of base models, increasing interaction between different data modalities, and reducing the gap between the decision and clinical action.

First, designing accurate base models is critical for final predictive accuracy. For example, in the majority voting scheme, we need to choose one or more base models for each data modality from different models, including decision tree, k nearest neighbors (kNN), support vector machine, logistic regression, and neural network. For each base model, there are also multiple parameters to be tuned. The selection of base models and parameter tuning can be tedious and also prone to overfitting, especially for small sample sizes, so inventing efficient hyper-parameter tuning algorithms such as Bayesian optimization can be an interesting direction.

Second, decision-level integration allows independent analysis of each data set, while the integration at this upper level also limits the possible interactions among different data modalities. How to use the knowledge and decision from a resource-rich modality to assist decision-making in another modality is a challenging but rewarding task. For example, we often have limited patient data in the biomedical domain. However, we have a large collection of biomedical domain knowledge, documented as research papers or medical ontologies. Transferring decisions from such knowledge or retrieving such knowledge for decision-making[153] can be helpful for physicians.

Third, bridging the huge gap in the decision-level integration can improve the final clinical action. After we design accurate predictive models with these integration methods, how to suggest viable actions (medications and procedures) for physicians remains a challenge because even though randomized trials are widely used in drug and clinical trials to determine the effect of new technology, in daily practice, physicians can only resort to observational study for such evaluations.[154] To solve the problem, it is necessary to understand the causal relationship between genomic data, medical conditions, and treatment to suggest and evaluate more reasonable actions. This causal inference modeling is a huge topic in biomedical big data analytics on its own and will not be covered in this chapter.

3. Case studies

3.1 Predicting length of stay after infant cardiac surgery by integrating proteomic data with genomic variations using majority voting

Immediately following newborn cardiac surgery, a profound inflammatory response may occur due to exposure of the blood to foreign materials during cardiopulmonary bypass (CPB).[167] It has been shown that the severity of systemic inflammatory responses correlates with prolonged length of stay (LOS) in either the intensive care unit (ICU) or hospital.[168] A full understanding of the mechanism for systemic inflammatory responses may help (1) identify high-risk patients and modify surgical procedures on a personalized basis to improve overall clinical outcomes; and (2) develop therapeutic strategies targeting specific biomarkers to alter clinical outcomes.

In this case study, we aim to integrate genomic data with proteomic data at the decision level for accurate prediction of clinical outcomes after infant cardiac surgery. Such study can not only facilitate biological interpretation of inflammatory response mechanisms but also provide an opportunity for biomarker translation to clinical settings. Specifically, combining postoperative serum cytokine levels and genome-wide variants, we investigate decision-level integration models that may predict the LOS of infants in the cardiac ICU after surgery.

For neonates who had undergone cardiac surgeries, blood samples were collected at admission back to the ICU immediately after the operation. We analyze 11 inflammatory markers from these postoperative blood samples. DNA of blood serum samples were sequenced at the Yerkes National Primate Research Center at Emory University using two Illumina HiSeq 1000 flowcells and 50 base-pair paired-end reads. Genomic variants, including SNPs and INDELs (i.e., insertions and deletions), were identified using The Genome Analysis Toolkit (GATK).[110] Fig. 5 summarizes the key modules and the workflow of the GATK pipeline. To reduce the number of

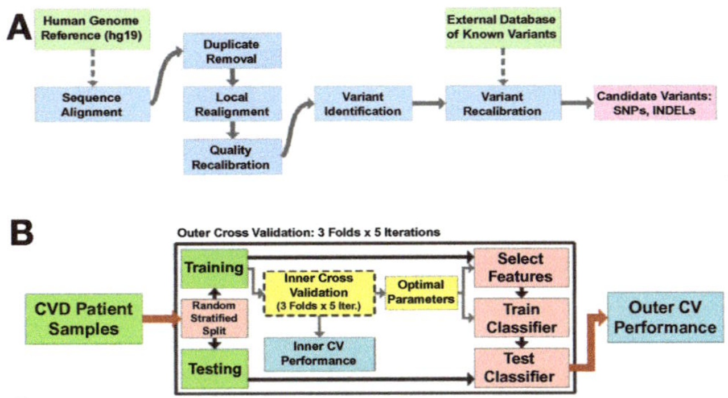

C

Majority voting:

$$H(x) = \begin{cases} c_j & \text{if } \sum_{i=1}^{T} h_i^j(x) > \frac{1}{2}\sum_{k=1}^{l}\sum_{i=1}^{T} h_i^k(x) \\ \text{rejection} & \text{otherwise.} \end{cases}$$

Weighted majority voting:

$$H(x) = c_{\arg\max_j \sum_{i=1}^{T} w_i h_i^j(x)}$$

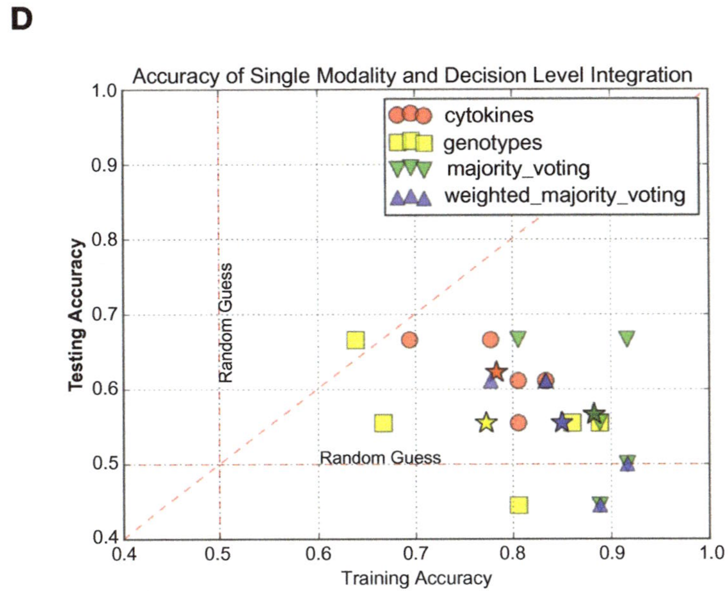

Fig. 5 Predicting length of stay (LOS) after infant cardiac surgery by integrating proteomic data with genomic variations. (A) Genome Analysis Toolkit (GATK) pipeline for extracting genomic variations. (B) Nested cross validation to estimate prediction performance. (C) Decision level integration using majority voting and weighted majority voting, where ***T*** is the number of classifiers, ***l*** is the number of labels, ***x*** is the input data of a sample, $h_i^j(x)$ is the probability of predicting ***x*** as label ***j*** by the ith classifier, and w_i is the weight for the ith classifier. (D) The accuracy of using single data modality and decision level integration for predicting the length of stay after infant cardiac surgery.

variants we discovered, we filtered the variants to include only variants that correspond to the inflammatory response (38 interferons and 92 interleukins (IL)) and identified 4411 variants. We binarized the LOS into long and short LOS with a threshold of 14 days. We applied nested cross-validation for prediction modeling of each data modality (Fig. 5B). We used mutual information for feature selection and k-nearest-neighbors (KNN) classifiers for prediction. We then used majority voting and weighted majority voting to combine prediction decisions of the model trained based on each data modality alone (Fig. 5C).

Fig. 5D shows the prediction results for LOS. With single data modality, accuracy using cytokines alone (test accuracy: 0.62 ± 0.042, train accuracy: 0.78 ± 0.048) outperforms accuracy using genotypes alone (test accuracy: 0.56 ± 0.070, train accuracy: 0.77 ± 0.10). After integrating cytokines and genotypes at the decision level, majority voting (test accuracy: 0.57 ± 0.089, train accuracy: 0.88 ± 0.041) and weighted majority voting (test accuracy: 0.56 ± 0.070, train accuracy: 0.85 ± 0.048) achieve similar prediction accuracies. Single data modality using cytokines alone achieves the best testing accuracy. Decision level integration improved the training accuracy but failed to improve the testing accuracy.

With decision level data integration, the training accuracy improved compared with that using a single modality alone. The improvement may be beneficial for realizing p-Health by accurate prediction of LOS after infant cardiac surgery. However, the testing accuracy with decision level integration is similar to that using genotypes alone and lower than that using cytokines alone, which may indicate overfitting when using majority voting and weighted majority voting at the decision level. The potential overfitting could be caused by the unbalanced number of features and the limited number of data modalities since we have only two data modalities with 4411 SNP genotypes while only 11 cytokines. To improve the predicting accuracy by integrating cytokines and genotypes, we may need more sophisticated data integration analytics to combine unbalanced data. In this case study, we present data integration application at the decision level for p-Health, where advanced analytics is important for making sense of multi-modality data.

4. Privacy and security

For translation to clinical applications, privacy and security preservation needs to be implemented through each step of the mutli-modal data integration pipeline (Fig. 1). The medical applications in mobile devices with embedded sensors provide superb services supporting healthcare interventions. Sophisticated data originating from the patients, from their medical records (EHR/PHR), and from the environments, can be used to assess users' health conditions and to generate alerts. The m-health apps often carry not only immediate illness-related data but also sensitive information, such as personal genetic condition, the results of any test and scan, medical histories, and personal information with tremendous value. If these records were not secured, the consequence could be severe, leading to financial losses and personal harm. Thus enforcing security and privacy through m-health apps on personal information is absolutely essential through data collection to predictive modeling (Fig. 1).

While these sensitive data are at stake, however, there is a lack of generally recognized security and privacy guidelines and legally binding data protection provisions to guarantee data privacy and safety. Through our in-depth medical data protection analysis of some popular m-health apps, we have evaluated the common data protection mechanisms, protection quality, and medical-related protection criteria.

In the past 20 years, the mobile "health and well-being" app (m-health app) outburst and is increasingly popular among patients due to its better health communication means with patients. Even though it is considered part of the IoT ecosystem, it has its uniqueness for entering into the patients' private space, where the goal is to achieve a flexible and low-cost healthcare service with a better clinical decision. However, along with the convenience, the improved accuracy, increased efficiency, and enhanced productivity need sharable Personal Health Records.

However, the unregulated software and hardware (mobile phone and wearables) m-health market only promises inadequate and shaggy protection of patients' personal data. Radical changes in medical data protection are needed.

Among others, the common technical privacy risks in mobile health are unencrypted traffic, embedded advertisements, and third-party analytics services. The privacy attacks could come from the internet, third-party service, Bluetooth connections, the logging system, SD card storage, and export capabilities. Potential damages include information leaks, information manipulation or information loss, and unauthorized information access from third parties.

The main security concerns of the m-health app embrace how the patient data were collected, managed, and shared, such as how to limit the information to its minimum requirement, just enough for medical services.

We have discussed the information concerning security from the following categories:

The privacy policy needs to avoid the dubious definition for protecting users from privacy issues, such as malicious or unintentional data leakage. The permission requests of the apps need to be specific with a sound reason, such as why tracking a user's location is necessary. In addition, some of the connections do not connect using HTTPS and have static code issues concerning connection components. The security and privacy attributes should be considered when transmitting health-related data, such as the transmit form and/or storage form, and how the health multimedia content is transmitted and stored (e.g., scans of x-rays images) must also be considered. These are very private medical information and are particularly important to the patients, as is the patients' geolocation information, which must be protected from third parties. A particularly important aspect of the m-Health app in China is how to use chat session transmission, as WeChat is very popular in mainland China. A lot of medical information uses WeChat to transmit sensitive medical information with exposure risk in the none-HTTPS WeChat connections, including email addresses, passwords, names, images, and health-related questions that could be leaked through this channel.

To address these potential security and privacy requirement of the m-Health app, we need to address the functional and content requirements of the m-Health app at the policy level.

5. Conclusion

In this chapter, we discussed state-of-the-art methods for data integration at three levels, viz., raw data, feature, and decision, for combining clinical informatics with genomics to enable evidence-based p-Health. For integration at the raw data level, the major challenge is data harmonization and data quality control. With the unified FHIR standards established by HL7, integrating different data resources with FHIR apps is the most promising solution. For integration at the feature level, the power of conventional hand-crafted features is limited in the big data era. Advanced AI methods such as deep learning can significantly increase the feature representation power with end-to-end training and are one key direction for feature-level integration. Interpretation of integrated features at the feature level is another opportunity for translating data integration to clinical practice. For integration at the decision level, the construction of base models, the lack of interactions between different data modalities, and the translation from decisions to final clinical actions are the three major challenges. Designing novel predictive models and efficient hyper-parameter tuning algorithms, enhancing interactions between different data modalities, and inferring causal relationships are opportunities for data integration at the decision level.

In conclusion, combing EHR data with clinical genomics requires advanced data integration analytics. These data integration enabled by advanced AI methods will predict mortality risk and re-admission frequency as well as provide diagnostic and prognostic information, thus reshaping p-Health care.

References

1. Tang PC, Ash JS, Bates DW, Overhage JM, Sands DZ. Personal health records: Definitions, benefits, and strategies for overcoming barriers to adoption. *J Am Med Inform Assoc*. 2006;13(2):121–126.
2. Buchanan BG, Shortliffe EH. *Rule-Based Expert Systems*. vol. 3. Reading, MA: Addison-Wesley; 1984.
3. Ledley RS, Lusted LB. Reasoning foundations of medical diagnosis. *Science*. 1959;130 (3366):9–21.
4. Jha AK, DesRoches CM, Campbell EG, et al. Use of electronic health records in US hospitals. *N Engl J Med*. 2009;360(16):1628–1638.
5. Charles D, Gabriel M, Furukawa MF. Adoption of electronic health record systems among US non-federal acute care hospitals: 2008–2013. *ONC Data Brief*. 2013;9:1–9.
6. Health UDo, Services H. *HITECH Act Enforcement Interim Final Rule*. US Department of Health Services; 2009.
7. Prevention USCfDCa. *National Electronic Health Records Survey Public Use File National Weighted Estimates. 2019*; 2019. https://www.cdc.gov/nchs/fastats/electronic-medical-records.htm. Accessed Feburary 28, 2022.
8. Kim JW, Ryu B, Cho S, et al. Impact of personal health records and wearables on health outcomes and patient response: three-arm randomized controlled trial. *JMIR Mhealth Uhealth*. 2019;7(1).
9. Omer AE, Shaker G, Safavi-Naeini S, et al. Low-cost portable microwave sensor for non-invasive monitoring of blood glucose level: novel design utilizing a four-cell CSRR hexagonal configuration. *Sci Rep-Uk*. 2020;10(1).
10. Deleted in review.
11. NHGRI. *Fact Sheets about Genetic and Genomic Science*; 2015. https://www.genome.gov/10000202/fact-sheets/. Accessed 09 November, 2015.
12. NHGRI. *National Human Genome Research Institute (NHGRI) Definition of 'Genomic Medicine'*; 2012. https://www.genome.gov/pages/about/nachgr/sept2012agendadocuments/genomic_medicine_definition_080112_rchisolm.pdf. Accessed August, 2012.
13. Nature. *Single Nucleotide Polymorphism*; 2014. https://www.nature.com/scitable.
14. Feuk L, Carson AR, Scherer SW. Structural variation in the human genome. *Nat Rev Genet*. 2006;7(2):85–97.
15. Matlin AJ, Clark F, Smith CWJ. Understanding alternative splicing: towards a cellular code. *Nat Rev Mol Cell Biol*. 2005;6(5):386–398.

16. Hunter DJ. Gene-environment interactions in human diseases. *Nat Rev Genet.* 2005;6 (4):287–298.
17. Gulbakan B, Ozgul RK, Yuzbasioglu A, Kohl M, Deigner HP, Ozguc M. Discovery of biomarkers in rare diseases: innovative approaches by predictive and personalized medicine. *EPMA J.* 2016;7:24.
18. Hayden EC. Technology: the $1,000 genome. *Nature.* 2014;507(7492):294–295.
19. AMIA.org. *Definition of Clinical Informatics*; 2017. https://www.amia.org/applications-informatics/clinical-informatics.
20. CMS.gov. *Electronic Health Records 2012*; 2017. https://www.cms.gov/Medicare/E-Health/EHealthRecords/index.html.
21. Procter RDE. Health Informatics Journal, Edinburgh, United Kingdom. Definition of health informatics [internet]. In: *Message to: Virginia van Horne.* Bethesda, MD: Content Manager HSR Information Central; 2009.
22. Russell S, Norvig P. Intelligence A. In: *A Modern Approach. Artificial Intelligence.* Vol. 25. Egnlewood Cliffs: Prentice-Hall; 1995:27.
23. Samuel AL. Some studies in machine learning using the game of checkers. *IBM J Res Dev.* 2000;44(1.2):206–226.
24. LeCun Y, Bengio Y, Hinton G. Deep learning. *Nature.* 2015;521(7553):436–444.
25. OmicsWiki. *Omes and Omics. 2013.* http://omics.org/index.php/Omes_and_Omics. Accessed 18 Aug, 2013.
26. Redon R, Ishikawa S, Fitch KR, et al. Global variation in copy number in the human genome. *Nature.* 2006;444(7118):444–454.
27. Schuster SC. Next-generation sequencing transforms today's biology. *Nat Methods.* 2008;5(1):16–18.
28. Greiwe J, Nyenhuis SM. Wearable technology and how this can be implemented into clinical practice. *Curr Allergy Asthma Rep.* 2020;20(8).
29. Bodenreider O. The unified medical language system (UMLS): integrating biomedical terminology. *Nucleic Acids Res.* 2004;32(Suppl 1):D267–D270.
30. Jensen PB, Jensen LJ, Brunak S. Mining electronic health records: towards better research applications and clinical care. *Nat Rev Genet.* 2012;13(6):395–405.
31. Berdahl CT, Nguyen AT, Diniz MA, et al. Using body temperature and variables commonly available in the EHR to predict acute infection: a proof-of-concept study showing improved pretest probability estimates for acute COVID-19 infection among discharged emergency department patients. *Diagnosis (Berlin, Germany).* 2021;8 (4):450–457.
32. Dagliati A, Malovini A, Tibollo V, Bellazzi R. Health informatics and EHR to support clinical research in the COVID-19 pandemic: an overview. *Brief Bioinform.* 2021;22 (2):812–822.
33. Garry EM, Weckstein AR, Quinto K, et al. Use of an EHR to inform a claims-based algorithm to categorize inpatient COVID-19 severity. *Pharmacoepidemiol Drug Saf.* 2021;30:93.
34. Holzer KJ, Lou SS, Goss CW, et al. Impact of changes in EHR use during COVID-19 on physician trainee mental health. *Appl Clin Inform.* 2021;12(03):507–517.
35. Icten Z, Watzker A, Friedman M, Menzin J. Thrombotic and cardiovascular events among patients hospitalized with COVID-19: findings from a large EHR database. *Pharmacoepidemiol Drug Saf.* 2021;30:366–367.
36. Nault K, Grgurich P, Stempek S, Dargin J, Gray A. Effect of EHR restriction on conservation of neuromuscular blocking agents during a COVID-19 surge. *Crit Care Med.* 2021;49(1):116.
37. Osborne TF, Veigulis ZP, Arreola DM, Roosli E, Curtin CM. Automated EHR score to predict COVID-19 outcomes at US Department of Veterans Affairs. *PLoS One.* 2020;15(7).

38. Poulos J, Zhu L, Shah AD. Data gaps in electronic health record (EHR) systems: an audit of problem list completeness during the COVID-19 pandemic. *Int J Med Inform*. 2021;150.
39. Middleton B, Bloomrosen M, Dente MA, et al. Enhancing patient safety and quality of care by improving the usability of electronic health record systems: recommendations from AMIA. *J Am Med Inform Assoc*. 2013;20(e1):e2–e8.
40. Yu SC, Hofford MR, Lai AM, Kollef MH, Payne PRO, Michelson AP. Respiratory support status from EHR data for adult population: classification, heuristics, and usage in predictive modeling. *J Am Med Inform Assoc JAMIA*. 2022.
41. Sun J, Wang F, Hu J, Edabollahi S. Supervised patient similarity measure of heterogeneous patient records. *ACM SIGKDD Explor Newsl*. 2012;14(1):16–24.
42. Victora CG, Habicht J-P, Bryce J. Evidence-based public health: moving beyond randomized trials. *Am J Public Health*. 2004;94(3):400–405.
43. Ashley EA. The precision medicine initiative: a new national effort. *JAMA*. 2015;313 (21):2119–2120.
44. Hammill BG, Curtis LH, Fonarow GC, et al. Incremental value of clinical data beyond claims data in predicting 30-day outcomes after heart failure hospitalization. *Circ Cardiovasc Qual Outcomes*. 2011;4(1):60–67.
45. Billings J, Dixon J, Mijanovich T, Wennberg D. Case finding for patients at risk of readmission to hospital: development of algorithm to identify high risk patients. *BMJ*. 2006;333(7563):327.
46. Krumholz HM, Chen Y-T, Wang Y, Vaccarino V, Radford MJ, Horwitz RI. Predictors of readmission among elderly survivors of admission with heart failure. *Am Heart J*. 2000;139(1):72–77.
47. Hasan O, Meltzer DO, Shaykevich SA, et al. Hospital readmission in general medicine patients: a prediction model. *J Gen Intern Med*. 2010;25(3):211–219.
48. Cai X, Perez-Concha O, Coiera E, et al. Real-time prediction of mortality, readmission, and length of stay using electronic health record data. *J Am Med Inform Assoc*. 2015;23(3):553–561.
49. Vedomske MA, Brown DE, Harrison JH. Random forests on ubiquitous data for heart failure 30-day readmissions prediction. In: *Paper presented at Machine Learning and Applications (ICMLA), 2013 12th International Conference on 2013*; 2013.
50. Ottenbacher KJ, Smith PM, Illig SB, Linn RT, Fiedler RC, Granger CV. Comparison of logistic regression and neural networks to predict rehospitalization in patients with stroke. *J Clin Epidemiol*. 2001;54(11):1159–1165.
51. Pollack MM, Ruttimann UE, Getson PR. Pediatric risk of mortality (PRISM) score. *Crit Care Med*. 1988;16(11):1110–1116.
52. Harrell F. *Regression Modeling Strategies: With Applications to Linear Models, Logistic and Ordinal Regression, and Survival Analysis*. Springer; 2015.
53. Thiébaut A, Bénichou J. Choice of time-scale in Cox's model analysis of epidemiologic cohort data: a simulation study. *Stat Med*. 2004;23(24):3803–3820.
54. Kononenko I. Machine learning for medical diagnosis: history, state of the art and perspective. *Artif Intell Med*. 2001;23(1):89–109.
55. Cruz JA, Wishart DS. Applications of machine learning in cancer prediction and prognosis. *Cancer Informat*. 2006;2:59.
56. Lipton ZC, Kale DC, Elkan C, Wetzell R. *Learning to Diagnose with LSTM Recurrent Neural Networks*. arXiv preprint arXiv:151103677; 2015.
57. Ridgway JP, Lee A, Devlin S, Kerman J, Mayampurath A. Machine learning and clinical informatics for improving HIV care continuum outcomes. *Curr HIV/AIDS Rep*. 2021;18(3):229–236.
58. Tyrer J, Duffy SW, Cuzick J. A breast cancer prediction model incorporating familial and personal risk factors. *Stat Med*. 2004;23(7):1111–1130.

59. Matsubara Y, Sakurai Y, Van Panhuis WG, Faloutsos C. FUNNEL: automatic mining of spatially coevolving epidemics. In: *Paper presented at: Proceedings of the 20th ACM SIGKDD International Conference on Knowledge Discovery and Data Mining 2014*; 2014.
60. Lee K, Agrawal A, Choudhary A. Real-time disease surveillance using twitter data: demonstration on flu and cancer. In: *Paper presented at: Proceedings of the 19th ACM SIGKDD International Conference on Knowledge Discovery and Data Mining 2013*; 2013.
61. Klein TE, Chang JT, Cho MK, et al. Integrating genotype and phenotype information: an overview of the PharmGKB project. *Pharm J.* 2001;1(3):167.
62. Wang Y, Chen R, Ghosh J, et al. Rubik: knowledge guided tensor factorization and completion for health data analytics. In: *Paper presented at: Proceedings of the 21th ACM SIGKDD International Conference on Knowledge Discovery and Data Mining*; 2015.
63. Sun L, Roesler J, Rösen-Wolff A, et al. CARD15 genotype and phenotype analysis in 55 pediatric patients with Crohn disease from Saxony, Germany. *J Pediatr Gastroenterol Nutr.* 2003;37(4):492–497.
64. Kale DC, Che Z, Bahadori MT, Li W, Liu Y, Wetzel R. In: *Causal Phenotype Discovery Via Deep Networks. Paper presented at: AMIA Annual Symposium Proceedings*; 2015.
65. Ho JC, Ghosh J, Steinhubl SR, et al. Limestone: high-throughput candidate phenotype generation via tensor factorization. *J Biomed Inform.* 2014;52:199–211.
66. Athey S, Imbens G. Recursive partitioning for heterogeneous causal effects. *Proc Natl Acad Sci.* 2016;113(27):7353–7360.
67. Chen Y, Dales R, Tang M, Krewski D. Obesity may increase the incidence of asthma in women but not in men: longitudinal observations from the Canadian National Population Health Surveys. *Am J Epidemiol.* 2002;155(3):191–197.
68. Johansson F, Shalit U, Sontag D. Learning representations for counterfactual inference. In: *Paper presented at: International Conference on Machine Learning*; 2016.
69. Heckerman D. A Bayesian approach to learning causal networks. In: *Paper presented at: Proceedings of the Eleventh Conference on Uncertainty in Artificial Intelligence*; 1995.
70. Duncan DT, Hickson DA, Goedel WC, et al. The social context of HIV prevention and care among black men who have sex with men in three US cities: the neighborhoods and networks (N2) cohort study. *Int J Env Res Pub He.* 2019;16(11).
71. Liu Y-Y, Moreno A, Li S, Li F, Song L, Rehg JM. Learning continuous-time hidden Markov models for event data. In: *Mobile Health.* Springer; 2017:361–387.
72. Che Z, Purushotham S, Kale D, et al. Time series feature learning with applications to health care. In: *Mobile Health.* Springer; 2017:389–409.
73. Dempsey WH, Moreno A, Scott CK, et al. iSurvive: an interpretable, event-time prediction model for mHealth. In: *Paper presented at: International Conference on Machine Learning*; 2017.
74. Paweletz CP, Charboneau L, Bichsel VE, et al. Reverse phase protein microarrays which capture disease progression show activation of pro-survival pathways at the cancer invasion front. *Oncogene.* 2001;20(16):1981–1989.
75. Johnson AE, Pollard TJ, Shen L, et al. MIMIC-III, a freely accessible critical care database. *Scientific Data.* 2016;3.
76. Jack CR, Bernstein MA, Fox NC, et al. The Alzheimer's disease neuroimaging initiative (ADNI): MRI methods. *J Magn Reson Imaging.* 2008;27(4):685–691.
77. Marek K, Jennings D, Lasch S, et al. The Parkinson progression marker initiative (PPMI). *Prog Neurobiol.* 2011;95(4):629–635.
78. Statistics NCfH. *Instructions for Classifying Multiple Causes of Death.* NCHS Instruction Manual; 2005.
79. Team HS. FHIRbase: open source storage based on the FHIR standard ready for use in production. In: *FHIR Standard Gives Specific Directions for Exchanging Structured Medical Data and FHIRbase Is Developed to Easily Store and Retrieve Medical Data in the FHIR Format*; 2017. Accessed 08/18, 2017. Available at http://fhirbase.github.io/index.html.

80. Labs C. *EHRServer: Open Source, Service-Oriented, Clinical Data Repository*; 2017. https://cabolabs.com/en/projects.
81. Ng K, Ghoting A, Steinhubl SR, Stewart WF, Malin B, Sun JM. PARAMO: a PARAllel predictive MOdeling platform for healthcare analytic research using electronic health records. *J Biomed Inform*. 2014;48:160–170.
82. Badgeley MA, Shameer K, Glicksberg BS, et al. EHDViz: clinical dashboard development using open-source technologies. *BMJ Open*. 2016;6(3).
83. Center T. *Open mHealth: The First and Only Open Standard for Mobile Health Data*; 2017. http://www.openmhealth.org/.
84. King MC, Marks JH, Mandell JB, Grp NYBCS. Breast and ovarian cancer risks due to inherited mutations in BRCA1 and BRCA2. *Science*. 2003;302(5645):643–646.
85. Esplin ED, Oei L, Snyder MP. Personalized sequencing and the future of medicine: discovery, diagnosis and defeat of disease. *Pharmacogenomics*. 2014;15(14):1771–1790.
86. Vogelstein B, Kinzler KW. Cancer genes and the pathways they control. *Nat Med*. 2004;10(8):789–799.
87. Strausberg RL, Simpson AJ, Old LJ, Riggins GJ. Oncogenomics and the development of new cancer therapies. *Nature*. 2004;429(6990):469–474.
88. Nguyen PL, Taghian AG, Katz MS, et al. Breast cancer subtype approximated by estrogen receptor, progesterone receptor, and HER-2 is associated with local and distant recurrence after breast-conserving therapy. *J Clin Oncol*. 2008;26(14):2373–2378.
89. Zheng SL, Sun JL, Wiklund F, et al. Cumulative association of five genetic variants with prostate cancer. *N Engl J Med*. 2008;358(9):910–919.
90. Chapman PB, Hauschild A, Robert C, et al. Improved survival with vemurafenib in melanoma with BRAF V600E mutation. *N Engl J Med*. 2011;364(26):2507–2516.
91. Altshuler D, Durbin RM, Abecasis GR, et al. A map of human genome variation from population-scale sequencing. *Nature*. 2010;467(7319):1061–1073.
92. Telenti A, Pierce LC, Biggs WH, et al. Deep sequencing of 10,000 human genomes. *Proc Natl Acad Sci U S A*. 2016;113(42):11901–11906.
93. Ecker JR. FORUM: genomics ENCODE explained. *Nature*. 2012;489(7414):52–53.
94. Allen N, Sudlow C, Downey P, et al. UK biobank: current status and what it means for epidemiology. *Health Policy Tech*. 2012;1(3):123–126.
95. Gerlinger M, Rowan AJ, Horswell S, et al. Intratumor heterogeneity and branched evolution revealed by multiregion sequencing. *N Engl J Med*. 2012;366(10):883–892.
96. Mueller SG, Weiner MW, Thal LJ, et al. The Alzheimer's disease neuroimaging initiative. *Neuroimaging Clin N Am*. 2005;15(4):869.
97. Horbal SR, Seffens W, Davis AR, et al. Associations of Apelin, Visfatin, and urinary 8-Isoprostane with severe hypertension in African Americans: the MH-GRID study. *Am J Hypertens*. 2016;29(7):814–820.
98. Liao XY, Lochhead P, Nishihara R, et al. Aspirin use, tumor PIK3CA mutation, and colorectal-cancer survival. *N Engl J Med*. 2012;367(17):1596–1606.
99. Pharoah PD, Tsai YY, Ramus SJ, et al. GWAS meta-analysis and replication identifies three new susceptibility loci for ovarian cancer. *Nat Genet*. 2013;45(4):362–370. 370e361–362.
100. Accurso FJ, Rowe SM, Clancy JP, et al. Effect of VX-770 in persons with cystic fibrosis and the G551D-CFTR mutation. *N Engl J Med*. 2010;363(21):1991–2003.
101. Okada Y, Wu D, Trynka G, et al. Genetics of rheumatoid arthritis contributes to biology and drug discovery. *Nature*. 2014;506(7488):376.
102. Vogel CL, Cobleigh MA, Tripathy D, et al. Efficacy and safety of trastuzumab as a single agent in first-line treatment of HER2-overexpressing metastatic breast cancer. *J Clin Oncol*. 2002;20(3):719–726.
103. Soda M, Choi YL, Enomoto M, et al. Identification of the transforming EML4-ALK fusion gene in non-small-cell lung cancer. *Nature*. 2007;448(7153):561–566.

104. Gire SK, Goba A, Andersen KG, et al. Genomic surveillance elucidates Ebola virus origin and transmission during the 2014 outbreak. *Science*. 2014;345(6202):1369–1372.
105. Harold D, Abraham R, Hollingworth P, et al. Genome-wide association study identifies variants at CLU and PICALM associated with Alzheimer's disease. *Nat Genet*. 2009;41(10):1088–U1061.
106. de Ligt J, Willemsen MH, van Bon BW, et al. Diagnostic exome sequencing in persons with severe intellectual disability. *N Engl J Med*. 2012;367(20):1921–1929.
107. Moffitt RA, Yin-Goen Q, Stokes TH, et al. caCORRECT2: improving the accuracy and reliability of microarray data in the presence of artifacts. *BMC Bioinformatics*. 2011;12:383.
108. Andrews S. *FastQC: A Quality Control Tool for High Throughput Sequence Data*; 2010. http://www.bioinformatics.babraham.ac.uk/projects/fastqc/. Accessed 26 Apr, 2010.
109. Li H, Durbin R. Fast and accurate short read alignment with burrows-wheeler transform. *Bioinformatics*. 2009;25(14):1754–1760.
110. DePristo MA, Banks E, Poplin R, et al. A framework for variation discovery and genotyping using next-generation DNA sequencing data. *Nat Genet*. 2011;43 (5):491.
111. Quinlan AR, Clark RA, Sokolova S, et al. Genome-wide mapping and assembly of structural variant breakpoints in the mouse genome. *Genome Res*. 2010;20(5):623–635.
112. Trapnell C, Roberts A, Goff L, et al. Differential gene and transcript expression analysis of RNA-seq experiments with TopHat and cufflinks. *Nat Protoc*. 2012;7(3):562–578.
113. Phan JH, Kothari S, Wang MD. omniClassifier: a desktop grid computing system for big data prediction modeling. *ACM BCB*. 2014;2014:514–523.
114. Phan JH, Young AN, Wang MD. omniBiomarker: a web-based application for knowledge-driven biomarker identification. *IEEE Trans Biomed Eng*. 2013;60 (12):3364–3367.
115. Stokes TH, Torrance JT, Li H, Wang MD. ArrayWiki: an enabling technology for sharing public microarray data repositories and meta-analyses. *BMC Bioinformatics*. 2008;9(Suppl 6):S18.
116. Wong L. Technologies for integrating biological data. *Brief Bioinform*. 2002;3 (4):389–404.
117. Goble C, Stevens R. State of the nation in data integration for bioinformatics. *J Biomed Inform*. 2008;41(5):687–693.
118. Gomez-Cabrero D, Abugessaisa I, Maier D, et al. Data integration in the era of omics: current and future challenges. *BMC Syst Biol*. 2014;8(Suppl 2):I1.
119. Ritchie MD, Holzinger ER, Li R, Pendergrass SA, Kim D. Methods of integrating data to uncover genotype-phenotype interactions. *Nat Rev Genet*. 2015;16(2):85–97.
120. Eid J, Fehr A, Gray J, et al. Real-time DNA sequencing from single polymerase molecules. *Science*. 2009;323(5910):133–138.
121. Rahm E, Do HH. Data cleaning: problems and current approaches. *IEEE Data Eng Bull*. 2000;23(4):3–13.
122. Shi LM, Reid LH, Jones WD, et al. The MicroArray quality control (MAQC) project shows inter- and intraplatform reproducibility of gene expression measurements. *Nat Biotechnol*. 2006;24(9):1151–1161.
123. Consortium SM-I. A comprehensive assessment of RNA-seq accuracy, reproducibility and information content by the sequencing quality control consortium. *Nat Biotechnol*. 2014;32(9):903–914.
124. Chin L, Andersen JN, Futreal PA. Cancer genomics: from discovery science to personalized medicine. *Nat Med*. 2011;17(3):297–303.
125. Vapnik VN, Vapnik V. *Statistical Learning Theory*. Vol. 1. New York: Wiley; 1998.
126. Dolin RH, Alschuler L, Boyer S, et al. HL7 clinical document architecture, release 2. *J Am Med Inform Assoc*. 2006;13(1):30–39.

127. Clifton C, Kantarcioğlu M, Doan A, et al. Privacy-preserving data integration and sharing. In: *Paper presented at: Proceedings of the 9th ACM SIGMOD Workshop on Research Issues in Data Mining and Knowledge Discovery*; 2004.
128. Bender D, Sartipi K. HL7 FHIR: an Agile and RESTful approach to healthcare information exchange. In: *Paper presented at: Computer-Based Medical Systems (CBMS), 2013 IEEE 26th International Symposium on 2013*; 2013.
129. Franz B, Schuler A, Kraus O. Applying FHIR in an integrated health monitoring system. *EJBI*. 2015;11(2):en56–en61.
130. Alterovitz G, Warner J, Zhang P, et al. SMART on FHIR genomics: facilitating standardized clinico-genomic apps. *J Am Med Inform Assoc*. 2015;22(6):1173–1178.
131. Meier L, Van De Geer S, Bühlmann P. The group lasso for logistic regression. *J R Stat Soc Ser B Stat Methodol*. 2008;70(1):53–71.
132. Ding C, Peng H. Minimum redundancy feature selection from microarray gene expression data. *J Bioinform Comput Biol*. 2005;3(02):185–205.
133. Gönen M, Alpaydın E. Multiple kernel learning algorithms. *J Mach Learn Res*. 2011;12:2211–2268.
134. Hardoon DR, Szedmak S, Shawe-Taylor J. Canonical correlation analysis: an overview with application to learning methods. *Neural Comput*. 2004;16(12):2639–2664.
135. Koller D, Friedman N. *Probabilistic Graphical Models: Principles and Techniques*. MIT Press; 2009.
136. Srivastava N, Salakhutdinov RR. Multimodal learning with deep Boltzmann machines. In: *Paper presented at: Advances in Neural Information Processing Systems*; 2012.
137. Andrew G, Arora R, Bilmes J, Livescu K. Deep canonical correlation analysis. In: *Paper presented at: International Conference on Machine Learning*; 2013.
138. Havaei M, Guizard N, Chapados N, Bengio Y. HeMIS: hetero-modal image segmentation. In: *Paper presented at: International Conference on Medical Image Computing and Computer-Assisted Intervention*; 2016.
139. Arya N, Saha S. Multi-modal advanced deep learning architectures for breast cancer survival prediction. *Knowl-Based Syst*. 2021;221.
140. De Silva T, Chew EY, Hotaling N, Cukras CA. Deep-learning based multi-modal retinal image registration for the longitudinal analysis of patients with age-related macular degeneration. *Biomed Opt Express*. 2021;12(1):619–636.
141. Islam KT, Wijewickrema S, O'Leary S. A deep learning based framework for the registration of three dimensional multi-modal medical images of the head. *Sci Rep-UK*. 2021;11(1).
142. Venugopalan J, Tong L, Hassanzadeh HR, Wang MD. Multimodal deep learning models for early detection of Alzheimer's disease stage. *Sci Rep*. 2021;11(1):3254.
143. Zhang F, Li Z, Zhang B, Dua H, Wang B, Zhang X. Multi-modal deep learning model for auxiliary diagnosis of Alzheimer's disease. *Neurocomputing*. 2019;361:185–195.
144. Lee G, Nho K, Kang B, et al. Predicting Alzheimer's disease progression using multi-modal deep learning approach. *Sci Rep-UK*. 2019;9.
145. Ribeiro MT, Singh S, Guestrin C. Why should i trust you? Explaining the predictions of any classifier. In: *Paper presented at: Proceedings of the 22nd ACM SIGKDD International Conference on Knowledge Discovery and Data Mining*; 2016.
146. Koh PW, Liang P. *Understanding Black-Box Predictions via Influence Functions. arXiv preprint arXiv:170304730*; 2017.
147. Dietterich TG. Ensemble methods in machine learning. *Multiple Classifier Syst*. 2000;1857:1–15.
148. Shen HB, Chou KC. Ensemble classifier for protein fold pattern recognition. *Bioinformatics*. 2006;22(14):1717–1722.
149. Breiman L. Bagging predictors. *Mach Learn*. 1996;24(2):123–140.

150. Friedman JH. Greedy function approximation: a gradient boosting machine. *Ann Stat.* 2001;1189–1232.
151. Akavia UD, Litvin O, Kim J, et al. An integrated approach to uncover drivers of cancer. *Cell.* 2010;143(6):1005–1017.
152. Gevaert O, Smet FD, Timmerman D, Moreau Y, Moor BD. Predicting the prognosis of breast cancer by integrating clinical and microarray data with Bayesian networks. *Bioinformatics.* 2006;22(14):e184–e190.
153. Rasiwasia N, Costa Pereira J, Coviello E, et al. A new approach to cross-modal multimedia retrieval. In: *Paper presented at: Proceedings of the 18th ACM International Conference on Multimedia*; 2010.
154. Bottou L, Peters J, Quiñonero-Candela J, et al. Counterfactual reasoning and learning systems: the example of computational advertising. *J Machine Learn Res.* 2013;14 (1):3207–3260.
155. Lam HYK, Marenco L, Clark T, et al. Research—AlzPharm: integration of neurodegeneration data using RDF. *BMC Bioinformatics.* 2007;8.
156. Sujansky W. Heterogeneous database integration in biomedicine. *J Biomed Inform.* 2001;34(4):285–298.
157. Holzinger ER, Dudek SM, Frase AT, Krauss RM, Medina MW, Ritchie MD. ATHENA: a tool for meta-dimensional analysis applied to genotypes and gene expression data to predict HDL cholesterol levels. *Pac Symp Biocomput.* 2013;385–396.
158. Fridley BL, Lund S, Jenkins GD, Wang L. A Bayesian integrative genomic model for pathway analysis of complex traits. *Genet Epidemiol.* 2012;36(4):352–359.
159. Mankoo PK, Shen R, Schultz N, Levine DA, Sander C. Time to recurrence and survival in serous ovarian tumors predicted from integrated genomic profiles. *PLoS One.* 2011;6(11):e24709.
160. Ovaska K, Laakso M, Haapa-Paananen S, et al. Large-scale data integration framework provides a comprehensive view on glioblastoma multiforme. *Genome Med.* 2010;2 (9):65.
161. Kovačević A, Dehghan A, Filannino M, Keane JA, Nenadic G. Combining rules and machine learning for extraction of temporal expressions and events from clinical narratives. *J Am Med Inform Assoc.* 2013;20(5):859–866.
162. Lanckriet GR, De Bie T, Cristianini N, Jordan MI, Noble WS. A statistical framework for genomic data fusion. *Bioinformatics.* 2004;20(16):2626–2635.
163. Daemen A, Gevaert O, De Moor B. Integration of clinical and microarray data with kernel methods. In: *Paper presented at: Engineering in Medicine and Biology Society, 2007. EMBS 2007. 29th Annual International Conference of the IEEE 2007*; 2007.
164. Kim D, Shin H, Song YS, Kim JH. Synergistic effect of different levels of genomic data for cancer clinical outcome prediction. *J Biomed Inform.* 2012;45(6):1191–1198.
165. Tsuda K, Shin H, Scholkopf B. Fast protein classification with multiple networks. *Bioinformatics.* 2005;21(Suppl 2):ii59–ii65.
166. Cheng F, Zhao Z. Machine learning-based prediction of drug–drug interactions by integrating drug phenotypic, therapeutic, chemical, and genomic properties. *J Am Med Inform Assoc.* 2014;21(e2):e278–e286.
167. Brix-Christensen V. The systemic inflammatory response after cardiac surgery with cardiopulmonary bypass in children. *Acta Anaesthesiol Scand.* 2001;45(6):671–679.
168. Leclerc F, Leteurtre S, Duhamel A, et al. Cumulative influence of organ dysfunctions and septic state on mortality of critically ill children. *Am J Resp Crit Care.* 2005;171 (4):348–353.

CHAPTER TWO

Precision diagnostics in cancer: Predict, prevent, and personalize

Judy S. Crabtree* and Lucio Miele*
Department of Genetics, Louisiana State University Health Sciences Center, New Orleans, LA, United States
*Corresponding authors: e-mail address: jcrabt@lsuhsc.edu; lmiele@lsuhsc.edu

Contents

1. Introduction 40
   1.1 Somatic oncology testing 40
   1.2 Approaches to precision oncology 41
   1.3 Whole genome sequencing (WGS) 41
   1.4 Whole exome sequencing (WES) 42
   1.5 Targeted panels 42
2. NGS-based technologies 43
   2.1 Step 1: Tumor sampling and extraction 43
   2.2 Step 2: Sequencing library generation 44
   2.3 Step 3: Sequencing 46
   2.4 Step 4: Data analysis 46
   2.5 Step 5: Variant interpretation 48
3. Therapy selection 49
   3.1 Targeted therapies 50
   3.2 Checkpoint inhibitors and immunotherapies 51
4. Clinical trial design 51
5. Conclusions 52
References 53

Abstract

In the past two decades since the first sequencing of the human genome, clinical genomics has undergone a renaissance with the advent of next generation sequencing (NGS) technologies. Whole genome sequencing is now increasingly available in the clinical setting at a fraction of the cost of the human genome reference sequence, and with turnaround times as short as 7–10 days. Clinical genetics-based management has become more tractable with improved payer reimbursement and increased availability of targeted therapies, especially in the oncology space. Precision diagnostics in cancer allow clinicians to assess a patient's risk of developing cancers, detect and classify tumors, predict prognoses, select targeted therapies when available, and monitor their patient's disease burden longitudinally. This approach ushers in a new era of predictive and preventive oncology care that can be personalized to the genetics of the individual patient, the patient's tumor, and their clinical presentation.

Progress in Molecular Biology and Translational Science, Volume 190
ISSN 1877-1173
https://doi.org/10.1016/bs.pmbts.2022.03.005

Within oncology, there are global collaborative efforts including the International Cancer Genome Consortium (ICGC), Pan-Cancer Analysis of Whole Genomes (PCAWG) and The Cancer Genome Atlas (TCGA) that have archived thousands of cancer genomes. In collaboration with databases including but not limited to cancer cataloging variant association with disease risk (ClinVar and others) and drug responses (PharmGKB and others), the study of somatic and germline genomic variation provides a wealth of information on cancer development, evolution, heterogeneity, and treatment response. Although many challenges still exist for defining all possible actionable genomic variants, we have unprecedented opportunity to gain mechanistic and clinical insights into human neoplastic diseases.

1. Introduction

1.1 Somatic oncology testing

There are many approaches to genomic testing that enable tumor biopsy, liquid biopsy, and single-cell analyses to assist in diagnosis, prognosis, and treatment selection. In cancers of unknown origin, molecular testing can offer a tissue-agnostic approach to assist in diagnosis and identifying appropriate therapies when combined with other clinical data. In one example a cancer of unknown primary origin pathologically mimicked a poorly differentiated soft tissue sarcoma, but did not respond to standard chemotherapy.[1] Molecular profiling identified the tumor as having a heterozygous KRAS G12S mutation, additional compound heterozygous mutations and deletions in TP53, high tumor mutational burden, and amplification of PD-L1, among other variants. Treatment with pembrolizumab, an immune checkpoint inhibitor with efficacy in tumors with this molecular signature, resulted in rapid clinical improvement and partial remission in this case,[1] demonstrating the power of molecular profiling in diagnosis and therapy selection.

Genomic testing has also identified morphologically identical tumors that often have variable molecular signatures.[2] For example, using transcriptomic data Sorlie, et al. identified different molecular subgroups in ER+ breast cancer including basal-like, luminal subtypes A, B and C, Her2+-enriched and normal breast-like. Similar studies have been performed in triple negative breast cancers resulting in molecular subtypes of luminal, basal-like, and claudin-low,[3] and basal-like1/2, mesenchymal, and luminal androgen receptor[4] that likely respond differently to therapies. Furthermore, genomic sequencing of an initial tumor can be used as a baseline for future comparisons upon recurrence, therapy resistance, and/or metastasis to understand the natural evolution of a tumor in response to therapy as well as to drive future drug discovery efforts. Several clinically available transcriptomic tests, such as Oncotype Dx,[5,6] PAM50[2,7] or MammaPrint/ Blueprint[8–10] offer information that is predictive of response to chemotherapy, allowing countless patients to avoid overtreatment with chemotherapy. While these tests were originally developed using older platforms such as microarray hybridization, they now use multiplex RT-PCR, nanostring or RNASeq technology.

1.2 Approaches to precision oncology

In addition to gene expression signatures, mutational profiling of solid and hematopoietic tumors can provide invaluable information for oncologists choosing treatment strategies. Such tests are now standard of care. Molecular testing in oncology includes a wide array of testing platforms from single gene assays to NGS-based approaches. Single gene analysis has been the mainstay of clinical testing for many years and encompasses PCR-based methods, Sanger sequencing, and/or fluorescent in situ hybridization (FISH). All of these single gene assays are relatively low throughput, but have faster turnaround times, lower cost, and lower complexity than NGS-based technologies. Furthermore, single gene assays are more easily customized and standardized across laboratories, and are extremely useful for querying problematic regions of the genome or variants that may be technically challenging from an NGS perspective (for example regions with high GC content, or large repeat regions). The primary disadvantage of single gene testing is that multiple single-gene queries become prohibitive with limited sample availability.

More recently, NGS-based approaches such as the Illumina TSO-500 or Foundation One CDx identify known germline and somatic variants as well as measuring tumor mutational burden (TMB) for immunotherapy guidance and markers of microsatellite instability (MSI) to identify DNA repair dysfunction. NGS-based approaches allow for simultaneous analysis of many genes in parallel, which is especially beneficial when sample is limited. Approaches may include whole genome sequencing (WGS), whole exome sequencing (WES), or targeted gene panels.

1.3 Whole genome sequencing (WGS)

Whole genome sequencing has become more tractable in the clinical setting with costs approaching \$1000 per genome and data availability in 7–10 days.

The potential advantage of WGS includes the availability of vast amounts of data to detect novel or rare variants present in any portion of the genome. However, WGS cannot reliably detect low-level variants in samples complicated by tumor heterogeneity or contaminating normal tissue. The large amount of data can also be an overwhelming disadvantage and challenges remain as to the best approaches for identifying clinically actionable variants, and distinguishing somatic mutations from germline polymorphisms. The distinction between germline and somatic variants is critical, and is best accomplished by analyzing tumor and non-tumor samples from the same patient as the recommended standard of care. If this is not an option, distinguishing between germline and somatic variants relies exclusively on database collections such as the 1000 Genomes Project, the Single Nucleotide Polymorphism Database dbSNP, the Database of Clinical Variation ClinVar, the Catalog of Somatic Mutations in Cancer (COSMIC) and others. Data interpretation based on these databases should be interpreted with caution since rare tumors are often poorly represented in these databases.

1.4 Whole exome sequencing (WES)

Whole exome sequencing is an NGS approach whereby only the 1.5% of the genome that encodes proteins and regulatory regions is analyzed. Roughly 85% of known pathogenic variants are located in the coding and regulatory regions of the genome, so this approach maximizes the amount of clinically relevant data, while generating significantly less data than WGS, an advantage from the data management perspective when linking with electronic medical records. Because only coding and regulatory regions are queried in this approach, WES fails to detect pathogenic variants resulting from intronic break points, such as structural rearrangements. Further, the depth of sequencing, while equivalent to WGS, is lower for WES than the targeted panels discussed below, and may result in missed identification of clinically actionable variants with low allele frequency or in samples with low tumor burden.

1.5 Targeted panels

Targeted panels are by far the most frequently used NGS-based approach. Laboratories performing these tests report lower sequencing costs, faster turnaround times (compared to WGS or WES), and high coverage of genomic regions of interest enabling detection of lower frequency alleles—a key component in samples with potentially mixed tumor/normal content or low tumor burden. Targeted panels return more focused data sets with lower complexity than WES or WGS facilitating variant interpretation. These panels are inherently centered on actionable variants, variants know to be predictors of recurrence, or variants associated with cancer risk and response to therapy. Targeted panels can also be optimized and designed to run with limited sample input (small biopsies, fine needle aspirates (FNA), or cell free DNA), allowing sufficient sensitivity to detect low level variations. One disadvantage is that novel, potentially informative variants in genes not included in the panels will not be detected and these test panels must be updated regularly as new clinically actionable variants emerge.

Targeted panels also offer the ability to identify patients with a germline susceptibility to cancer. For example, patients with Lynch Syndrome (also known as hereditary nonpolyposis colorectal cancer, HNPCC) develop a spectrum of cancers including endometrial and colorectal cancers as a result of germline mutations in genes involved in genome integrity such as MLH1, MSH2, MSH6 or PMS2. Other examples include screening for known cancer driver mutations such as MEN1 or BRCA1/2. Mutations in the MEN1 gene are associated with elevated risk for parathyroid, pituitary and endocrine pancreatic tumors, while variants present in BRCA1 or 2 are correlated with an increased risk for hormone-driven cancers including breast, ovarian, and prostate.

2. NGS-based technologies

The details of NGS-based technologies vary depending on the particular assay in use, but all of these technologies rely on five fundamental steps: (1) tumor sampling/enrichment of tumor cells and extraction of nucleic acid (DNA and/or RNA), (2) preparation of sequencing libraries, (3) sequencing, (4) data analysis, and (5) interpretation.

2.1 Step 1: Tumor sampling and extraction

Optimal performance of NGS-based technologies requires high quality DNA and/or RNA that is ideally obtained from fresh frozen samples when available, or from formalin-fixed paraffin embedded (FFPE) specimens. Extraction of DNA from FFPE can result in artifactual C->T mutations as a result of DNA fragmentation and crosslinking that contribute to the mutational "noise" in NGS results.[11] Intact RNA is historically challenging to isolate from FFPE, but evidence suggests this concern has been minimized with recent advances in RNA extraction technology.[12] Appropriate preservation of nucleic acids and the use of targeted panels in lieu of WES or WGS overcomes some limitations of FFPE.[11] NGS-based strategies use DNA and RNA as complementary approaches: DNA analyses can detect single nucleotide variants (SNV), insertions/deletions (indels), and copy number variants (CNVs) whereas RNA analysis detects gene fusion and changes in gene expression.

Primary sources of DNA and/or RNA include tumor biopsies at resection, FNA of solid tumors in patients with advance stage cancers for whom resection is not an option, or liquid biopsies which utilize circulating tumor DNA (ctDNA) or circulating tumor cells (CTCs) in blood. Tumor biopsies, and especially FNA, require further enrichment for tumor cells prior to sequencing library generation typically by manual dissection from FFPE tissue as guided by staining from a neighboring section stained for hematoxylin and eosin. This "macrodissection" approach is relatively low precision compared to microdissection and is most successful in cases with large regions of tumor versus surrounding normal tissue. Digitally guided or laser capture microdissection further improve precision and are more often used when sample is limited. Using CTCs as input not only provides RNA or DNA for molecular analyses, but the absolute number of CTCs is also used to monitor response to therapy and disease progression. Additionally, peripheral blood from patients with solid tumors can be analyzed to monitor disease burden using cell free DNA (cfDNA) in plasma. As mentioned previously, it is recommended to analyze tumor and non-tumor from the same patient in parallel to identify germline variants and better inform therapy selection.

2.2 Step 2: Sequencing library generation

Once nucleic acids are extracted from the sample, sequencing libraries must then be prepared by hybrid capture or PCR-based amplification of the regions of interest (Fig. 1). For RNA-based transcriptomic tests that measure the abundance of specific transcripts by NGS, a reverse transcription step to produce cDNAs must precede library generation. For DNA-based tests that detect mutations or copy number aberration, libraries are prepared directly from genomic DNA. For large-scale whole exome and panel sequencing, most laboratories utilize hybrid capture strategies for target enrichment. This involves fragmentation of nucleic acids by sonication, nebulization, transposase/tagmentation, or chemical means followed by hybridization to complementary oligonucleotides either in solution or on an array. Tagmentation refers to the transposome-based initial step in the Illumina library generation process that allows for selection of fragments of optimal

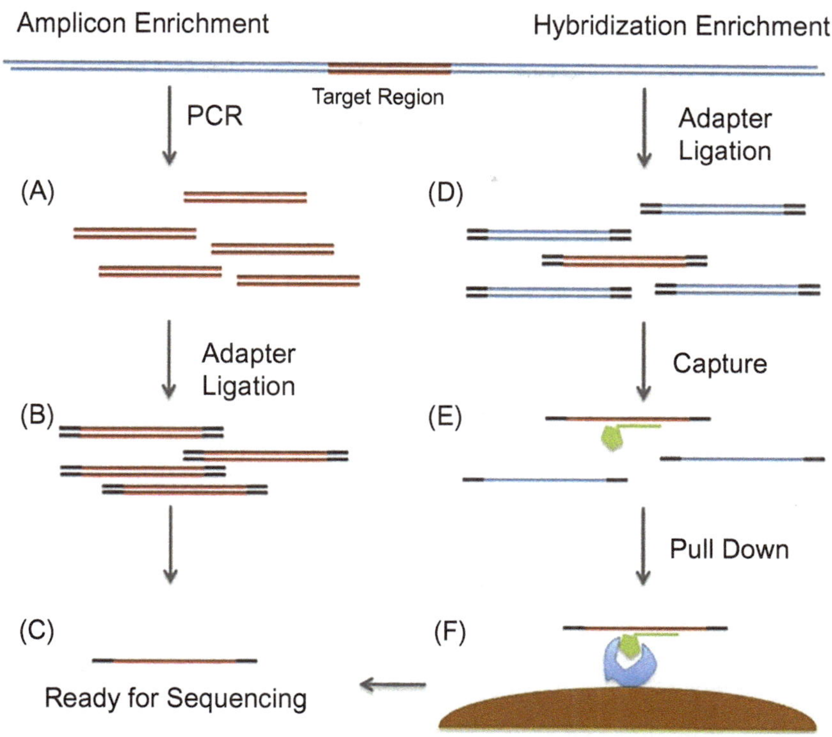

Fig. 1 A comparison of two popular sequence enrichment methods. (A) For amplicon enrichment, PCR primers specific to the region of interest are used to amplify the target area. (B) These PCR products are then prepared for sequencing via ligation with sequencer-specific DNA molecules (adapters). (C) Molecules are then ready for sequencing. (D) For hybridization enrichment, the entire genome is sheared into small fragments which are subsequently ligated to sequencer-specific adapter DNA molecules. (E) Biotinylated oligomers that have been designed to be complementary to the region of interest are incubated with the previously generated sequencing library. (F) Captured molecules from the region of interest are pulled down using streptavidin-coated magnetic beads. DNA molecules are then eluted and ready for sequencing (C). *Reprinted with permission from Wiley GB, Kelly JA, Gaffney PM. Use of next-generation DNA sequencing to analyze genetic variants in rheumatic disease. Arthritis Res Ther (2014);16(6):490. doi:10.1186/s13075-014-0490-4.*

size for sequencing.[13] Once samples are enriched for the targets of interest, molecular barcodes and sequencing adapters are ligated to the fragments in order to identify individual samples and facilitate adherence to beads or a flow cell for direct sequencing.

Amplification methods utilize a set of primers to enrich specific regions of the genome and can use as little as 10 ng DNA or RNA, but there is the risk of a variant or deletion being present under a primer binding site leading to a false negative results. Gene fusions can be detected with specific primer sets developed for each fusion, but novel fusions will not be detected. Alternatively, hybrid capture-based libraries are able to detect fusions since only one of the two genes in a fusion needs to anneal to a capture probe for enrichment and detection.

2.3 Step 3: Sequencing

Two major chemistries exist for NGS-based sequencing (Fig. 2). The first is the Illumina "sequencing by synthesis" platform[14] that utilizes a flow cell on which each individual library fragment is amplified into a cluster using solid phase bridge amplification. Sequencing occurs as fluorescently-labeled nucleotides are incorporated and detected by imaging the flow cell at each round of chain extension. Fluorescent emissions from each cluster are identified by wavelength and result in a called nucleotide. The cycle of incorporation, imaging, and quenching is repeated to generate millions of short DNA sequences that are aligned in the data analysis phase of the process. Incorrect substitutions (most particularly A > C errors) are more prevalent than indel errors in the Illumina platform.[15]

The second platform frequently used is ThermoFisher's Ion Torrent pH-sensing technology.[16] In this platform, each of the fragments from the library generation phase is amplified on an individual bead. The beads are then placed into individual wells of the detection plate, with one bead per well. The bead-containing wells are then flooded with each nucleotide in turn. Incorporation of a nucleotide releases hydrogen ions (due to the release of pyrophosphate from the nucleotide triphosphate substrate of DNA polymerase) and changes the pH of the well. This change in voltage is recorded and converted to a nucleotide base call. Repeated cycles of this process generate millions of short DNA sequences for alignment in the data analysis phase. Ion Torrent technology works equivalently well as the Illumina platform in terms of accuracy, but the Ion Torrent does tend to exhibit a higher error rate in sequences with long stretches of homopolynucleotides.[17] Recent improvements in base-calling algorithms have begun to abrogate these concerns.[18]

2.4 Step 4: Data analysis

The large amounts of raw data must be converted into an informative format to support clinical decisions. Both of the platforms identified above contain associated bioinformatics workflows for quality assurance, quality control

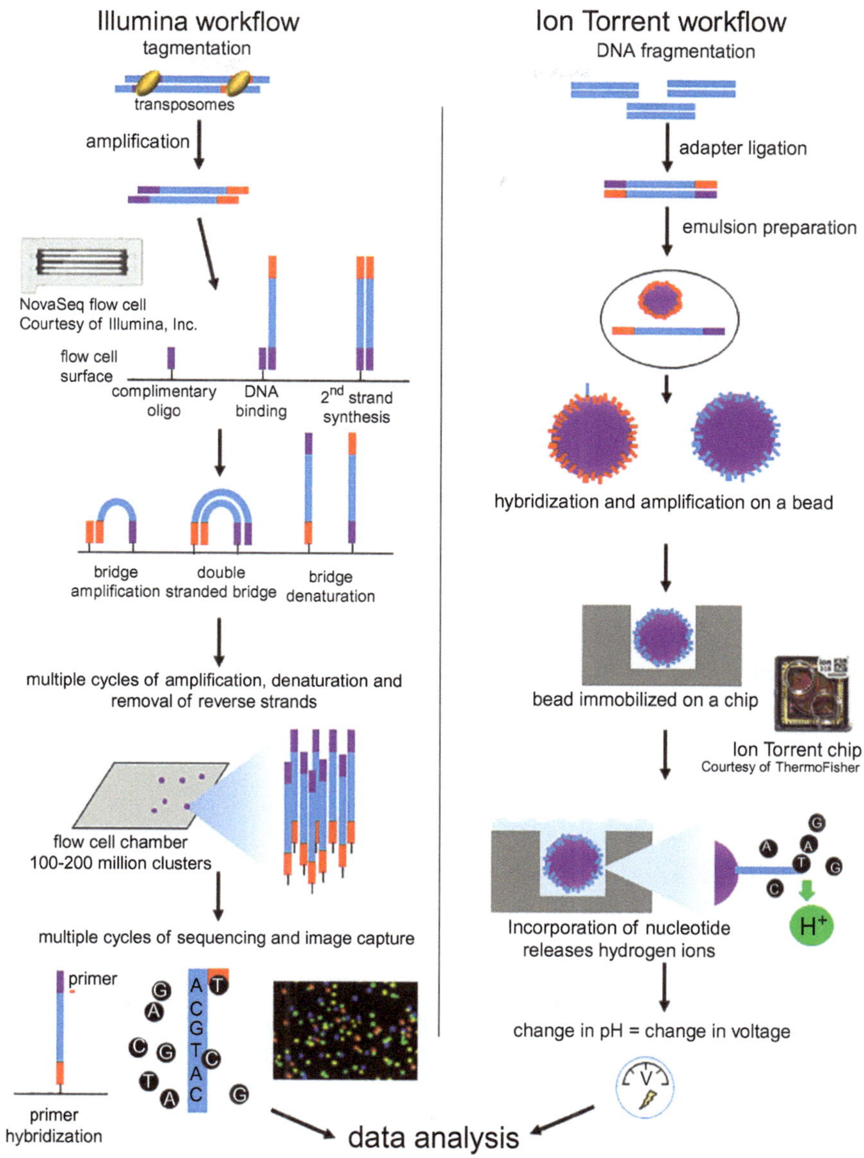

Fig. 2 A comparison of two popular NGS sequencing platforms. The *left side* is the Illumina platform for next generation sequencing that utilizes transposomes—a combination of transposases and accessory proteins on a proprietary bead. The bead-linked transposomes help generate even insert sizes of ~350 bp. Once the fragments are generated, adapters are added to allow binding to the flow cell and bridge amplification generates clusters on the flow cell—each cluster corresponding to one fragment. Each flow cell, depending on the size and instrument in use, can accommodate

(Continued)

and base calling algorithms to convert raw image/pH data into a text file format called FASTQ. This process often assigns a Phred quality score (Q score) to each base as a measure of error probability.[19] Following base calling, the sequencing reads are aligned with the reference sequence of the human genome, resulting in a binary alignment map, or a BAM file. This mapping process is critical as misalignments/mapping errors can result in missed variant identification, or the false identification of artifacts as variants. Incorrect mapping of DNA fragments can mask the presence of indels, translocations, or variants present in genes located in multiple regions of the genome, while RNA sequence mapping has the additional challenges of identifying reads that span splice junctions, or gene fusions.[20–23] Mapped sequences are then used to identify variants and result in a variant call format or VCF file. Identified SNVs are evaluated by multiple criteria to ensure accurate positive calls. These criteria are based on variant allele frequency, quality scores, strand bias, read number, sequencing depth in the gene region, read length, and the percentage of base calls greater than Q20. Indels are identified using one of a number of algorithms such as Pindel, or the Genome Analysis Toolkit Suite.[24,25] Identified variants are annotated as to the gene and location (aVCF files) and are often manually curated to ensure accuracy, quality, and adherence to nomenclature and naming conventions.

2.5 Step 5: Variant interpretation

Variants have been identified by the MD Anderson Cancer Center's Knowledge Base for Precision Oncology to fall on a spectrum that includes benign/likely benign, variants of unknown significance (VUS), likely pathogenic, and pathogenic based on the availability of supporting clinical evidence.[26] The majority of variants identified fall into the category of VUS because there is simply not enough data yet to understand the role of the

Fig. 2—Cont'd up to 200 million clusters. Subsequent cycles of sequencing and image capture result in the raw data available for data analysis. The *right side* of the figure is the ThermoFisher Ion Torrent workflow that uses an emulsion technology to amplify one fragment per bead, with the binding to beads facilitated by adapters. The microwells of the Ion Torrent chip can accommodate only one bead/well. As the sequencing cycles progress, the sensor located at the bottom of the well converts the change in pH to a voltage signal proportional to the number of incorporated bases. *Adapted from Fig. 2 in Ballard D, Winkler-Galicki J, Wesoły J. Massive parallel sequencing in forensics: advantages, issues, technicalities, and prospects. Int J Leg Med (2020);134 (4):1291–1303. doi:10.1007/s00414-020-02294-0*

detected variant in human health, i.e., the variant is not observed at significant allele frequency in the general population and no convincing evidence of cancer association exists.

Clinically relevant variants are classified according to the availability of well-powered, supporting clinical evidence, FDA-approved therapies, and guidance for patient management from professional clinical societies. To this end, variants are classified as Tier I, II, III or IV with each tier having different therapeutic, diagnostic and prognostic significance. Tier I includes variants with strong clinical significance that are supported by published professional guidance, FDA approved therapies, and large, well-powered studies. Tier II variants have potential clinical significance and are supported by smaller clinical trials with some consensus among researchers and practitioners, with the limited availability of experimental therapies. Tier III variants are VUS that, as discussed above, are not observed at a significant allele frequency in the general population, and lack studies conclusively indicating a link to disease. Tier IV variants are benign or likely benign variants with no significant correlation to disease and having significant allele frequency present in the general population.[27] Distillation of the many identified variants into a clinically useful and relevant form requires generation and careful maintenance of a database for interpretation. These extensive databases contain current research on identified variants, supporting evidence of cancer association, clinical trial availability, successful therapeutic management, and patient outcome information all of which supports an iterative algorithm to prioritize variants most critical to patient care. Several commercial groups offer assistance to laboratories in the interpretation of variants and the generation of clinical reports. Obviously, the quality of the information in the report is directly correlated to the quality of the knowledge base on which it is founded, including linked evidence to recommended therapies and available clinical trials in the geographic vicinity of the patient. The most relevant clinical information is displayed on the face page of the report, with additional details available in subsequent pages for use as needed by pathology and oncology specialists, and molecular tumor boards.

3. Therapy selection

The four main uses of genomic tests in oncology include: 1. Risk stratification (e.g., the transcriptomic tests predicting risk of recurrence and therefore the need for chemotherapy in breast cancer and other solid tumors, or the detection of ESR1 mutations that predict endocrine resistance); 2.

Choice of targeted therapy, with agents shown to be effective in tumors carrying specific mutations or gene rearrangements; 3. Determination of eligibility for immunotherapy based on tumor mutational load; and 4. Determination of clinical trial eligibility for investigational agents in basket or umbrella trials.[28]

3.1 Targeted therapies

Molecular data is now used to screen patients prior to therapy selection since many pathologically identical tumors have very different molecular signatures. Further, within the same tumor there may be intratumoral genetic heterogeneity, and tumors with different tissue origins may carry identical genetic alterations. In 2017, pembrolizumab was granted accelerated approval for ANY solid tumor with high microsatellite instability (MSI-H) or deficient mismatch repair (dMMR).[29] This landmark decision was the first approval based on a molecular signature and not on tissue/location-based tumor type. Now almost a decade later, implementation of molecular screening and targeted therapies is standard of care for certain tumor types, with promising research for future targeted therapies ongoing. For example, the recent accelerated approval of mobocertinib in patients with EGFR exon 20 insertion mutations in patients with non-small cell lung cancer (NSCLC) that progressed on platinum-based therapy, or infigratinib for patients with fibroblast growth receptor 2 (FGFR2) gene fusions or other rearrangements in cholangiocarcinoma.[30] Selpercatinib (LOXO-292), a specific RET receptor tyrosine kinase inhibitor, targets the RET V803M/L gatekeeper mutation and based on the results of the LIBRETTO-001 trial (NCT03157128) was approved by the US FDA in May 2020 for patients with RET-altered NSCLC and thyroid cancers.[31–33] Finally, sotorasib (AMG-510), the first targeted KRAS G12C inhibitor, showed efficacy in patients with KRAS G12C driver mutations.[34,35] As a result of the CodeBreaK100 trial (NCT03600883), sotorasib was granted accelerated approval by the US FDA in May 2021 for the treatment of KRAS G12C-driven NSCLC.[36] While resistance to therapy remains a concern, longitudinal disease monitoring before and after first line therapy often detects therapy-acquired mutations that allow for specificity in second and third line targeted therapies. Further, the ability to screen prior to therapy selection provides justification for the use of FDA approved medications for off label indications. For example, RET rearrangements have been identified in breast cancers and treatment with cabozantinib which is FDA approved only for locally advanced or metastatic differentiated thyroid cancer, demonstrated both radiographic and clinical responses.[37]

3.2 Checkpoint inhibitors and immunotherapies

Many targeted oncology panels include markers indicative of tumor mutational burden (TMB) and microsatellite instability (MSI). These results are often extremely informative for immunotherapies such as anti-programmed cell death ligand 1 (anti-PDL1; pembrolizumab or nivolumab) and anti-cytotoxic T lymphocyte-associated protein – 4 (anti-CTLA-4; ipilimumab). Cancers with high TMB and/or MSI are predicted to have strong, durable responses to these immune-enhancing therapies. Presumably, the basis for the clinical usefulness of TMB is that tumors with high mutational loads are more likely to express neoantigens resulting from mutations in protein-coding genes. At least some of these neoantigens can be presented to the patient's immune system and trigger T- (and sometimes B-) cell responses. Unfortunately, the unaided immune system is usually unable to eradicate tumors even in the presence of anti-tumor immune responses. This is due to "immune editing,"[38] the process whereby tumor cells decrease their immunogenicity and acquire the ability to produce immune-suppressive mediators that directly or indirectly dampen the immune response against cancer cells. A vast array of approved and investigational immunotherapeutics aim to stimulate cancer immune responses.[39–41] The largest class of standard of care agents at the time of this writing includes immune checkpoint inhibitors. These are generally engineered monoclonal antibodies that block ligand-receptor interactions on T-cells responsible for termination of T-cell activation.[41–43] These agents non-specifically enhance CD4 and CD8 T-cell activation, resulting in prolonged and intensified immune responses. In cancers that carry numerous neoantigens, this can result in complete remissions. Squamous NSCLC, a cancer previously incurable, has shown remarkable responses with checkpoint inhibitors.[44,45] These cancers usually carry significant TMBs.

4. Clinical trial design

The advent of molecular oncology and targeted therapies has required a necessary adjustment in clinical trial design. Recent years have seen a significant shift in the design of master protocols, now consisting primarily of basket and umbrella trials, with enrollment based on genetic variants present in the tumors instead of tumor location/tissue site of origin. Enrollees in umbrella trials are recruited based on tumor type, then following WES of tumor and germline, are assigned to a trial group for targeted therapies that match the genetic alterations in individual tumors. The goal of umbrella trials is to identify the subset of mutations and matched therapies that have efficacy within a particular cancer subtype. Basket trials, on the other hand, recruit patients with tumors from multiple different organ and tissue sites of origin, but which all contain a common oncogenic driver mutation. These patients are enrolled and treated with therapy targeted to the driver mutation with the goal of determining whether targeted therapy has efficacy across a wide range of tumor types based on the driver mutation.

To support the revised trial designs, several platforms are now available to match patients with clinical trials based on their unique genetic signature. The Molecular Analysis for Therapy Choice (NCI-MATCH) by the National Cancer Institute,[46] and the Targeted Agent and Profiling Utilization Registry (TAPUR) trial conducted by the American Society of Clinical Oncology (ASCO)[47] are two such platforms. The goal of NCI-MATCH is to determine the objective response rates and length of progression-free survival in patients with any advanced solid tumor, lymphoma or myeloma that has progressed on standard of care therapy, or patients with rare cancers for which there is no standard of care.[46] The TAPUR trial has similar goals to collect information on trial efficacy, particularly for drugs chosen and used off-label.

5. Conclusions

Molecular profiling has changed the practice of oncology, and it will continue to do so for the foreseeable future. Precision oncology is no longer a distant promise but a growing reality. That said, there remain multiple challenges that will require additional research as well as policy adjustments. From a scientific standpoint, tumor heterogeneity and clonal evolution are a major obstacle to molecularly driven targeted therapeutics. It is now well established that many advanced human tumors contain a mixture of different clones with different driver mutations. Targeted therapies can potentially eradicate clones carrying the targeted mutation only to allow recurrences fueled by clones driven by different mutations. This implies that basing therapeutic strategies exclusively on a single molecular test performed at diagnosis runs the risk of "treating yesterday's tumor". Non-targeted immunotherapy does not suffer from this limitation, as it doesn't require a single mutant target. However, clonal evolution can nevertheless result in loss of immunogenicity and immune-resistant relapses.[48,49] Longitudinal follow-up with repeat testing is likely to become routine, and single-cell sequencing may be necessary to better define the clonal architecture of cancers and their ranges of targetable mutations. This raises another practical problem, namely, the accessibility of tumor tissue for repeated biopsies, and the associated discomfort and risk for patients. Liquid biopsies, i.e., analysis of nucleic acids in biological fluids and/or in exosomes can potentially solve this problem. However, the amount of nucleic acids released by different tumors depends on tumor burden and anatomical location. Increasingly sensitive methods for isolation and enrichment of CTC, tumor-derived exosomes, and other vesicles will be necessary to minimize repeat biopsies.

It must be remembered that not all targetable mutations are associated with an approved drug. However, recent successes are encouraging. Targets that were previously thought to be "undruggable" (c-Myc, K-Ras, Bcl-2[50–52]) have been successfully targeted. The future of cancer targeted drug development (and the clinical trials resulting from such efforts) will by necessity have to be driven by genomics.

Finally, insurance coverage for genomic testing (and in the future, for other "omics") as well as potentially coverage for off-label use of therapeutics in cancers shown by molecular tests to possess a validated target must be expanded. The cost of such coverage will be more than balanced by the savings associated with avoidance of overtreatment with ineffective medications, avoidance of treatment for the side effects of toxic chemotherapy, prevention of unnecessary loss of life, and increased productive years. As the science progresses and cancer care becomes increasingly guided by molecular evidence, health care policy must follow the evidence if we are to reap the full benefits of precision oncology.

References

1. Groschel S, Bommer M, Hutter B, et al. Integration of genomics and histology revises diagnosis and enables effective therapy of refractory cancer of unknown primary with PDL1 amplification. *Cold Spring Harb Mol Case Stud.* 2016;2(6), a001180. https://doi.org/10.1101/mcs.a001180.
2. Sorlie T, Perou CM, Tibshirani R, et al. Gene expression patterns of breast carcinomas distinguish tumor subclasses with clinical implications. *Proc Natl Acad Sci U S A.* 2001;98 (19):10869–10874. https://doi.org/10.1073/pnas.191367098.
3. Sakach E, O'Regan R, Meisel J, Li X. Molecular classification of triple negative breast cancer and the emergence of targeted therapies. *Clin Breast Cancer.* 2021;21(6):509–520. https://doi.org/10.1016/j.clbc.2021.09.003.
4. Lehmann BD, Colaprico A, Silva TC, et al. Multi-omics analysis identifies therapeutic vulnerabilities in triple-negative breast cancer subtypes. *Nat Commun.* 2021;12(1):6276. https://doi.org/10.1038/s41467-021-26502-6.
5. Nguyen TTA, Postlewait LM, Zhang C, et al. Utility of Oncotype DX score in clinical management for T1 estrogen receptor positive, HER2 negative, and lymph node negative breast cancer. *Breast Cancer Res Treat.* 2022. https://doi.org/10.1007/s10549-022-06530-6.

6. Tesch ME, Speers C, Diocee RM, et al. Impact of TAILORx on chemotherapy prescribing and 21-gene recurrence score-guided treatment costs in a population-based cohort of patients with breast cancer. *Cancer.* 2022;128(4):665–674. https://doi.org/10.1002/cncr.33982.
7. Perou CM, Sørlie T, Eisen MB, et al. Molecular portraits of human breast tumours. *Nature.* 2000;406(6797):747–752. https://doi.org/10.1038/35021093.
8. van't Veer LJ, Dai H, van de Vijver MJ. Gene expression profiling predicts clinical outcome of breast cancer. *Nature.* 2002;415:530–536.
9. Rutgers E, Piccart-Gebhart MJ, Bogaerts J. The EORTC 10041/BIG 03-04 MINDACT trial is feasible: results of the pilot phase. *Eur J Cancer.* 2011;47:2742–2749. https://doi.org/10.1016/j.ejca.2011.09.016.
10. Cardoso F, Piccart-Gebhart M, van't Veer L, Rutgers E. The MINDACT trial: the first prospective clinical validation of a genomic tool. *Mol Oncol.* 2007;1:246–251. https://doi.org/10.1016/j.molonc.2007.10.004.
11. Do H, Dobrovic A. Sequence artifacts in DNA from formalin-fixed tissues: causes and strategies for minimization. *Clin Chem.* 2015;61(1):64–71. https://doi.org/10.1373/clinchem.2014.223040.
12. Lien TG, Ohnstad HO, Lingjaerde OC, et al. Sample preparation approach influences PAM50 risk of recurrence score in early breast cancer. *Cancers (Basel).* 2021;13(23). https://doi.org/10.3390/cancers13236118.
13. Illumina. *Tagmentation.* Accessed March 7, 2022. https://www.illumina.com/techniques/sequencing/ngs-library-prep/tagmentation.html#:~:text=Tagmentation%20is%20the%20initial%20step,to%20in%2Dsolution%20tagmentation%20reactions.
14. Illumina. *Illumina Sequencing by Synthesis Video*; 2022. https://www.youtube.com/watch?v=fCd6B5HRaZ8. Accessed 7 March 2022.
15. Minoche AE, Dohm JC, Himmelbauer H. Evaluation of genomic high-throughput sequencing data generated on Illumina HiSeq and genome analyzer systems. *Genome Biol.* 2011;12(11), R112. https://doi.org/10.1186/gb-2011-12-11-r112.
16. ThermoFisher. *Ion Torrent Next-Generation Sequencing*; 2022. https://www.youtube.com/watch?v=zBPKj0mMcDg. Accessed 7 March 2022.
17. Huse SM, Huber JA, Morrison HG, Sogin ML, Welch DM. Accuracy and quality of massively parallel DNA pyrosequencing. *Genome Biol.* 2007;8(7):R143. https://doi.org/10.1186/gb-2007-8-7-r143.
18. Wommack KE, Bhavsar J, Ravel J. Metagenomics: read length matters. *Appl Environ Microbiol.* 2008;74(5):1453–1463. https://doi.org/10.1128/AEM.02181-07.
19. Ewing B, Hillier L, Wendl MC, Green P. Base-calling of automated sequencer traces using phred. I Accuracy assessment. *Genome Res.* 1998;8(3):175–185. https://doi.org/10.1101/gr.8.3.175.
20. Grant GR, Farkas MH, Pizarro AD, et al. Comparative analysis of RNA-Seq alignment algorithms and the RNA-Seq unified mapper (RUM). *Bioinformatics.* 2011;27(18):2518–2528. https://doi.org/10.1093/bioinformatics/btr427.
21. Kim D, Salzberg SL. TopHat-fusion: an algorithm for discovery of novel fusion transcripts. *Genome Biol.* 2011;12(8), R72. https://doi.org/10.1186/gb-2011-12-8-r72.
22. Kinsella M, Harismendy O, Nakano M, Frazer KA, Bafna V. Sensitive gene fusion detection using ambiguously mapping RNA-Seq read pairs. *Bioinformatics.* 2011;27(8):1068–1075. https://doi.org/10.1093/bioinformatics/btr085.
23. McPherson A, Hormozdiari F, Zayed A, et al. deFuse: an algorithm for gene fusion discovery in tumor RNA-Seq data. *PLoS Comput Biol.* 2011;7(5), e1001138. https://doi.org/10.1371/journal.pcbi.1001138.
24. Ghoneim DH, Myers JR, Tuttle E, Paciorkowski AR. Comparison of insertion/deletion calling algorithms on human next-generation sequencing data. *BMC Res Notes.* 2014;7:864. https://doi.org/10.1186/1756-0500-7-864.

25. Ye K, Schulz MH, Long Q, Apweiler R, Ning Z. Pindel: a pattern growth approach to detect break points of large deletions and medium sized insertions from paired-end short reads. *Bioinformatics*. 2009;25(21):2865–2871. https://doi.org/10.1093/bioinformatics/btp394.
26. Center MAC. *Personalized Cancer Therapy: knowledge Base for Precision Oncology. MD Anderson Cancer Center*; 2022. https://pct.mdanderson.org. Accessed 5 February 2022.
27. Li MM, Datto M, Duncavage EJ, et al. Standards and guidelines for the interpretation and reporting of sequence variants in cancer: a joint consensus recommendation of the Association for Molecular Pathology, American Society of Clinical Oncology, and College of American Pathologists. *J Mol Diagn*. 2017;19(1):4–23. https://doi.org/10.1016/j.jmoldx.2016.10.002.
28. Park JJH, Hsu G, Siden EG, Thorlund K, Mills EJ. An overview of precision oncology basket and umbrella trials for clinicians. *CA Cancer J Clin*. 2020;70(2):125–137. https://doi.org/10.3322/caac.21600.
29. FDA. *FDA Grants Accelerated Approval to Pembrolizumab for First Tissue/Site Agnostic Indication*; 2022. https://www.fda.gov/drugs/resources-information-approved-drugs/fda-grants-accelerated-approval-pembrolizumab-first-tissuesite-agnostic-indication. Accessed 7 March 2022.
30. FDA. *Ongoing Cancer Accelerated Approvals*; 2022. https://www.fda.gov/drugs/resources-information-approved-drugs/ongoing-cancer-accelerated-approvals. Accessed 6 February 2022.
31. Drilon A, Oxnard GR, Tan DSW, et al. Efficacy of Selpercatinib in RET fusion-positive non-small-cell lung cancer. *N Engl J Med*. 2020;383(9):813–824. https://doi.org/10.1056/NEJMoa2005653.
32. Wirth LJ, Sherman E, Robinson B, et al. Efficacy of Selpercatinib in RET-altered thyroid cancers. *N Engl J Med*. 2020;383(9):825–835. https://doi.org/10.1056/NEJMoa2005651.
33. Thein KZ, Velcheti V, Mooers BHM, Wu J, Subbiah V. Precision therapy for RET-altered cancers with RET inhibitors. *Trends Cancer*. 2021;7(12):1074–1088. https://doi.org/10.1016/j.trecan.2021.07.003.
34. Hong DS, Fakih MG, Strickler JH, et al. KRAS(G12C) inhibition with Sotorasib in advanced solid tumors. *N Engl J Med*. 2020;383(13):1207–1217. https://doi.org/10.1056/NEJMoa1917239.
35. Skoulidis F, Li BT, Dy GK, et al. Sotorasib for lung cancers with KRAS p.G12C mutation. *N Engl J Med*. 2021;384(25):2371–2381. https://doi.org/10.1056/NEJMoa2103695.
36. Nakajima EC, Drezner N, Li X, et al. FDA approval summary: sotorasib for KRAS G12C-mutated metastatic NSCLC. *Clin Cancer Res*. 2021. https://doi.org/10.1158/1078-0432.Ccr-21-3074.
37. Paratala BS, Chung JH, Williams CB, et al. RET rearrangements are actionable alterations in breast cancer. *Nat Commun*. 2018;9(1):4821. https://doi.org/10.1038/s41467-018-07341-4.
38. Shaaban M, Othman H, Ibrahim T, et al. Immune checkpoint regulators: a new era toward promising cancer therapy. *Curr Cancer Drug Targets*. 2020;20(6):429–460. https://doi.org/10.2174/1568009620666200422081912.
39. Aghajanian H, Rurik JG, Epstein JA. CAR-based therapies: opportunities for immuno-medicine beyond cancer. *Nat Metab*. 2022;4(2):163–169. https://doi.org/10.1038/s42255-022-00537-5.
40. Sun Q, Melino G, Amelio I, Jiang J, Wang Y, Shi Y. Recent advances in cancer immunotherapy. *Discov Oncol*. 2021;12(1):27. https://doi.org/10.1007/s12672-021-00422-9.
41. Walia HK, Sharma P, Singh N, Sharma S. Immunotherapy in small cell lung cancer treatment: a promising headway for future perspective. *Curr Treat Options Oncol*. 2022. https://doi.org/10.1007/s11864-022-00949-1.

42. Sochacka-Ćwikła A, Mączyński M, Regiec A. FDA-approved drugs for hematological malignancies-the last decade review. *Cancers (Basel)*. 2021;14(1). https://doi.org/10.3390/cancers14010087.
43. Chen BJ, Zhao JW, Zhang DH, Zheng AH, Wu GQ. Immunotherapy of cancer by targeting regulatory T cells. *Int Immunopharmacol*. 2022;104, 108469. https://doi.org/10.1016/j.intimp.2021.108469.
44. Mamdani H, Matosevic S, Khalid AB, Durm G, Jalal SI. Immunotherapy in lung cancer: current landscape and future directions. *Front Immunol*. 2022;13, 823618. https://doi.org/10.3389/fimmu.2022.823618.
45. Seegobin K, Majeed U, Wiest N, Manochakian R, Lou Y, Zhao Y. Immunotherapy in non-small cell lung cancer with actionable mutations other than EGFR. *Front Oncol*. 2021;11, 750657. https://doi.org/10.3389/fonc.2021.750657.
46. NCI. *NCI-MATCH (Molecular Analysis for Therapy Choice)*; 2022. http://www.cancer.gov/about-cancer/treatment/clinical-trials/nci-supported/nci-match/. Accessed 6 February 2022.
47. ASCO. *Targeted Agent and Profiling Utilization Registry (TAPUR) Study*; 2022. https://www.asco.org/research-data/tapur-study. Accessed 6 February 2022.
48. Kwok M, Wu CJ. Clonal evolution of high-risk chronic lymphocytic leukemia: a contemporary perspective. *Front Oncol*. 2021;11, 790004. https://doi.org/10.3389/fonc.2021.790004.
49. Wang S, Xie K, Liu T. Cancer immunotherapies: from efficacy to resistance mechanisms - not only checkpoint matters. *Front Immunol*. 2021;12, 690112. https://doi.org/10.3389/fimmu.2021.690112.
50. Madden SK, de Araujo AD, Gerhardt M, Fairlie DP, Mason JM. Taking the Myc out of cancer: toward therapeutic strategies to directly inhibit c-Myc. *Mol Cancer*. 2021;20(1):3. https://doi.org/10.1186/s12943-020-01291-6.
51. Uprety D, Adjei AA. KRAS: from undruggable to a druggable cancer target. *Cancer Treat Rev*. 2020;89, 102070. https://doi.org/10.1016/j.ctrv.2020.102070.
52. Zhang Z, Bai L, Hou L, et al. Trends in targeting Bcl-2 anti-apoptotic proteins for cancer treatment. *Eur J Med Chem*. 2022;232, 114184. https://doi.org/10.1016/j.ejmech.2022.114184.

CHAPTER THREE

Artificial intelligence and machine learning in precision medicine: A paradigm shift in big data analysis

Mehar Sahu†, Rohan Gupta†, Rashmi K. Ambasta, and Pravir Kumar*

Molecular Neuroscience and Functional Genomics Laboratory, Delhi Technological University (Formerly Delhi College of Engineering), Shahbad Daulatpur, Delhi, India
*Corresponding author: e-mail addresses: pravirkumar@dtu.ac.in; kpravir@gmail.com

Contents

1. Introduction 58
2. Artificial intelligence and big data: Changes in the landscape of precision healthcare and medicine 61
3. The emergence of AI in precision medicine 63
4. Synergies between artificial intelligence and precision medicine 64
   4.1 Therapy planning 64
   4.2 Risk prediction and diagnosis 67
5. Promises of implementing AI/ML in healthcare 69
   5.1 Solving the human resource crisis in healthcare 69
   5.2 Improving workflow and reducing medical errors 70
   5.3 Enhancing healthcare delivery 71
6. The integration of nanomaterials in AI-based precision medicine 73
7. The perspective of big data analytics in precision medicine 75
8. AI and personalized medicine: A case of neurological disease 76
9. Obstacles in translating big data into healthcare 79
10. Opportunities for translating big data into healthcare 81
11. Artificial intelligence and machine learning in precision medicine: An industry perspective 82
    11.1 Multi-omics profiling 82
    11.2 Digital biomarkers and biomarker-based clinical trials 83
    11.3 Precision dosing and safety measures 85
12. Future challenges and perspectives of artificial intelligence in precision medicine 87

† Equal contribution.

Progress in Molecular Biology and Translational Science, Volume 190
ISSN 1877-1173
https://doi.org/10.1016/bs.pmbts.2022.03.002

Conflict of interest 89
Acknowledgment 89
Author's contribution 89
References 89

Abstract

The integration of artificial intelligence in precision medicine has revolutionized healthcare delivery. Precision medicine identifies the phenotype of particular patients with less-common responses to treatment. Recent studies have demonstrated that translational research exploring the convergence between artificial intelligence and precision medicine will help solve the most difficult challenges facing precision medicine. Here, we discuss different aspects of artificial intelligence in precision medicine that improve healthcare delivery. First, we discuss how artificial intelligence changes the landscape of precision medicine and the evolution of artificial intelligence in precision medicine. Second, we highlight the synergies between artificial intelligence and precision medicine and promises of artificial intelligence and precision medicine in healthcare delivery. Third, we briefly explain the promise of big data analytics and the integration of nanomaterials in precision medicine. Last, we highlight the challenges and opportunities of artificial intelligence in precision medicine.

Highlights

- Therapy planning, risk prediction, and diagnostic approach are three constituents of AI in precision medicine.
- AI improves healthcare delivery and reduces medical errors.
- Nanomaterials enhance AI-based precision medicine.
- Data handling and replication complications are the limitations of AI-based precision medicine.
- Electronic health records and imaging analysis are opportunities for AI-based precision medicine.

1. Introduction

The implementation of artificial intelligence (AI) algorithms in the field of drug discovery and design offers tremendous opportunities. AI and machine learning (ML) enhance drug design and development by improving our knowledge of disease pathology, identifying dysregulated molecular pathways, predicting novel therapeutic targets, and analyzing *in silico* clinical efficacy of drugs.[1] Through its enhancement of knowledge of both patient specificity and drug properties, AI and ML enable the design and prevention of disease at the individual patient level. Thus, AI and

ML algorithms are modernizing several aspects of our lives through their involvement in precision medicine, including medical imaging, risk analysis, phenotype prediction, and analysis of gene expression patterns.[2] The term "precision medicine" refers to the identification of a subpopulation within a large population, which is different from their disease risk, prognosis, detection, and response to treatment due to differences in their habitat, physiology, and other characteristics.[3] The emphasis of personalized medicine is placed on tailored prevention, diagnosis, and treatment for each individual based on individual genetics, phenotype, epigenetics, and lifestyle.[4] Deep phenotyping of every individual is necessary before interventions can be applied. The availability of multimodal datasets could improve the our ability to find disease phenotypic and genotypic characteristics, which help predict dose-response, risk prediction, prognosis, treatment response, and patient outcomes.[5–8] Recently, new AI/ML algorithms for precision medicine for neurological disease, cancer, and cardiovascular disease have shown promising results with respect to disease risk prediction, phenotypic prediction, dose-treatment response, with a high degree of precision and accuracy.[9–14] Further, an AI-based decision-making process allows analysis of three different factors of big data, multimodal examination, scientific knowledge, and decision support systems. In addition, AI/ML-based precision medicine enables the identification of different single-cell parameters, namely genomic/epigenomic profiling, proteomics signature analysis, immune markers profiling, and morphology studies, which guide selection of target-based medicines. However, a significant drawback of computational precision medicine is the requirement of advance computing systems to process multimodal datasets. Volume, velocity, variety, veracity, variability, and value are the "V's" of big data, and these are the core challenges and hurdles of effective and efficient biomedical and health informatics.[15] In summary, AI, ML, precision medicine, and big data together enhance healthcare delivery through their high degree of precision and accuracy (Fig. 1).

Here, we discuss the role of AI and big data in changing the landscape of precision medicine, the potential synergies between precision medicine and AI, and We also explain the recent applications and future prospectives that AI and big data holds for precision medicine. As an exemplar of the AI/ML/precision medicine approach, we examine its role in the context of neurological diseases and consider obstacles in translating big data into effective healthcare. Lastly, we shed light on the role of AI/ML and precision medicine from an industry perspective.

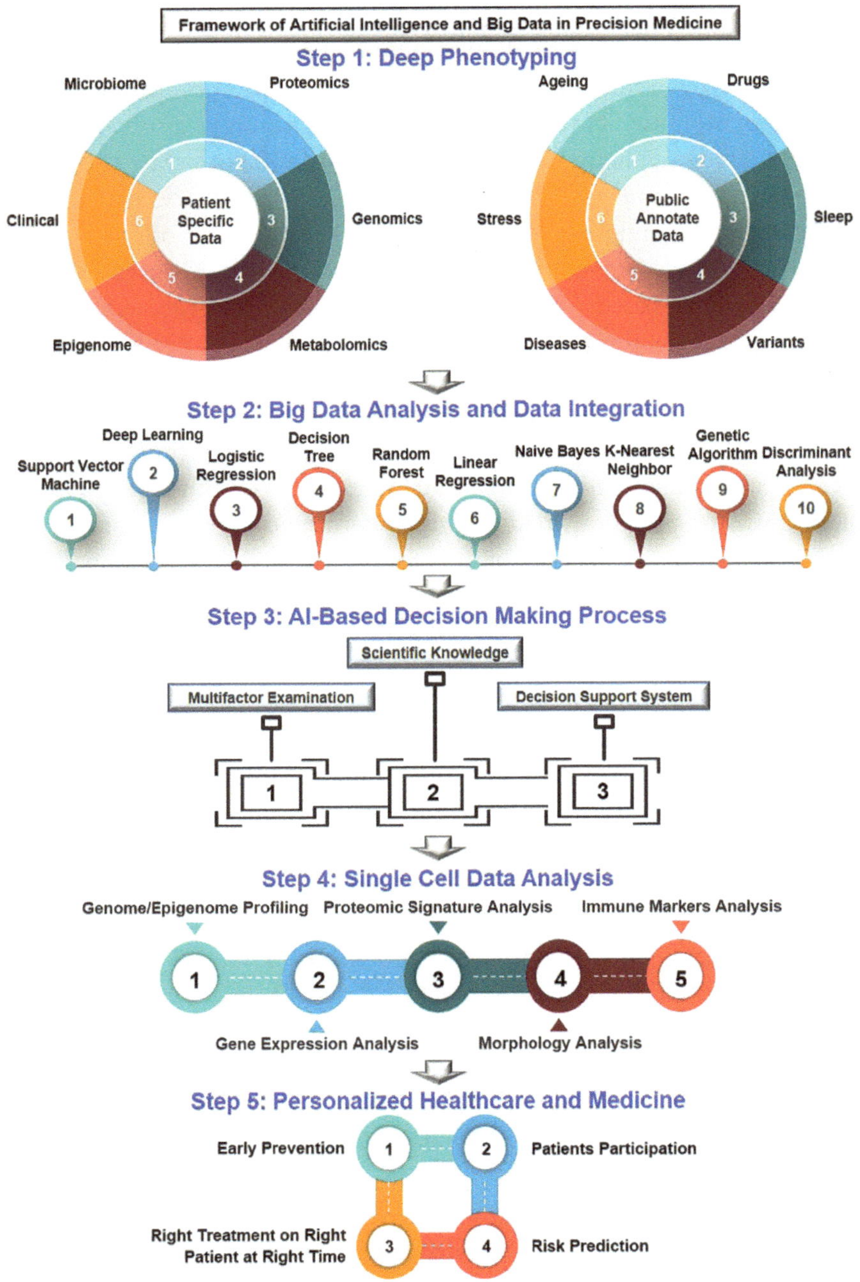

Fig. 1 Framework of artificial intelligence and big data in precision medicine: deep phenotyping data from public annotation database, such as variants, sleep pattern, drugs data, aging, stress, diseases and patient-specific data, such as genomics, proteomics,

2. Artificial intelligence and big data: Changes in the landscape of precision healthcare and medicine

AI is central to precision medicine, where big data play an essential role. In particular situations, there is no precision medicine without AI, as precision medicine requires high-performance computing systems, ML techniques, and deep learning algorithms. Further, precision medicine uses the cognitive abilities of the physician to a great extent.[16,17] AI is not one technology; rather, it combines several technologies, such as ML, neural networks, deep learning, natural language processing, rule-based expert systems, physical robots, and robotic process automation.[18] ML algorithms predict the most appropriate treatment protocols based on patients' unique characteristics. However, the ML algorithms require an enormous amount of data for training, termed "supervised learning."[19] The most complex forms of ML are neural networks and deep learning algorithms, which recognize potential lesions in radiology images.[20,21] Similarly, natural language processing and rule-based expert systems are other AI methods that involve accuracy of recognition and clinical decision support, respectively.[22,23] Physical robots and robotic process automation are two other AI-based technologies that are heavily incorporated in personalized healthcare, especially in precision medicine. Surgical robots approved in the USA in early 2000 were considered the first non-human surgeon that improved the ability to create precise and minimally invasive incisions and also aided in other aspects of surgical interventions. However, decisions of utmost importance are still made by human surgeons.[24] Likewise, robotic process automation based on computer programs on servers enables the clinicians to bill and update patients' records.[25] AI provides decision-making abilities at a scale that can transform the healthcare system, and thus physicians are moving faster towards using modern technologies to analyze big healthcare data.[26,27] Currently, big data is employed in almost every government

microbiome, clinical, epigenomic, and metabolomic data were collected. The data were analyzed through different machine learning algorithms, namely support vector machines, deep learning, logistic regression, decision tree, random forest, and others, which are used for the artificial intelligence-based decision-making process. There are three types of artificial intelligence-based decision-making processes, namely multifactor examination, scientific knowledge, and decision support systems. Further, the data is used in single-cell data analysis, such as genome profiling assay, proteomic signature analysis, and immune markers analysis.

and non-government institution, especially in hospitals, pharmacies, and care-giving institutions. However, to transform current clinical and medical practice, big data must employ AI-generated algorithms that have the potential to process big data efficiently and effectively.[28,29] To be effective, big data analytics involving AI/ML algorithms must enable the analysis of data from millions of patients. However, before implementing big data analytics and AI/ML algorithms in precision medicine, it is of utmost importance to elucidate the relationships among AI, ML, big data, and precision medicine.[30] Currently, IBM, Google, Apple, and Amazon are investing heavily in health care data analytics to promote precision medicine.[17,30] There are three crucial areas of precision medicine based on clinical areas, namely complex algorithms, digital health applications, and the "OMICS" test. "OMICS" tests are further classified into two groups: genetics-based and drug discovery based on selected therapeutic applications.[31] First, complex algorithms include ML/AI algorithms that can be used with large complex datasets to analyze genetic information, demographic data, and electronic health records to predict optimal treatment strategies and responses.[32] Secondly, digital health applications should include health data, such as oxygen levels, dietary intake, cardiac data, and blood sugar levels obtained using health monitoring devices such as wearables and mobile sensors.[33,34] Thirdly, OMICS-based tests are encouraged where genetic information from a particular population subtype is taken by integrating it with ML algorithms that collects patient's response to particular treatment and helps to find a correlation amongst them. The "OMICS" test is also used to predict other characteristic features, including biomarker expression, epigenetics, metabolic profiles, and protein activity that comprises of study of protein expression, modification, structure, and functional analysis.[35,36] Apart from this, AI/ML enables machine vision for diagnosis and surgery. For instance, the smart tissue autonomous robot (STAR) from Johns Hopkins University outperforms human surgeons in several areas. Similarly, a group at the Institute of Information Technology at the Alpen-Adria Universität Klagenfurt used a surgery video as a training sample to identify a particular intervention made by the human surgeon.[37] Further, clinicians at every hospital will go through only a relatively small number of data out of enormous data available worldwide due to the limited number of patients. Thus, worldwide sharing data is of utmost importance. AI enables the clinicians and scientists to share data across different platforms to improve the treatment and identification of particular patient's response to particular treatments.[38] The Canadian Distributed Infrastructure for Genomics

(CanDIG) enables data sharing across the country in Canada. It is an organization led by different clinicians, scientists, AI specialists, and geneticists to provide a solution to the people of Canada facing healthcare challenges.[39] Likewise, Flatiron health is a data-sharing organization in New York that uses big data and analytics through its cloud-based software platform and thus enables scientists, clinicians, and physicians around the country to overcome hurdles in cancer diagnosis and prevention.[40] Additionally, a number of start-ups, e.g., Embleema, K Health, Prognos Health, LunaDNA, and Humanity.co have been set up to share clinical and research data worldwide.[41–44]

3. The emergence of AI in precision medicine

The term "personalized medicine" was introduced in ancient times (~1500 BC) to face the challenges of the healthcare system and the research community that aims at different hurdles, such as prevention, diagnosis, and treatment of disease. Advancements in science and technology have contributed significantly to improving healthcare outcomes.[45] From 2700 BC until the Hippocratic period, where the term "personalized medicine" was re-introduced, many medicines were developed, and at a level that we face hurdles to attain.[46] However, after the Hippocratic period, a significant research gap was seen, and the theory of "one size fits all" predominated healthcare delivery. However, after the 1950s, with the great discoveries in scientific and healthcare research, such as the discovery of biomarkers and the introduction of the morphogenesis model, personalized medicine regained attention.[47] The post-1950s to early 2000s witnessed impactful discoveries in the healthcare system, including understanding of the genetic code (1980s), the human genome project (1990s), fMRI images (1991), identification of novel therapeutic targets (1993), introduction of Biochips (1994), and introduction of ML algorithms in medicine (the early 2000s) have benefitted the medicine developmental process.[47] However, the applications of personalized medicine in healthcare still struggle to be fully executed.

The "OMICS" era began in the 21st century, which enabled clinicians and researchers to analyze the molecular phenotypes of the individual patient. In addition, advancements in sequencing technologies with "OMICS" research increase the possibility of early diagnosis and treatment in different diseases.[48] The discovery of the human genome and human Inhalative PK profiling in early 2001 laid the foundation of precision medicine that improves the understanding of molecular characteristics of the individual patient.[49] Despite potential of the human genome to enhance the discovery of precision medicine, it possesses several hurdles. The introduction of the "HapMap project" and publication of "GWAS" studies have the ability to overcome the limitations.[50,51] The "HapMap" project adopted a hierarchical mapping strategy and contributed to the advancement of human genetics and pharmacogenomics.[52] Later on, during 2010–2011, the introduction of the "developmental human brain atlas" and the "research domain criteria (RDoC)" project were considered as the major outcome in precision medicine to combat neurological manifestations and identification of cells that are important in development.[53,54] However, a major advance in precision medicine came in 2012 with the simultaneous introduction of ENCODE data and deep learning algorithms in medicine.[55,56] Deep learning algorithms change the entire scenario of the healthcare industry because they enable both clinicians and researchers to identify novel therapeutic targets and several other parameters of disease of an individual patient.[57] Introduction of the human proteome map and cell atlas in 2014 and 2017, respectively, enabled clinicians, researchers, and physicians to identify novel drug targets, deep knowledge of immunological lineages, and altered signaling cascades that lead to disease.[58,59] Likewise, the application of deep learning algorithms in radiology, imaging, genomics, diagnosis, and evolution makes it possible to identify different disease characteristics at an individual patient level.[60–62] Currently, AI/ML algorithms have been successfully employed in the development of prediction tools, which increase the possibility of targeted therapy.[63,64] Additional major advances are expected in the coming years (Fig. 2A).

4. Synergies between artificial intelligence and precision medicine

Today, the integration of AI/ML algorithms with precision medicine assists clinicians, physicians, and researchers in solving the most complex problems in personalized healthcare.[65] The synergies between AI/ML and precision medicine are divided into two groups, therapy planning and risk prediction. Therapy planning is further divided into three sub-groups, genomic considerations, environmental considerations, and clinical considerations, whereas, risk prediction is divided into two sub-groups, genomic approaches and non-genomics approaches (Fig. 2B).

4.1 Therapy planning

4.1.1 Genomic considerations

Genomic information is one of the most crucial areas of precision medicine as it enables prediction of disease parameters at an individual level. However, the

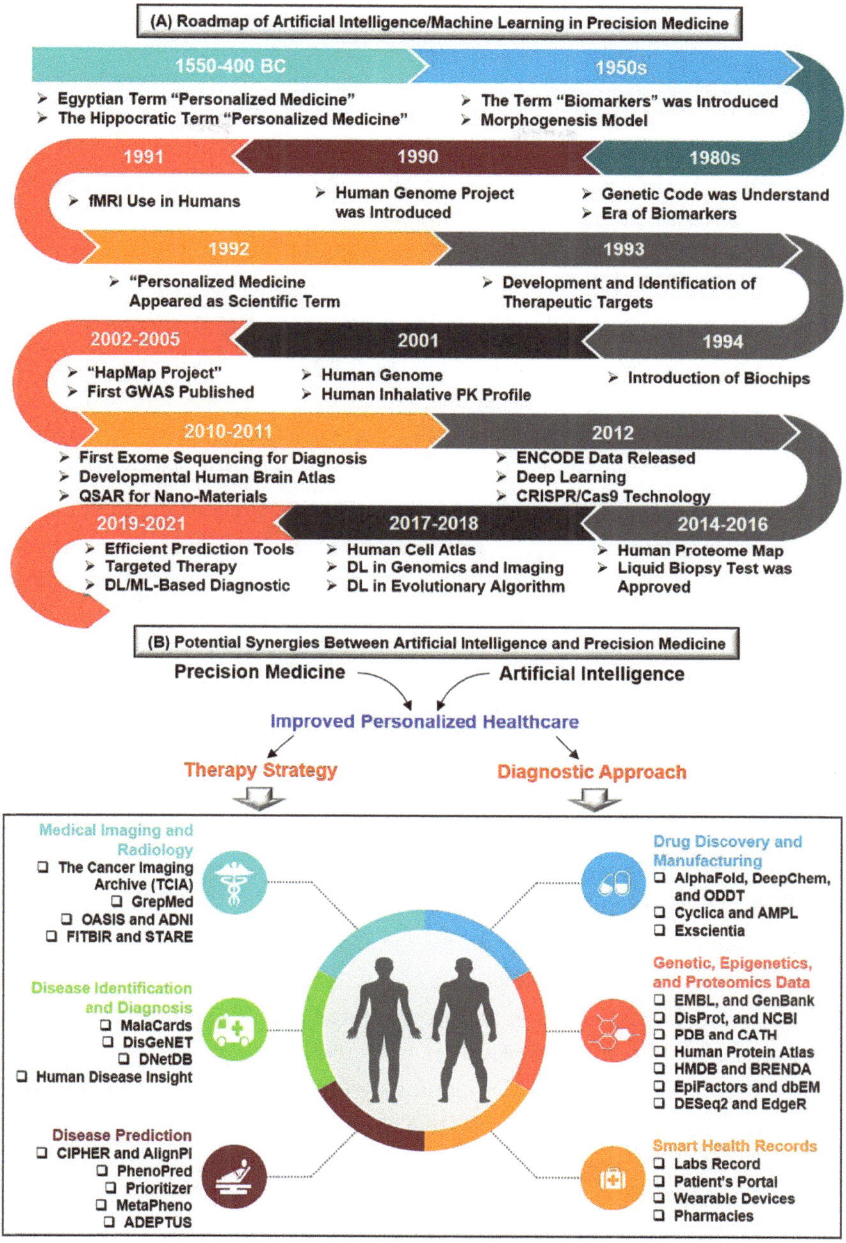

Fig. 2 (A) Artificial intelligence in healthcare was started in early 1550 BC with the invention of the term "Personalized medicine." Later in the 1950s, the term biomarkers were introduced, followed by the introduction of the morphogenesis model. Afterward, the study of genetic code in the 1980s and the human genome project in the 1990s marked the historical landmark in the intervention of artificial intelligence in healthcare. The discovery of deep learning, ENCODE data, and CRISPR/Cas9 technology in 2012 marked the revolution in precision medicine. Later, deep learning in genomics, imaging, and evolutionary algorithms proved to be a landmark feature of artificial intelligence in precision medicine. (B) Potential synergies between artificial intelligence and precision medicine are divided into two categories: therapy strategy and diagnostic approach.

data collected in real-time is a hindrance for developing AI/ML-based algorithms to predict the relationship between patient, medication, and genomic information. Thus, deep-genotyping of the patient is of utmost importance to identify personalized dosage and medications.[66,67] The first-ever synergy between AI and precision medicine occurred with efficient and high-throughput genome sequencing. For example, Zou et al., incorporated deep learning algorithms to combine scientific literature data with sequence to protein 3D structural data, which identifies the transcription sites, regulatory elements, gene expression, and single nucleotide polymorphisms from genotype data.[68] Likewise, Erijman et al., incorporated a deep learning algorithm to identify known acidic activation domains within transcription factors and predict the consequences of mutations.[69] In addition, Guo et al., developed a deep learning system, Deep Metabolism, to predict the individual cell phenotype from genome sequencing.[70] Thus, these interpretations and discoveries laid the foundation for identifying links between genomic sequence and disease prognosis. Further, AI in exome sequence analysis has facilitated the administration of the right treatment at the right dosage to the right cohort of patients.[71] For example, Vega et al., proposed the diagnostic performance of Fabric GEM, an AI-based clinical decision support tool, which accelerates the genome interpretation process that will decrease the treatment cost.[72] Further, the introduction of AI in image recognition led to the successful application of radiogenomics in personalized treatment to establish a relationship between imaging data and gene expression.[73,74] Liao et al., applied four ML, algorithms-based prediction models to determine the interaction between image features and genomics data through Pearson's correlation analysis.[75] Similarly, Iwatate et al., incorporated an ML algorithm to predict p53 status and PD-L1 expression in pancreatic cancer. In the study, the authors demonstrated that p53 mutations were significantly associated with poor prognosis, whereas the predicted abnormal expression of PD-L1 was not significant.[76] Knowing the effect of a particular therapy on a particular patient could help clinicians and physicians to choose an effective and precise treatment plan. Abajian et al., predicted the treatment response to intraarterial therapies for hepatocellular carcinoma through logistic regression and random forest models. The authors predicted transarterial chemoembolization treatment response with an accuracy of 78%.[77] Similarly, Facciorusso et al., demonstrated the response of repetitive echoendoscopic celiac plexus neurolysis in pancreatic cancer though a ML model involving artificial neural network and logistic regression models. The authors predicted the pain response following repeat neurolysis is generally less pronounced than after initial treatment.[78]

Taliaz et al., predicted the response to antidepressant medications through integration of genetic, clinical, and demographic data using ML algorithms.[79] However, more prospective and retrospective clinical research still needs to be conducted to provide training for ML purposes.

4.1.2 Environmental considerations

Integrating environmental considerations in treatment plans is a significant challenge in personalized healthcare and precision medicine as it requires an enormous amount of environmental data. One such application of AI algorithms came from the study of Wille et al., where the authors predicted the barriers to healthcare in American Indian patients that experiencing homelessness.[80] Similarly, Truong et al., performed survey analysis of dermatologic health-related quality of life and resource access in homeless patients. The authors concluded that resource limitations were associated with female patients and high dermatology life quality index (DLQI) scores with non-white participants.[81] The major environmental consideration is the availability of experts and trained professionals in a remote area at the point of need. AI provides several examples that augment diagnostic capabilities in resource-poor locations. For example, Li et al., developed an ML algorithms-based system for on-time and efficient identification of heart diseases and Wroge et al., developed a system using voice datasets of patients and unsupervised deep neural networks for the diagnosis of Parkinson's disease.[82,83]

4.2 Risk prediction and diagnosis

4.2.1 Genomic approaches

At present, genomic information is central to disease risk and prevention, and AI algorithms play an essential role in identifying gene inheritance and mutations in the disease progression. Thus, integrating genomic information and disease risk prediction promotes personalized healthcare or precision medicine. For example, Peng et al., developed an ML model that predicts genes involved in the progression of Parkinson's disease. The authors proposed a model, N2A-SVM, which is divided into three parts, extracting features of a gene by analyzing the gene regulatory network, reducing the dimensions using deep neural networks, and predicting Parkinson's disease genes.[84] Likewise, Barman et al., incorporated deep neural networks with 16 selected features for pseudo-amino acid composition and network properties to predict unknown infectious disease-associated host genes.[85] Complex diseases often involve environmental and lifestyle factors and AI algorithms can discover this complex relationship individually. For instance, a study by Malki et al., predicts the association between weather data and COVID-19 mortality rate, and Alaa et al., use automated ML algorithms for cardiovascular disease prediction considering both traditional and non-traditional variables.[86,87] Further, AI-enabled phenotype recognition through image matching with genetic variations could allow faster disease prediction.[88] Qin and Li in 2020 identified facemask-wearing conditions using super image resolution with a classification network to prevent COVID-19.[89] Similarly, Yang et al., developed and validated a deep learning algorithm for scoliosis screening using unclothed back images. The authors found that the algorithms' accuracy was 20% more accurate than human specialists in detecting disease conditions.[90] Another example came from the study of Mishima et al., who evaluated Face2Gene using facial images of patients with congenital dysmorphic syndromes in Japan.[91]

4.2.2 Nongenomic approaches

Automated speech analysis now is being used in the early detection and diagnosis of several movement disorders, including Parkinson's disease, cognitive impairment, and mental disorders.[92,93] Campbell et al., augmented Mini-Mental state examination tool for early detection of Alzheimer's disease that evaluates functional performance in several fields, including language; the model consists of two algorithms, AcceXible and ADReSS, which tested two databases on two languages, namely Spanish and English.[94] Likewise, Karaman et al., proposed a convolution neural network-based algorithm to identify Parkinson's disease through voice features. The proposed model has over 91% accuracy, and thus it could be implemented for personal usage as personalized healthcare.[95] Further, AI-driven processes of surgery and colonoscopy decrease complications and adverse effects. For instance, Lundberg et al., developed a system that predicts the risk of hypoxemia and provides explanations of the risk factors in real-time during anesthesia.[96] Similarly, Memarian et al., developed a multimodal data and ML algorithm-based system for surgery outcome prediction in temporal lobe epilepsy. The model accurately predicted the post-surgical outcome in epileptic patients who showed resistance against drugs.[97] Bihorac et al., developed a system called as MySurgeryRisk to predict complications and death after surgery.[98] The widespread involvement of smart devices and wearables collecting patient data have helped in monitoring and diagnosis of the disease.[99–102] Rashid et al., developed a binary convolutional neural network-based system that facilitates the early detection of myocardial infarction through wearable devices.[103] AI thus is useful in the early detection and diagnosis of disease condition, irrespective of genomic signatures and facilitates personalized medicine.

5. Promises of implementing AI/ML in healthcare

The merger of artificial intelligence and precision medicine has revolutionized patients' healthcare. Precision medicine helps identify a unique healthcare approach for patients who respond less to conventional treatment, whereas artificial intelligence provides computation that influences the generated data and trains the system to learn, which further empowers clinicians' decision-making. The application of artificial intelligence in precision medicine also aims to simplify personalized diagnosis and prognostication.[104]

5.1 Solving the human resource crisis in healthcare

One of the most critical problems in achieving health and development goals is the shortage of healthcare workers. This has created major problems in some countries. There is no adequate staffing in rural areas in comparison to metropolises. The majority of the priority programs initiated by the government and big agencies suffer a from scarcity of staff. Even in developed countries, increased demand more health workers exist because of their aging population and the aging of their workforce. The spread of chronic diseases, changes in climate threatening global security and frequent alterations in work patterns all combine to indicate the ever-growing demand for health workers worldwide.[105] The WHO's "Global Strategy on Human Resources for Health: Workforce 2030" report states that by 2030 the shortage of healthcare workers could be up to 9.9 million physicians, nurses and midwives worldwide.[106] In a Joint Learning Initiative, where a syndicate of more than 100 health leaders proposed that human resources in the healthcare sector should be mobilized and strengthened, poor countries should be backed by international reinforcement in combating health crises. Thus, a sustainable health system should be encouraged in all countries. As a human resource crisis is a shared problem, it should be dealt with by taking shared responsibilities and cooperative actions.[107] Moreover, for artificial intelligence to provide an effective system, it depends on the availability, accessibility, acceptability and quality of its health workers. It also facilitates easy diagnostics, decision-making, big data analytics and administration amidst health workers. However, some ethical challenges like informed consent to use, safety and transparency, algorithmic fairness and biases and data privacy, whereas legal hindrances such as safety and effectiveness, liability, data protection, cybersecurity and intellectual property law prevent AI technology from being used to its full potential. Moreover, the human resource crisis cannot be solved by providing this technology only to physicians; there must be an overall involvement of all those in the healthcare profession.[108] Smart algorithms could assist healthcare workers in designing treatment plans and providing personalized treatment for every patient. They could also perform repetitive and monotonous tasks, allowing clinicians and nurses to focus more on patient care. Engineers and data scientists should monitor the precision of artificial intelligence algorithms by training them with reliable big data as well as provide training to personnel so they work confidently alongside the intelligent machine.

5.2 Improving workflow and reducing medical errors

AI has an impact on three levels of medicine: (1) for clinicians, it eases many tasks and gives better image interpretations; (2) for the healthcare system, it helps in enhancing workflow and reducing medical errors; and (3) for patients, they can keep a track of their own data by wearing holter monitors that captures their information time to time. AI tools give leverage to the health system to deliver complex information without any need for trained personnel. In Geisinger Health in Pennsylvania, the Artificial intelligence chatbox (Clear Genetics) has provided results for exome sequencing for more than 100,000 patients, minimizing the need for genetic counselors.[109] As the potential of precision medicine is high in the field of genomics, however healthcare organization should focus more on how to execute precision medicine plans. At the individual patient level, the AI technology should enable patient to get a lab test done at inpatient side and get the result and discharge planner on the outpatient side. However, it becomes important to understand whether integrating this technology will be cost-effective for the patient and the healthcare facility or not. Moreover, significant partnering with the right lab such as radiology or pathology lab and the overall workflow of the tests becomes a necessity. The potential of AI is not restricted to solving human resource issues but also enables provision of care for large numbers of patients. Neural networks are mathematical models that use various algorithms to find complex relationships within large data sets (analytics). Machines learn from the errors encountered in response to slight changes in the algorithms and correct themselves (training), which further improves the accuracy of models (confidence).[110] The ambiguity regarding patients' diagnoses versus predicted response to treatment at times may lead to cognitive and judgment errors. Nonetheless, reinforcement learning, a subfield of machine learning, helps to identify the sequence of action that aims to reach a more accurate and predetermined goal.[111] Many people die because of prescription errors, and these errors come from either flawed electronic health records (EHRs) or when the doctors choose the wrong drugs or dosing unit from the drop-down menu or they miss out potential injurious drug interactions. This problem can be solved by machine learning models that compare new prescriptions with historic EHR data and flag prescriptions that deviate from the past. This allows further review by the doctor[112] (Fig. 3A).

5.3 Enhancing healthcare delivery

A shift in the traditional public health decision-making standards has been observed with the incoming of precision medicine. The healthcare delivery model in precision medicine focuses on customized healthcare tailored for a subgroup of patients, whereas historically, funds have been allocated to safety, efficacy and efficiency. To achieve broad public health goals and minimize healthcare disparities, public health policy and decision-making must enhance healthcare delivery.[113] When the human genome was first sequenced, no one could imagine that precision medicine would be creating a huge impact on patients, healthcare delivery systems and research contributors. There has been a substantial increase in identification of disease-causing and drug-response genetic variants, whereas developments in clinical medicine have lagged.[114] To deliver this innovation in the best manner, precision medicine needs to be made readily accessible to patients and clinicians. These genetically targeted therapies will extensively be utilized if payers opt to cover them; thus, a collaboration among payers, clinicians and scientists become essential to accelerate the application of precision medicine. Moreover, scientists and clinicians need to be sensitive to payers' perspectives or there will be delays and substantial loss of resources.[115] Despite many efforts to move towards personalized medicine, there are some major challenges, including education, informatics, patient engagement, internal funding, and ensuring high-value testing and data collection practices.[116] Although leaders in the field of medicine have addressed some of the barriers, health care delivery organizations have to do substantial work to guarantee patients benefit from the clinical and economic returns of personalized medicine.[117] Moreover, a recent public survey stated that only 4

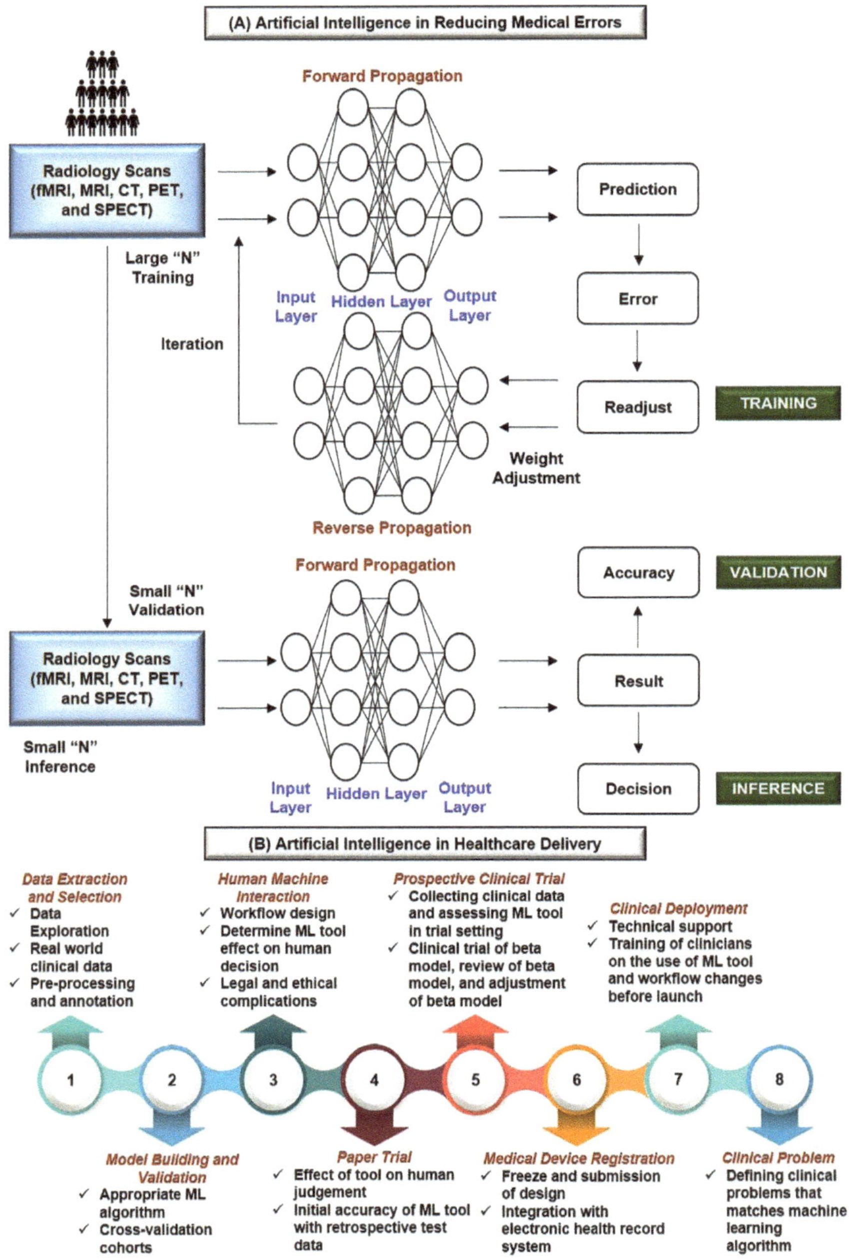

Fig. 3 (A) AI in reducing medical errors: radiology scans were taken as input for forwarding propagation that predicts the outcomes, and if the error was present, the data was readjusted for reverse propagation. (B) AI in healthcare delivery: there are nine different areas in which artificial intelligence could help to enhance healthcare delivery, such as data extraction and selection, human-machine interface, prospective clinical trials, clinical deployment, model building and validation, paper trial, medical device registration, and clinical problems.

out of 10 patients are aware of personalized medicine. Further, only 11% of patients were made aware of it.[118] Precision medicine has revolutionized healthcare delivery by providing innovative insights that give health workers an opportunity to decide the best treatment option for their patients. To enhance healthcare delivery of personalized medicine, there is a requirement of new knowledge, emphasizing patient's perspectives, identifying new molecular pathways in guiding healthcare, development of new infrastructure, management setting to process data and improving delivery services to ensure the accessibility of personalized medicine to all. Overcoming these barriers will require short-term and long-term strategies in order to prevent systemic and cultural change. Moreover, opting best strategies will help in creating a roadmap for healthcare systems to advance the personalized medicine paradigm[119] (Fig. 3B).

6. The integration of nanomaterials in AI-based precision medicine

AI and nanotechnology promote the goal of precision medicine. AI benefits nanomedicine through improved material properties, which could increase drug targeting and therapeutic efficacy (Fig. 4A). Moreover, computational methods, such as ML and AI, improve the results of *in situ* nano-based screening.[120] For instance, Yang et al., developed a novel tool, iDriver, which uses Bayesian frameworks and multivariate statistical modeling to identify driver genes. The authors applied the model on 4 cancer types and identified highly enriched driver genes which acquires mutation and are liked with cancer progression.[121] Further, owing to the structural variations of nanomedicines and lack of relevant high-throughput synthesis methods, only a little of the nanomedicine-design space has been explored.[122] Thus, integrating AI/ML algorithms into nanoparticle design is another major aspect for improving targeted therapies. Alafeef et al., combined ML algorithms, including artificial neural networks and characteristics of cellular uptake responses for individual cancer cell types, to classify various cancer cell types. The authors successfully classified 36 unknown cancer samples with an accuracy of >98%, demonstrating the potential application of AI/ML in nanomedicine-based target therapies.[123] In addition, computational biology through AI/ML algorithms are of utmost importance for predicting the ability of nanoparticles to cross barriers on their way to target organs.[124,125] Different models have been constructed to predict the ability of nanoparticles to permeate the blood-brain barrier and their potential

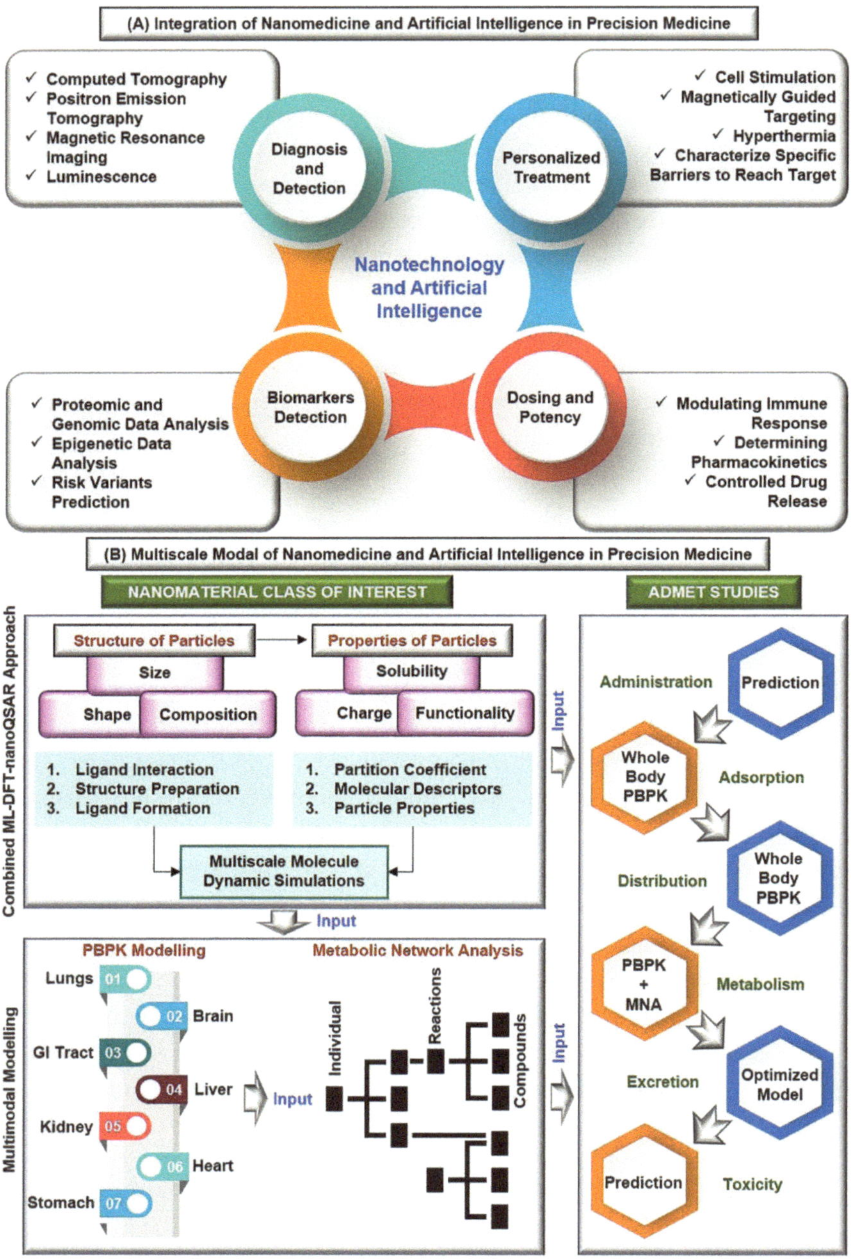

Fig. 4 (A) Integration of nanomedicine and artificial intelligence in precision medicine: Involvement of artificial intelligence in nanotechnology is helpful in different areas of precision medicine, such as biomarkers detection, diagnosis and detection, personalized treatment, and dosing and potency. (B) Artificial intelligence in PBPK modelling: artificial intelligence and machine learning algorithms are involved in improving drug dosage and potency through the enhanced structure of particles and properties of particles, which are used in PBPK modeling. Further, artificial intelligence is involved in ADMET prediction that improves precision medicine.

toxicity.[126] Kingston et al., assessed micrometastases as a potential target for nanoparticles using ML algorithms and 3D microscopy, whereas, Liu et al., predicted the properties of hydroxyapatite nanoparticles datasets with different morphologies and sizes.[127,128] Priyadarshani et al., detected the chromium toxicity of diethylenetriaminepentaacetic acid-functionalized gold nanoparticles through a support vector regression model. The authors predicted the accurate concentration of chromium (III and IV) in water and in the aqueous solution of ferrochrome industrial waste with mean squared errors of 0.00039 and 404, respectively.[129] Moreover, ML/AI algorithms have been successfully employed in predicting the optimal shape and size for nanoparticle binding and drug release.[130] Pellegriono et al., applied ML algorithms for elucidating and predicting the role of synthesis parameters on the shape and size of TiO_2 nanoparticles.[131] Similarly, Mekki-Berrada et al., integrated the Gaussian process-based Bayesian optimization with a deep neural network to predict the color palette accessible with the reaction synthesis.[132] Drug dosage is a critical aspect of personalized medicine, and nanoparticles have the potential to achieve this goal. Different studies have demonstrated the potential value of AI/ML algorithms on drug dosing and therapeutic efficacy.[133,134] For example, the implementation of a convolutional neural network with a long short-term memory network and time-series feature extraction improved the temporal dynamics significantly when HepG2 cells were exposed to lipid nanoparticles loaded with mRNA for expression of GFP.[135] Moreover, physiologically based pharmacokinetic (PBPK) modeling is an important for accelerating the development of drugs in solid dispersion formulations. Recently, studies have demonstrated that AI/ML algorithms have the potential to predict the accuracy and precision of PBPK modeling[136,137] (Fig. 4B). Gao et al., constructed a random forest algorithm based on a dataset consisting of 674 dissolution profiles of solid dispersion to distinguish two types of dissolution profiles. Further, the authors predicted time-dependent dissolution profiles. The authors concluded that the integrated computational algorithms could improve pharmaceutical formulation development relative to traditional method.[138]

7. The perspective of big data analytics in precision medicine

One of the key enablers of the use precision medicine is to digitally collect specific, accurate and timely data of an individual. Big data containing information can help precision medicine applications make decisions about precision therapies that can be delivered at the population level. Nowadays, medical information is captured electronically, therefore analytics and machine learning can be applied to find clinical pathways, better decision-making support and fewer variations. Combinations of chronic disease management and an engaged patient will boost population health, thereby providing a maximum health benefit to all at the lowest price.[139]

There is immense potential in big data for the use of artificial intelligence and machine learning tools in three aspects of precision medicine. For clinicians, it promises more efficient and standardized image interpretation to improve diagnosis, predict health risk and guide treatment options. For the health system, it has the potential to improve workflow and minimize medical errors. Lastly, for patients, it provides the means to process their data and take necessary precautions.[140] However, big data requires interdisciplinary efforts that will make the most of the information and medical data of newly formed groups fusing different backgrounds and expertise.[141] Rapid advances in high throughput technologies and implementation of electronic health records (EHRs) have led to the fast growth of -omic and EHR data. These multifaceted data contain abundant information for precision medicine, and with the help of big data analytics, one can extract knowledge and improve the quality of healthcare. However, most of the -omic and EHR data are high-dimensional and requires more computational power, which at times affects the accuracy of the analysis. Therefore, it becomes important to reduce data dimensionality without affecting its informational content. There are two possible strategies: (1) feature selection that aims to select an optimal subset of existing features; and (2) feature extraction that aims to transform existing features into a more compressed set of dimensions.[142,143] Another important barrier that big data in healthcare must overcome are ethical and legal challenges because of the personal nature of the information enclosed. These challenges include risk to privacy, personal autonomy, along with effects on public demand for transparency, trust and fairness.[144] Moreover, some technical and infrastructural matters might also endanger big data driven-healthcare due to data heterogeneity, data protection, analytical flows in investigating data and the lack of proper infrastructures for data storage[145] (Fig. 5).

8. AI and personalized medicine: A case of neurological disease

Patients with Alzheimer's disease (AD) require extended medical treatment because of the disease's slow rate of progression. It has been predicted that early diagnosis of AD could lead to savings of $7 trillion in

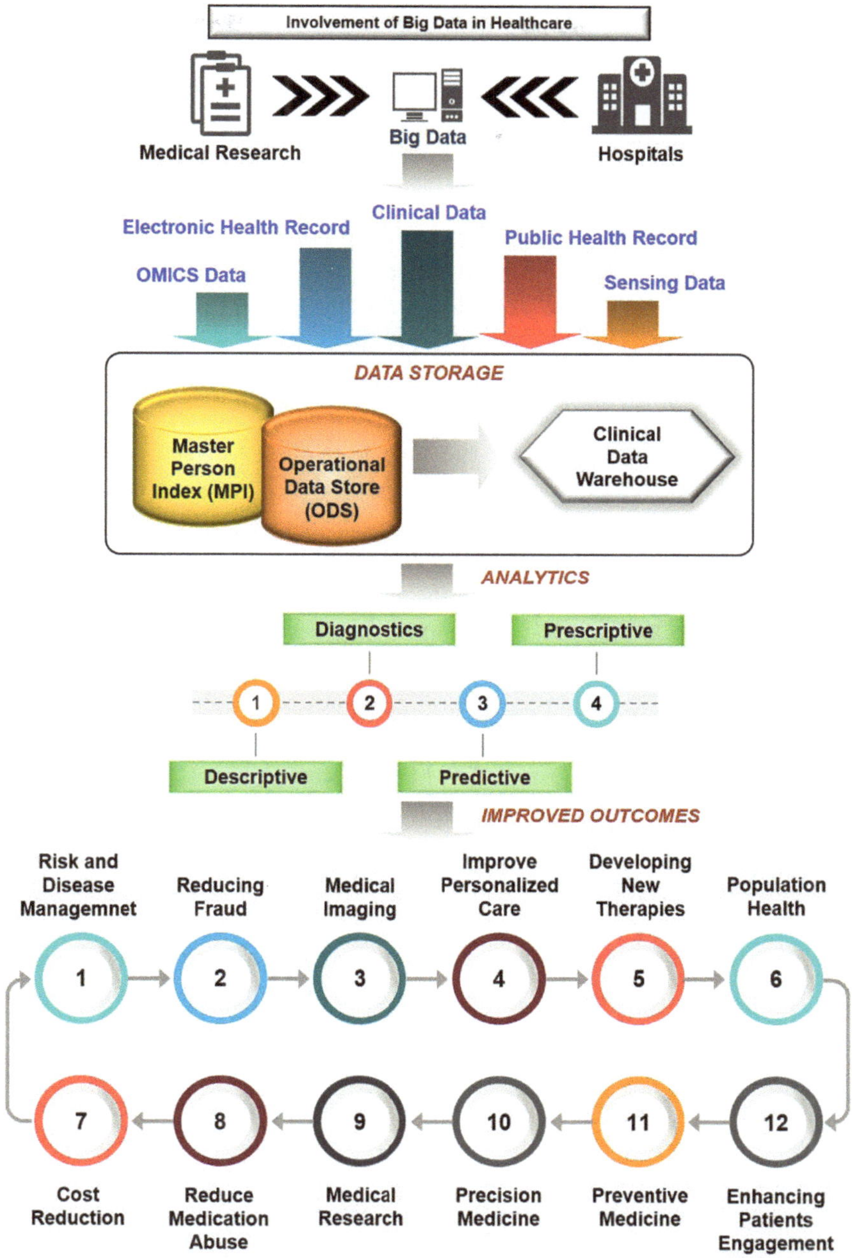

Fig. 5 Recently big data from hospitals and medical research institutes enhances the healthcare delivery through integration of artificial intelligence and machine learning algorithms. OMICS data, electronic health record, clinical data, public health record, and sensing data were stored in data storage through master person index (MPI) and operational data store (ODS), which is analyzed for their diagnostic, perspective, descriptive, and predictive values.

the United States alone.[146] Precision medicine aims to target issues underlying molecular and clinical heterogeneity by detecting patients' risk factors, pathophysiologic processes, and targeting a therapeutic intervention that is explicitly personalized to detect disease progression.[147] Apolipoprotein E genotype and age are the two largest risk factors for AD. Hence, a precision medicine strategy customized for each individual with different genotypes might be advantageous.[148] O'Bryant and his colleagues proposed a model based on precision medicine that targets non-steroidal anti-inflammatory drugs (NSAIDs) therapy to slow down the progression of AD. Here, they found that a blood proteomic companion diagnostic (CDx) is able to predict response to NSAIDs treatment.[149] The study of the proteome in plasma samples from the Alzheimer's Disease Cooperative Studies (ADCS) anti-inflammatory clinical trial with the inclusion of 1-year administration of rofecoxib and naproxen showed that NSAID and CDx can detect treatment response with 87% accuracy.[150] Drug-specific CDx, Rofecoxib-CDx and Naproxen-CDx generate high accuracy of 98% with rofecoxib and 97% with naproxen. Hence, it suggests the usefulness of a precision medicine-based model to treat AD.[151] A 9-month study has been done on AD patients in which a personalized, precision medicine protocol using Montreal Cognitive Assessment (MoCA) scores was evaluated for different markers. The results revealed prevention of cognition decline by giving a highly significant improvement in MoCA scores and CNS Vital Signs Neurocognitive Index. Thus, this approach of precision medicine therapeutics is warranted.[152] Moreover, computerized cognitive testing of some risk biomarkers can help predict the high risk of late-onset of AD in adults.[153,154] Few clinical interventions have been studied for precision medicine to delay cognition decline in AD patients. However, there are many publications on another multidomain approach called "one-size-fits-all." A common risk factor that has been observed in nearly 10% of the patients with Parkinson's disease (PD) is the mutation in the glucocerebrosidase (GBA) gene that is known to cause Gaucher disease. Likewise, mutation in LRRK2 has also been observed that is known to cause autosomal dominant PD. Depending upon the mutation in both genes, the severity of the disease can be predicted. Moreover, polymorphisms in the gene are strong risk factors. The mechanism behind these mutations causing PD has not yet been fully understood, however, it appears as a result of increased LRRK2 kinase function, indicating the probable advantage of kinase inhibitors.[155] Treatment can be directed towards repairment of this enzyme function such as enzyme-replacement therapy (ERT), gene therapy or by investigating early glucocerebrosidase chaperones that consequentially increase GBA activity in the brain.[156]

Recently symptomatic treatment of PD through drugs, targeting the dopamine system has become even more personalized with multiple drugs delivery systems that are used according to the need of a specific individual. However, the precision approach at the molecular level has not yet been fully tested. A "one-size-fits-only-one-or-a-few" approach determines pathogenic cascades concerning different molecular pathways in different PD patients, thereby suggesting that the best way to move ahead is with precision medicine.[157] Advancements in the field of cellular reprogramming and personalized medicine promise cell-based therapies and patient-specific modeling of PD through induced pluripotent stem cells (iPSCs). iPSC models, in contrast to other artificially-induced models, have well-maintained endogenous cellular machinery and transcriptional feedback. Moreover, iPSC cell lines can be established with specific genetic risk factors that will help to evaluate the differing response of different populations to treatment.[158] Many successful examples of digital biomarkers have been identified for PD through traditional clinical outcomes that are sparse, highly variable and rater-dependent. Smartphone-based measurement has been developed and introduced in clinical trials to monitor various symptoms of PD.[159,160] For example, inertial measurement unit features have been gauged to predict PD severity. In PD, the symptoms are heterogenous as different regions of the brain are affected resulting in different motor and behavioral symptoms. Thus, with the help of imaging techniques, optimal personalized medicine strategies can be developed that will help differentiate PD from Lewy body dementia, PD-AD and frontotemporal dementia.[161] In the case of Huntington's disease (HD), silencing expression of both expanded CAG-repeat and normal copies of the Huntingtin protein gene (HTT) may be therapeutically useful, but normal HTT expression must be maintained. Allele-specific strategies can be developed that can either silence the expanded CAG repeat or single nucleotide polymorphisms (SNPs). Based on PD patients' genotypes, both approaches require personalized treatment strategies.[162] Moreover, clinically relevant gait parameters have been assessed using wearable inertial devices.[163] Thus, for patients exhibiting both movement disorder and cognition decline, personalized and properly targeted treatment is required. If not personalized, then even the accurate diagnosis and treatment option selected may prove to be inappropriate, ineffective, and even deleterious.

9. Obstacles in translating big data into healthcare

A number of areas need to be addressed for better translation of existing data in healthcare solutions such as collection and standardization of heterogeneous datasets, curation of the resultant data, prior informed consent for use of de-identified data, and the ability to deliver these data back to the healthcare and research communities for further use. A significant barrier in the translation of big data is the incorporation of new techniques in clinical practices as it will demand traditional clinical practices to undergo changes to incorporate these data-driven techniques. Those techniques that might replace human actions require testing time, thereby speeding up the process. Whereas, those techniques that help in prioritizing work are likely to be applied faster. Moreover, techniques that were earlier not possible because of massive data set are now likely to gain early adoption. The reason is improved incorporation of hardware and software to support big data assisting in managing workflow.[55] There are technical hurdles that are related to the construction of prediction models of health risks, diagnoses and outcomes from integrated biomedical databases. At times, the medical record data are inherently biased so that even the most advanced deep learning de-noising auto-encoders cannot overcome the bias. Moreover, societal challenges include the evaluation of ethically actionable risk factors at the individual and population levels.[4] The development of new, data-driven hypotheses based on interpretable models will always require strict validation and experimental testing. Therefore, the experimental model working on a large dataset must follow traditional hypotheses with new techniques for easy incorporation and improvement of clinical practices. Overfitting is another insidious obstacle in the translation of big data. Overfitting happens easily and tends to be more dangerous because it is a self-fulfilling prophesy. We see what we want to see, not what is there. The risk of overfitting is likely to increase with increasing data complexity, for instance, as *P*-values increase, an increase in the number of independent samples (n) becomes fundamental.[164] To process the data of an independent cohort, generating sufficient samples and funds poses a major challenge at the level of an individual academic lab. Thus, it becomes essential to have a network of labs or even bigger consortia in order to generate reliable and robust big genomic datasets. It is recommended to have large training datasets to compare test datasets available for independent validation. Upgrading and increasing collaboration and communication in the biomedical community also will facilitate the validation of models built on Big data.[165] Today, software developers have universally made source code freely available through open-source tools that are used and improved by the scientific community. This initiative by the Open Bioinformatics Foundation and the International Society for Computational Biology has had a huge impact on the progress rate of harnessing big data[55] Moreover, many methods have been established to safeguard standardization of extracted genomic material because non-standardized data will produce flawed experimental results. Further, data standardization of a sample also permits it to be used in other studies and obviates the need to generate the same result over and over again.

10. Opportunities for translating big data into healthcare

The rise of electronic medical records (EMRs) and robust information technology (IT) systems in support of both research and healthcare delivery has brought patients and cohorts to agree upon sharing their data for precision medicine research.[114] EHRs can be structured or unstructured. Dictated or written clinical notes describing a patient's condition are not an efficient method of documentation. Such data are more challenging for computational analysis because unstructured data are heterogeneous, including typing and spelling errors, prone to grammatical errors and are rich in abbreviations, acronyms, and idiosyncrasies.[166] Whereas, structured data can be differentiated into two classes. Firstly, administrative data, which remains unchanged during the entire course of treatment or are kept updated over time. Secondly, ancillary clinical data are frequently changed during the entire course of treatment, whether be discrete or continuous.[143] A descriptive study found that doctors spend 16 min per patient just to fill out EHR forms. However, with the use of artificial intelligence workflow optimization, one can navigate medical records with voice commands. In addition, artificial intelligence technology still requires thorough expertise, process, planning, workflow transformation, and assessment of outcomes. To ensure that precision medicine is tailored according to the genetic make-up of an individual, it becomes important to gather as much information as possible from individual patients. However, there are risks associated with widespread data sharing. Thus, health counselors and researchers need to encourage patients to actively take part in the studies and also teach them the risks and benefits of data sharing. Therefore, there are two forms of consent prepared, viz. consent for receiving medical treatment that focuses on benefiting the individual and consent for releasing data and samples for research that focuses on generalizing knowledge.[167] Healthcare at the level of individuals suggests that if there is discharge summary or document of reimbursement from an insurance standpoint, then it becomes a situation of operationalization, where one can look into different ways to integrate their data into the healthcare system through the EHR. The size and the quality of data drive data-driven personalized medicine. Moreover, to understand the complexity and identify specific patterns hidden in big data, personalized models are necessary. Data-intensive algorithms such as deep learning models need to be made understandable for precision medicine and healthcare.[168] If a sample size is small, analyzing biomarkers in translational medicine becomes challenging. A possible solution for this is to use longitudinal model-based data integration. As these models are time-dependent, time courses of PK and PD/biomarker changes can be determined, which further helps in better understanding the experimental outcomes.[169] Moreover, opting for machine learning is better for finding integrative solutions for multi-view data, e.g., generalized linear models (GLMs), where the outcome is linearly related to factors and covariates through a link function that permits the estimation of the model parameters (typically by maximum likelihood or Bayesian techniques).[170] Along with GLMs, common machine learning models for multi-view data are Bayesian models such as the naïve Bayes classifier, ensemble-learning models such as random forest, neural networks and more recently deep learning.[171]

11. Artificial intelligence and machine learning in precision medicine: An industry perspective

11.1 Multi-omics profiling

Technology is now shifting from single omics to multi-omics profiling platforms that include genomics, epigenomics, transcriptomics, proteomics, lipidomics, metabolomics, microbiomics and many others. There are several biological layers in multi-omics profiling that help researchers to better understand the interplay amongst genetics, gene regulation and protein synthesis. This will not only assist in identifying the molecular pattern and key nodes involved in a diseased condition but will also increase the chances of finding novel drug targets, endotypes and potential biomarkers.[169] Despite this potential, some major challenges stand ahead. For instance, omics technologies evaluate big data in a semi-quantitative manner, therefore it makes it highly sensitive to batch effects. Thus, it becomes essential to include appropriate controls in the experimental design and validate top hits by orthogonal quantitative methods.[172] Assimilation of different multi-omics datasets into a single data repository also faces the problem of data transfer, integration and harmonization of different data formats and fragmentation. Moreover, analyzing single-cell multi-omics data might increase the chances of false results which demands appropriate data processing, normalization and analysis with appropriate statistical methods.[173] Another challenge is clinical trial feasibility, particularly for multi-center and tissue-derived omics. To generate and manage robust data, there are requirements for well-selected clinical sites, strict standard operating procedures (SOPs), cross-site controls, qualified analytical core facilities and the development of biobanks that will help understand disease characteristics and clinical trial outcomes.[174] Moreover, for a successful multi-omics collaboration network, alignment on human biosample accessibility, FAIR data principles and dissemination policies are also required.[175] Hence, multi-omics technology enables the identification of key nodes of a disease, where multiple layers of converging profiles maximize the opportunity to identify novèl drug targets, endotypes and biomarkers (Fig. 6A).

11.2 Digital biomarkers and biomarker-based clinical trials

Digital biomarkers have more of a role to play than just identifying the key nodes affecting an individual or providing guidance for suitable treatment options. Digital biomarkers give the opportunity to identify clinical endpoints in a remote, objective and unbiased manner. However, the use of omics technology and the availability of a massive amount of data creates a barrier as it requires sophisticated computational and statistical methods.[65] Digital biomarkers are demarcated as objective, quantifiable, physiological and behavioral data that is collected through digital devices. The collected data helps to understand and predict health-related outcomes. Nowadays, many applications help to manage health and can be accessed through smartphones which may or may not be attached to a sensor device.[176] There are ~300,000 health apps and 340 + sensor devices available today and their number double every 2 years. Nowadays, wearable sensors like smartwatches has become a trend, where people can monitor their health status anytime without any limitations. With the fast upgrading in these wearable sensors and their corresponding apps, there is also a requirement for digital exerts that can access and evaluate different healthcare applications and find the limitations and risks involved behind data sharing.[65,177] Many medical device companies like Philips, GE and Leica are working on developing new imaging technologies that could aid in better detection of digital biomarkers. Concurrently, there are many information technology companies like Google, IBM, Microsoft and PathAI that are developing tools, models and methods using machine learning and artificial intelligence for better analysis of big data.[65] Several clinical trials are currently in place to bring precision medicine to the bedside. Researchers are working toward

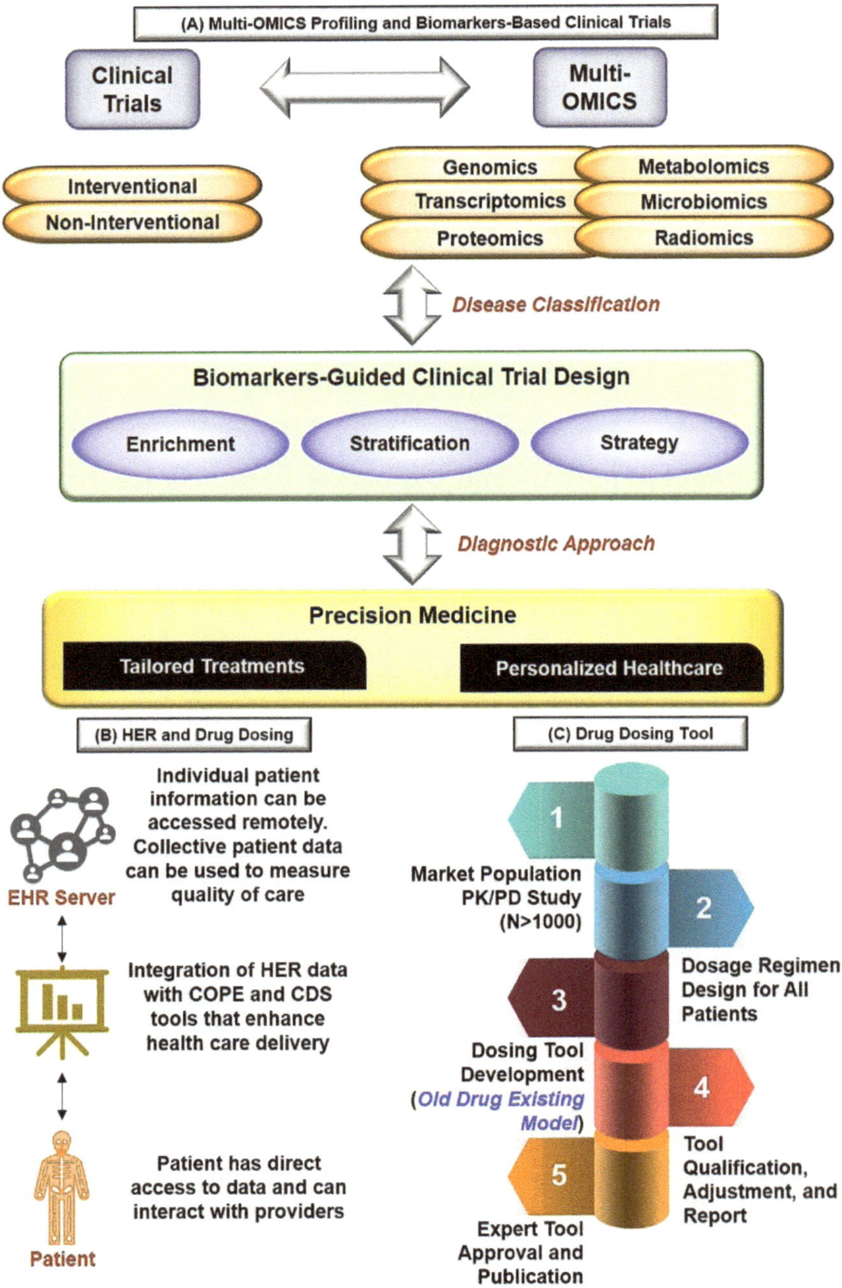

Fig. 6 (A) artificial intelligence is involved in precision medicine through multi-OMICS profiling and biomarkers-based clinical trials. Clinical trials and multi-OMICS are two inter-connected branches. Clinical trials involve interventional and non-interventional, where multi-OMICS include genomics, transcriptomics, proteomics, metabolomics,

the use of digital markers, allowing tests to be performed using smartphones or wearable sensors. The Clinical Trials Transformation Initiative (CTTI) has put forth the guidelines and detailed instructions for the development of digital biomarkers. It also emphasizes the use of digital biomarkers in clinical trials that are patient-centric enables faster decision-making, time savings and reduced costs. Diagnostics and therapeutic strategies targeting subpopulations are now at our doorsteps, demanding better ways to integrate artificial intelligence and machine learning tools into clinical trial design. As the treatment regime is individualized, we must move beyond fixed p-value based evaluation to further help to implement patient-centric clinical trials, incorporate Bayesian decision analysis (BDA) and patient preference into the clinical trial design.[178] Moreover, it is important to describe the context-of-use (CoU) for each biomarker used in clinical trial settings.[179] The CoU covers a broad range of factors, from diagnostic, safety monitoring, pharmacodynamic response, to predictive and prognostic biomarker applications. One needs to refer to the FDA-NIH biomarker working group and its related online resource BEST (Biomarkers, Endpoints and other Tools) before taking it to clinical trials.[180] For a smooth flow of clinical trials, one needs to select the right method along with a master protocol such as basket, umbrella or adaptive platform trials.[181] Currently, physiological biomarkers like blood pressure, clinically-established protein biomarkers such as C-reactive protein or imaging readouts are being accessed. However, if there is convincing evidence from independent epidemiological studies and clinical trials that biomarkers correlate closely with clinical outcome assessments, then it will substitute for traditional clinical endpoints.[169]

11.3 Precision dosing and safety measures

Precision dosing interchangeably used as individualized dosing works on drug attributes such as narrow therapeutic index (NTI), pharmacokinetic/pharmacodynamic (PK/PD) variability, disease state characterization and

microbiomics, and radiomics. Multi-OMICS and clinical trials further involved in disease classification, which helps in biomarkers-guided clinical trial design in three different ways, such as enrichment, stratification, and strategy. This will help in diagnostic approach through tailored treatments and personalized healthcare. (B) human electronic record integrated with drug dosing through the application of artificial intelligence algorithms. Electronic health record server integrated with COPE and CDS tools that enhance health care delivery. (C) artificial intelligence is also involved in drug dosing tools in five different ways, namely PK/PD study, dosage regimen design, dosing tool development, tool qualification, and tool approval.

patient-specific factors to optimize drug treatment. The goal of choosing the right drug and dose for the right patient still stands as a challenge for clinicians.[182] Precision dosing ensures great benefits and the least side effects except that one need to adjust dosages based on age, ethnicity, organ failure, or the use of interacting drugs.[183] Precision dosing depends on pharmacological models to predict a patient's response to a given dose and with data on individuals improving, the predictive results become better. According to the industrial perspective, precision dosing is an aspect of precision medicine that can be applied to a drug maker's portfolio. Moreover, the inclusion of precision dosing early in trials can eliminate probable risk factors of a drug candidate by improving drug efficacy and reducing toxicity that will further result in less patient dropout even if it is not genetically targeted. Moreover, it will not only give the opportunity to understand individual dose-response relationships but will also identify true non-responders to therapy. Nonetheless, the potential of getting a drug approved increases if the dosing is personalized.[184] Many drugs cannot follow precision dosing. However, it is needed in clinical fields with narrow therapeutic index (TI) such as antiarrhythmics, anticoagulants, antiepileptics, antineoplastics, aminoglycoside antibiotics and immunosuppressants, and also for those patients who belong to a special population.[185] Research is heading towards biosimulation, which can predict the dose for a given patient more accurately.[186] Moreover, model-informed approaches have become increasingly important to identify unique physiological and molecular characteristics of individual patients. It's important to shift to precision dosing from the currently ongoing approach, "one dosage for all," because precision medicine is better and will be increasingly feasible. Its impact is likely to be substantial for some drugs, but not necessarily for all. Thus, it is important to identify factors that make drugs good targets for precision dosing. Another feature of precision medicine is safety in order to avoid side effects of the drug. The FDA Table of Pharmacogenomic Biomarkers in Drug Labels (96 drugs in total) confirmed that 62% of the biomarkers are used to exclude patients or adjust dose based on safety-related concerns, and 31% are used for efficacy.[187,188] Some of the safety biomarkers measure polymorphisms in genes encoding metabolizing enzymes and most of the drug transporters evaluate adequate drug exposure. Other safety biomarkers focus on idiosyncratic safety issues which assess enzyme deficiencies and immune response.[189] Moreover, drug dosing modifications on undefined genomic data might compromise patients' safety. To solve this issue further, advancements are required in the field of pharmacogenomics to elucidate genotype-phenotype relationships.[190] Many drug development programs have been initiated by the government to fill these unmet needs, including the orphan drug program implemented in 1983 that focuses on products designed to treat rare diseases and the Food and Drug Administration Safety and Innovation Act (FDASIA) of 2012 that introduced the breakthrough therapy designation, which supports the expedited development of qualifying products. If precision medicine could be applied to such programs, it would enhance the flexibility for designing and conducting clinical trials for approval.[191]

12. Future challenges and perspectives of artificial intelligence in precision medicine

The proposed initiative of precision medicine has two components i.e., short-term focus on disease and the long-term goal to generate knowledge that can be applied to improve patients' health. With the advancements in basic research including molecular biology, genomics and bioinformatics; these two components are now well-attainable. Precision medicine will have a firm impact on clinicians, health systems and patients. However, the current limitations include data bias, privacy and security issues, and lack of transparency.[109] Regions with poor resources, e.g., underdeveloped countries, will face challenges in adopting artificial intelligence, pushing them further behind in improving healthcare. It is, however, well accepted that incorporation of new technologies in healthcare will be cost-effective in the long run. If such countries can invest in AI technology, it would help physicians to improve decision-making and reduce unnecessary hospitalizations, further reducing costs.[108] A review by 40 experts proposed that the key barrier in precision medicine is not science and technology, but policy-making and governance. To maximize the potential of precision medicine, it becomes mandatory to unlock the data, prioritize inclusivity and rethink pricing models. The Global Precision Medicine Council's Vision Statement for Precision Medicine presented a document representing models that provide directions in a variety of areas such as building trust and involvement of consumers in genetic testing, make availability on precision medicine approach in fast-track situations and constructing a regulatory system that could maintain privacy and ownership of genomic data. This step by the council was needed to understand the robust international collaborative environment needed to comprehend the challenges and therefore find a solution that can bridge the policy and governance gaps.[192] Achieving the goals of precision medicine in a nation also demands an improved regulatory framework. A collaboration between the NIH and the Department of Health and Human Services is working to bring a common rule that brings people willing to participate in research in innovative ways. Further, the Food and Drug Administration is working with many scientific communities such that the genomic technology supports innovation and brings trust in people that the technology is safe and effective. One needs to evolve with the experience by accepting advancements in healthcare systems, supporting data-sharing approaches and continuously improving infrastructure learning. This would help to add testing methodologies for new hypotheses, where real-time evaluation is done against a limited set of treatment options for a given condition that further help to determine what treatments provide the best results for different patient subgroups. This would partially offset cost concerns by dropping expenditures associated with unnecessary or ineffective care. It might take time to build an improved infrastructure that fully supports extensive sharing and effective use of genomic data among the population. However, the ultimate result would transform the healthcare system such that it will make the clinical-care system less dependent on externally funded research. The establishment of open data-sharing resources will not only reduce the cost of conducting genetic research but will also limit the number of cases related to conflict of interests that occurs when academic centers partner with commercial activities. Moreover, more research is needed to accelerate the delivery of precision medicine, such as the development of guidelines for -omic data integration that will enable better prediction of bio-system behavior. Nowadays, wearable sensors are used in a clinical settings that demand real-time streaming of data analytics so that fitness and preventative medicine can be monitored and alarm fatigue is reduced.[193,194] Additionally, due to high patient variability for any disease, the precise subtypes of many diseases remain unknown. Therefore, systematic big data analytics to model physician knowledge to validate the reliability of patient subgrouping based on EHR mining in necessary.[143] Two well-known efforts of IBM Watson and Arivale's wellness project reported that the data collected through their artificial intelligence model could not deal with complex and real-time personalized data. Thus, the healthcare industries need more well-labeled data, knowledge-guided models and experienced human resources.[195,196] Artificial intelligence may optimize the care trajectory of chronic disease patients, thereby suggesting precision therapies for complex illnesses, reducing medical errors and improving subject enrollment into clinical trials. Although AI technology was not specifically developed for healthcare, it is poised to address some rigid medical errors because AI technology is neither astute nor intuitive. Thus, human intelligence will always remain an essential component while using AI in medical practices.

Conflict of interest

The authors declare no conflict of interest

Acknowledgment

We would like to thanks the senior management of Delhi Technological University for their constant support and guidance.

Author's contribution

RG, MS, RK and PK conceived and designed this project. Data collected and analyzed by RG, MS, and RK. Art work is done by RG and PK. Manuscript has been written by RG, MS and PK.

References

1. Moingeon P, Kuenemann M, Guedj M. Artificial intelligence-enhanced drug design and development: toward a computational precision medicine. *Drug Discov Today*. 2022;27:215–222. https://doi.org/10.1016/J.DRUDIS.2021.09.006.
2. Mak KK, Pichika MR. Artificial intelligence in drug development: present status and future prospects. *Drug Discov Today*. 2019;24:773–780. https://doi.org/10.1016/J.DRUDIS.2018.11.014.
3. Uddin M, Wang Y, Woodbury-Smith M. Artificial intelligence for precision medicine in neurodevelopmental disorders. *NPJ Digit Med*. 2019;21(2):1–10. https://doi.org/10.1038/s41746-019-0191-0.
4. Prosperi M, Min JS, Bian J, Modave F. Big data hurdles in precision medicine and precision public health. *BMC Med Inform Decis Mak*. 2018;18:1–15. https://doi.org/10.1186/S12911-018-0719-2/FIGURES/7.
5. Xiong HY, Alipanahi B, Lee LJ, et al. The human splicing code reveals new insights into the genetic determinants of disease. *Science*. 2015;347:1254806. https://doi.org/10.1126/SCIENCE.1254806.
6. Uddin M, Tammimies K, Pellecchia G, et al. Brain-expressed exons under purifying selection are enriched for de novo mutations in autism spectrum disorder. *Nat Genet*. 2014;467(46):742–747. https://doi.org/10.1038/ng.2980.
7. Alipanahi B, Delong A, Weirauch MT, Frey BJ. Predicting the sequence specificities of DNA- and RNA-binding proteins by deep learning. *Nat Biotechnol*. 2015;338(33):831–838. https://doi.org/10.1038/nbt.3300.
8. Kermany DS, Goldbaum M, Cai W, et al. Identifying medical diagnoses and treatable diseases by image-based deep learning. *Cell*. 2018;172:1122–1131.e9. https://doi.org/10.1016/J.CELL.2018.02.010.
9. Kosmicki JA, Sochat V, Duda M, Wall DP. Searching for a minimal set of behaviors for autism detection through feature selection-based machine learning. *Transl Psychiatry*. 2015;5. https://doi.org/10.1038/TP.2015.7, e514.
10. Poplin R, Varadarajan AV, Blumer K, et al. Prediction of cardiovascular risk factors from retinal fundus photographs via deep learning. *Nat Biomed Eng*. 2018;23(2):158–164. https://doi.org/10.1038/s41551-018-0195-0.

11. Bone D, Goodwin MS, Black MP, Lee CC, Audhkhasi K, Narayanan S. Applying machine learning to facilitate autism diagnostics: Pitfalls and promises. *J Autism Dev Disord.* 2015;45:1121. https://doi.org/10.1007/S10803-014-2268-6.
12. Bello GA, Dawes TJW, Duan J, et al. Deep-learning cardiac motion analysis for human survival prediction. *Nat Mach Intell.* 2019;12(1):95–104. https://doi.org/10.1038/s42256-019-0019-2.
13. Ainscough BJ, Barnell EK, Ronning P, et al. A deep learning approach to automate refinement of somatic variant calling from cancer sequencing data. *Nat Genet.* 2018;5012(50):1735–1743. https://doi.org/10.1038/s41588-018-0257-y.
14. Esteva A, Kuprel B, Novoa RA, et al. Dermatologist-level classification of skin cancer with deep neural networks. *Nature.* 2017;5427639(542):115–118. https://doi.org/10.1038/nature21056.
15. Leff DR, Yang GZ. Big data for precision medicine. *Engineering.* 2015;1:277–279. https://doi.org/10.15302/J-ENG-2015075.
16. Mesko B. The role of artificial intelligence in precision medicine. *Exp Rev Precis Med Drug Develop.* 2017;2:239–241. https://doi.org/10.1080/23808993.2017.1380516.
17. Rodriguez F, Scheinker D, Harrington RA. Promise and perils of big data and artificial intelligence in clinical medicine and biomedical research. *Circ Res.* 2018;123:1282–1284. https://doi.org/10.1161/CIRCRESAHA.118.314119.
18. Davenport T, Kalakota R. The potential for artificial intelligence in healthcare. *Futur Healthc J.* 2019;6:94. https://doi.org/10.7861/FUTUREHOSP.6-2-94.
19. Lee SI, Celik S, Logsdon BA, et al. A machine learning approach to integrate big data for precision medicine in acute myeloid leukemia. *Nat Commun.* 2018;9. https://doi.org/10.1038/S41467-017-02465-5.
20. Soffer S, Ben-Cohen A, Shimon O, Amitai MM, Greenspan H, Klang E. Convolutional neural networks for radiologic images: a radiologist's guide. *Radiology.* 2019;290:590–606. https://doi.org/10.1148/RADIOL.2018180547/ASSET/IMAGES/LARGE/RADIOL.2018180547.TBL5.JPEG.
21. Montagnon E, Cerny M, Cadrin-Chênevert A, et al. Deep learning workflow in radiology: a primer. *Insights Imaging.* 2020;11:1–15. https://doi.org/10.1186/S13244-019-0832-5/TABLES/2.
22. Rajput A. *Natural Language Processing, Sentiment Analysis, and Clinical Analytics.* Innov Heal Informatics A Smart Healthc Prim; 2020:79–97. https://doi.org/10.1016/B978-0-12-819043-2.00003-4.
23. Abdullah U, Ligęza A, Zafar K. Performance evaluation of rule-based expert systems: an example from medical billing domain. *Expert Syst.* 2017;34. https://doi.org/10.1111/EXSY.12218, e12218.
24. Davenport TH, Glaser J. Just-in-time delivery comes to knowledge management. *Harv Bus Rev.* 2002;80:107. https://pubmed.ncbi.nlm.nih.gov/12140850/. [accessed January 25, 2022].
25. Hussain A, Malik A, Halim MU, Ali AM. The use of robotics in surgery: a review. *Int J Clin Pract.* 2014;68:1376–1382. https://doi.org/10.1111/IJCP.12492.
26. Noorbakhsh-Sabet N, Zand R, Zhang Y, Abedi V. Artificial intelligence transforms the future of healthcare. *Am J Med.* 2019;132:795. https://doi.org/10.1016/J.AMJMED.2019.01.017.
27. Krittanawong C. The rise of artificial intelligence and the uncertain future for physicians. *Eur J Intern Med.* 2018;48:e13–e14. https://doi.org/10.1016/J.EJIM.2017.06.017.
28. Krittanawong C, Zhang HJ, Wang Z, Aydar M, Kitai T. Artificial intelligence in precision cardiovascular medicine. *J Am Coll Cardiol.* 2017;69:2657–2664. https://doi.org/10.1016/J.JACC.2017.03.571.

29. Xu J, Yang P, Xue S, et al. Translating cancer genomics into precision medicine with artificial intelligence: applications, challenges and future perspectives. *Hum Genet.* 2019;1382(138):109–124. https://doi.org/10.1007/S00439-019-01970-5.
30. Afzal M, Riazul Islam SM, Hussain M, Lee S. Precision medicine informatics: principles, prospects, and challenges. *IEEE Access.* 2020;8:13593–13612. https://doi.org/10.1109/ACCESS.2020.2965955.
31. Fröhlich H, Balling R, Beerenwinkel N, et al. From hype to reality: data science enabling personalized medicine. *BMC Med.* 2018;16:1–15. https://doi.org/10.1186/S12916-018-1122-7/FIGURES/5.
32. Ahmed Z, Mohamed K, Zeeshan S, Dong XQ. Artificial intelligence with multi-functional machine learning platform development for better healthcare and precision medicine. *Database J Biol Database Curat.* 2020;2020. https://doi.org/10.1093/DATABASE/BAAA010.
33. Kodali RK, Swamy G, Lakshmi B. In: *An Implementation of IoT for Healthcare. 2015 IEEE Recent Advances in Intelligent Computational Systems (RAICS).* 2016; 2015:411–416. https://doi.org/10.1109/RAICS.2015.7488451.
34. Dias D, Cunha JPS. Wearable health devices—vital sign monitoring, systems and technologies. *Sensors (Basel).* 2018;18. https://doi.org/10.3390/S18082414.
35. Tebani A, Afonso C, Marret S, Bekri S. Omics-based strategies in precision medicine: toward a paradigm shift in inborn errors of metabolism investigations. *Int J Mol Sci.* 2016;17. https://doi.org/10.3390/IJMS17091555.
36. Stankovic B, Kotur N, Nikcevic G, Gasic V, Zukic B, Pavlovic S. Machine learning modeling from omics data as prospective tool for improvement of inflammatory bowel disease diagnosis and clinical classifications. *Genes (Basel).* 2021;12:1438. https://doi.org/10.3390/GENES12091438.
37. Petscharnig S, Schöffmann K. Learning laparoscopic video shot classification for gynecological surgery. *Multimed Tools Appl.* 2018;77:8061–8079. https://doi.org/10.1007/S11042-017-4699-5/FIGURES/8.
38. Hee Lee D, Yoon SN. Application of artificial intelligence-based technologies in the healthcare industry: opportunities and challenges. *Int J Environ Res Public Health.* 2021;18:1–18. https://doi.org/10.3390/IJERPH18010271.
39. Dursi LJ, Bozoky Z, de Borja R, et al. CanDIG: federated network across canada for multi-omic and health data discovery and analysis. *Cell Genom.* 2021;1:100033. https://doi.org/10.1016/J.XGEN.2021.100033.
40. Ma X, Long L, Moon S, Blythe JSA, Baxi SS. Comparison of population characteristics in real-world clinical oncology databases in the US: Flatiron Health, SEER, and NPCR. *medRxiv.* 2020. 2020.03.16.20037143 https://doi.org/10.1101/2020.03.16.20037143.
41. Crow D. A new wave of genomics for all. *Cell.* 2019;177:5–7. https://doi.org/10.1016/J.CELL.2019.02.041.
42. Halamka JD, Chu R, Hanna K, Kuzmeskas K. Early experiences with blockchain and EHRs. *Blockchain Healthc Today.* 2018;0. https://www.blockchainhealthcaretoday.com/index.php/journal/article/view/75 (accessed January 25, 2022).
43. Köhler W, Linde K, Halbach S, et al. Prognos®in the diagnosis of amalgam hypersensitivity—a diagnostic case-control study. *Complement Med Res.* 2007;14:18–24. https://doi.org/10.1159/000097998.
44. Doerr M, Guerrini C. Who are the people in your neighborhood? Personas populating unregulated mhealth research. *J Law Med Ethics.* 2020;48:37–48. https://doi.org/10.1177/1073110520917028.
45. Nimmesgern E, Benediktsson I, Norstedt I. Personalized medicine in Europe. *Clin Transl Sci.* 2017;10:61. https://doi.org/10.1111/CTS.12446.

46. Yeo IS. Hippocrates in the context of Galen: Galen's commentary on the classification of fevers in Epidemics VI. *Stud Anc Med*. 2005;31:433–443. https://doi.org/10.1163/9789004377271_027.
47. Visvikis-Siest S, Theodoridou D, Kontoe MS, Kumar S, Marschler M. Milestones in personalized medicine: from the ancient time to nowadays—the provocation of COVID-19. *Front Genet*. 2020;11:1442. https://doi.org/10.3389/FGENE.2020.569175/BIBTEX.
48. Visvikis-Siest S, Gorenjak V, Stathopoulou MG. Personalised medicine: the Odyssey from hope to practice. *J Pers Med*. 2018;8:31. https://doi.org/10.3390/JPM8040031.
49. Sorich MJ, McKinnon RA. Personalized medicine: potential, barriers and contemporary issues. *Curr Drug Metab*. 2012;13:1000–1006. https://doi.org/10.2174/138920012802138615.
50. O'Shaughnessy KM. HapMap, pharmacogenomics, and the goal of personalized prescribing. *Br J Clin Pharmacol*. 2006;61:783. https://doi.org/10.1111/J.1365-2125.2006.02683.X.
51. Van Der Wijst MGP, De Vries DH, Brugge H, Westra HJ, Franke L. An integrative approach for building personalized gene regulatory networks for precision medicine. *Genome Med*. 2018;101(10):1–15. https://doi.org/10.1186/S13073-018-0608-4.
52. Deloukas P, Bentley D. The HapMap project and its application to genetic studies of drug response. *Pharmacogenom J*. 2004;4:88–90. https://doi.org/10.1038/sj.tpj.6500226.
53. Fins JJ, Shapiro ZE. Deep brain stimulation, brain maps and personalized medicine: lessons from the human genome project. *Brain Topogr*. 2014;27:55–62. https://doi.org/10.1007/S10548-013-0297-7.
54. Insel TR. The NIMH research domain criteria (RDoC) project: precision medicine for psychiatry. *Am J Psychiatry*. 2014;171:395–397. https://doi.org/10.1176/appi.ajp.2014.14020138.
55. Hulsen T, Jamuar SS, Moody AR, et al. From big data to precision medicine. *Front Med*. 2019;6. https://doi.org/10.3389/FMED.2019.00034.
56. Azencott CA. Machine learning and genomics: precision medicine versus patient privacy. *Philos Trans R Soc A Math Phys Eng Sci*. 2018;376. https://doi.org/10.1098/RSTA.2017.0350.
57. Martorell-Marugán J, Tabik S, Benhammou Y, et al. *Deep Learning in Omics Data Analysis and Precision Medicine*. Exon Publ; 2019:37–53. https://doi.org/10.15586/COMPUTATIONALBIOLOGY.2019.CH3.
58. Regev A, Teichmann SA, Lander ES, et al. The human cell atlas. *Elife*. 2017;6. https://doi.org/10.7554/ELIFE.27041.
59. Eckhard U, Marino G, Butler GS, Overall CM. Positional proteomics in the era of the human proteome project on the doorstep of precision medicine. *Biochimie*. 2016;122:110–118. https://doi.org/10.1016/J.BIOCHI.2015.10.018.
60. Hosny A, Parmar C, Quackenbush J, Schwartz LH, Aerts HJWL. Artificial intelligence in radiology. *Nat Rev Cancer*. 2018;18:500. https://doi.org/10.1038/S41568-018-0016-5.
61. Lundervold AS, Lundervold A. An overview of deep learning in medical imaging focusing on MRI. *Z Med Phys*. 2019;29:102–127. https://doi.org/10.1016/J.ZEMEDI.2018.11.002.
62. Giger ML. Machine learning in medical imaging. *J Am Coll Radiol*. 2018;15:512–520. https://doi.org/10.1016/J.JACR.2017.12.028.
63. Ho DSW, Schierding W, Wake M, Saffery R, O'Sullivan J. Machine learning SNP based prediction for precision medicine. *Front Genet*. 2019;10:267. https://doi.org/10.3389/FGENE.2019.00267/BIBTEX.
64. Tuena C, Semonella M, Fernández-Álvarez J, Colombo D, Cipresso P. Predictive precision medicine: towards the computational challenge, P5 EHealth an agenda heal. *Technol Futur*. 2020;71–86. https://doi.org/10.1007/978-3-030-27994-3_5.

65. Seyhan AA, Carini C. Are innovation and new technologies in precision medicine paving a new era in patients centric care? *J Transl Med.* 2019;17:1–28. https://doi.org/10.1186/S12967-019-1864-9/FIGURES/4.
66. Pulley JM, Denny JC, Peterson JF, et al. Operational implementation of prospective genotyping for personalized medicine: the design of the Vanderbilt PREDICT project. *Clin Pharmacol Ther.* 2012;92:87. https://doi.org/10.1038/CLPT.2011.371.
67. Schildcrout JS, Shi Y, Danciu I, et al. A prognostic model based on readily available clinical data enriched a preemptive pharmacogenetic testing program. *J Clin Epidemiol.* 2016;72:107. https://doi.org/10.1016/J.JCLINEPI.2015.08.028.
68. Zou J, Huss M, Abid A, Mohammadi P, Torkamani A, Telenti A. A primer on deep learning in genomics. *Nat Genet.* 2019;51:12–18. https://doi.org/10.1038/S41588-018-0295-5.
69. Erijman A, Kozlowski L, Sohrabi-Jahromi S, et al. A high-throughput screen for transcription activation domains reveals their sequence features and permits prediction by deep learning. *Mol Cell.* 2020;78:890–902.e6. https://doi.org/10.1016/J.MOLCEL.2020.04.020.
70. Guo W, Xu YE, Feng X. Deep metabolism: a deep learning system to predict phenotype from genome sequencing. *arXiV.* 2017. https://doi.org/10.1101/135574.
71. O'Brien TD, Campbell NE, Potter AB, Letaw JH, Kulkarni A, Richards CS. Artificial intelligence (AI)-assisted exome reanalysis greatly aids in the identification of new positive cases and reduces analysis time in a clinical diagnostic laboratory. *Genet Med.* 2022;24:192–200. https://doi.org/10.1016/J.GIM.2021.09.007.
72. De La Vega FM, Chowdhury S, Moore B, et al. Artificial intelligence enables comprehensive genome interpretation and nomination of candidate diagnoses for rare genetic diseases. *Genome Med.* 2021;13:1–19. https://doi.org/10.1186/S13073-021-00965-0/FIGURES/5.
73. Infante T, Cavaliere C, Punzo B, Grimaldi V, Salvatore M, Napoli C. Radiogenomics and artificial intelligence approaches applied to cardiac computed tomography angiography and cardiac magnetic resonance for precision medicine in coronary heart disease: A systematic review. *Circ Cardiovasc Imaging.* 2021;14:1133–1146. https://doi.org/10.1161/CIRCIMAGING.121.013025.
74. Hoshino I, Yokota H. Radiogenomics of gastroenterological cancer: the dawn of personalized medicine with artificial intelligence-based image analysis. *Ann Gastroenterol Surg.* 2021;5:427–435. https://doi.org/10.1002/AGS3.12437.
75. Liao X, Cai B, Tian B, Luo Y, Song W, Li Y. Machine-learning based radiogenomics analysis of MRI features and metagenes in glioblastoma multiforme patients with different survival time. *J Cell Mol Med.* 2019;23:4375–4385. https://doi.org/10.1111/JCMM.14328.
76. Iwatate Y, Hoshino I, Yokota H, et al. Radiogenomics for predicting p53 status, PD-L1 expression, and prognosis with machine learning in pancreatic cancer. *Br J Cancer.* 2020;1238(123):1253–1261. https://doi.org/10.1038/s41416-020-0997-1.
77. Abajian A, Murali N, Savic LJ, et al. Predicting treatment response to intra-arterial therapies for hepatocellular carcinoma with the use of supervised machine learning—an artificial intelligence concept. *J Vasc Interv Radiol.* 2018;29:850–857.e1. https://doi.org/10.1016/J.JVIR.2018.01.769.
78. Facciorusso A, Del Prete V, Antonino M, Buccino VR, Muscatiello N. Response to repeat echoendoscopic celiac plexus neurolysis in pancreatic cancer patients: a machine learning approach. *Pancreatology.* 2019;19:866–872. https://doi.org/10.1016/J.PAN.2019.07.038.
79. Taliaz D, Spinrad A, Barzilay R, et al. Optimizing prediction of response to antidepressant medications using machine learning and integrated genetic, clinical, and demographic data. *Transl Psychiatry.* 2021;111(11):1–9. https://doi.org/10.1038/s41398-021-01488-3.

80. Wille SM, Kemp KA, Greenfield BL, Walls ML. Barriers to healthcare for American Indians experiencing homelessness. *J Soc Distr Homelessness*. 2017;26:1–8. https://doi.org/10.1080/10530789.2016.1265211.
81. Truong A, Secrest AM, Zhang M, et al. A survey of dermatologic health-related quality of life and resource access in patients experiencing homelessness. *J Am Acad Dermatol*. 2021;85:775–778. https://doi.org/10.1016/J.JAAD.2020.08.085.
82. Li JP, Haq AU, Din SU, Khan J, Khan A, Saboor A. Heart disease identification method using machine learning classification in e-healthcare. *IEEE Access*. 2020;8:107562–107582. https://doi.org/10.1109/ACCESS.2020.3001149.
83. Wroge TJ, Özkanca Y, Demiroglu C, Si D, Atkins DC, Ghomi RH. In: *Parkinson's Disease Diagnosis Using Machine Learning and Voice. 2018 IEEE Signal Process. Med. Biol. Symp. SPMB 2018 - Proc*; 2019. https://doi.org/10.1109/SPMB.2018.8615607.
84. Peng J, Guan J, Shang X. Predicting Parkinson's disease genes based on node2vec and autoencoder. *Front Genet*. 2019;10:226. https://doi.org/10.3389/FGENE.2019.00226/BIBTEX.
85. Barman RK, Mukhopadhyay A, Maulik U, Das S. Identification of infectious disease-associated host genes using machine learning techniques. *BMC Bioinform*. 2019;20:1–12. https://doi.org/10.1186/S12859-019-3317-0/FIGURES/4.
86. Alaa AM, Bolton T, Di Angelantonio E, Rudd JHF, van der Schaar M. Cardiovascular disease risk prediction using automated machine learning: a prospective study of 423,604 UK Biobank participants. *PLoS One*. 2019;14:e0213653. https://doi.org/10.1371/JOURNAL.PONE.0213653.
87. Malki Z, Atlam ES, Hassanien AE, Dagnew G, Elhosseini MA, Gad I. Association between weather data and COVID-19 pandemic predicting mortality rate: machine learning approaches. *Chaos Solitons Fractals*. 2020;138:110137. https://doi.org/10.1016/J.CHAOS.2020.110137.
88. Gurovich Y, Hanani Y, Bar O, et al. Identifying facial phenotypes of genetic disorders using deep learning. *Nat Med*. 2019;251(25):60–64. https://doi.org/10.1038/s41591-018-0279-0.
89. Qin B, Li D. Identifying facemask-wearing condition using image super-resolution with classification network to prevent COVID-19. *Sensors*. 2020;20:5236. https://doi.org/10.3390/S20185236.
90. Yang J, Zhang K, Fan H, et al. Development and validation of deep learning algorithms for scoliosis screening using back images. *Commun Biol*. 2019;21(2):1–8. https://doi.org/10.1038/s42003-019-0635-8.
91. Mishima H, Suzuki H, Doi M, et al. Evaluation of Face2Gene using facial images of patients with congenital dysmorphic syndromes recruited in Japan. *J Hum Genet*. 2019;648(64):789–794. https://doi.org/10.1038/s10038-019-0619-z.
92. Martínez-Nicolás I, Llorente TE, Martínez-Sánchez F, Meilán JJG. Ten years of research on automatic voice and speech analysis of people with Alzheimer's disease and mild cognitive impairment: a systematic review article. *Front Psychol*. 2021;12:645. https://doi.org/10.3389/FPSYG.2021.620251/BIBTEX.
93. Jeancolas L, Petrovska-Delacrétaz D, Mangone G, et al. X-vectors: new quantitative biomarkers for early Parkinson's disease detection from speech. *Front Neuroinform*. 2021;15:4. https://doi.org/10.3389/FNINF.2021.578369/BIBTEX.
94. Campbell EL, Mesía RY, Docío-Fernández L, García-Mateo C. Paralinguistic and linguistic fluency features for Alzheimer's disease detection. *Comput Speech Lang*. 2021;68:101198. https://doi.org/10.1016/J.CSL.2021.101198.
95. Karaman O, Çakın H, Alhudhaif A, Polat K. Robust automated Parkinson disease detection based on voice signals with transfer learning. *Expert Syst Appl*. 2021;178:115013. https://doi.org/10.1016/J.ESWA.2021.115013.

96. Lundberg SM, Nair B, Vavilala MS, et al. Explainable machine-learning predictions for the prevention of hypoxaemia during surgery. *Nat Biomed Eng.* 2018;210(2):749–760. https://doi.org/10.1038/s41551-018-0304-0.
97. Memarian N, Kim S, Dewar S, Engel J, Staba RJ. Multimodal data and machine learning for surgery outcome prediction in complicated cases of mesial temporal lobe epilepsy. *Comput Biol Med.* 2015;64:67–78. https://doi.org/10.1016/J.COMPBIOMED.2015.06.008.
98. Bihorac A, Ozrazgat-Baslanti T, Ebadi A, et al. MySurgeryRisk: development and validation of a machine-learning risk algorithm for major complications and death after surgery. *Ann Surg.* 2019;269:652. https://doi.org/10.1097/SLA.0000000000002706.
99. Iqbal SMA, Mahgoub I, Du E, Leavitt MA, Asghar W. Advances in healthcare wearable devices. *NPJ Flex Electron.* 2021;51(5):1–14. https://doi.org/10.1038/s41528-021-00107-x.
100. Sparaco M, Lavorgna L, Conforti R, Tedeschi G, Bonavita S. The role of wearable devices in multiple sclerosis. *Mult Scler Int.* 2018;2018:1–7. https://doi.org/10.1155/2018/7627643.
101. Patel MS, Asch DA, Volpp KG. Wearable devices as facilitators, not drivers, of health behavior change. *J Am Med Assoc.* 2015;313:459–460. https://doi.org/10.1001/JAMA.2014.14781.
102. Mizuno A, Changolkar S, Patel MS. Wearable devices to monitor and reduce the risk of cardiovascular disease: evidence and opportunities. 2021;72:459–471. https://doi.org/10.1146/annurev-med-050919-031534. https://doi.org/10.1146/ANNUREV-MED-050919-031534.
103. Rashid N, Al Faruque MA. Energy-efficient real-time myocardial infarction detection on wearable devices. *Proc Annu Int Conf IEEE Eng Med Biol Soc EMBS.* 2020; 4648–4651. https://doi.org/10.1109/EMBC44109.2020.9175232.
104. Johnson KB, Wei WQ, Weeraratne D, et al. Precision medicine, AI, and the future of personalized health care. *Clin Transl Sci.* 2021;14:86–93. https://doi.org/10.1111/CTS.12884.
105. WHO. The human resources for health crisis. *WHO.* 2011.
106. WHO. *World Health Organization, Global Strategy on Human Resources for Health: Workforce 2030.* WHO; 2016:64.
107. Chen L, Evans T, Anand S, et al. Human resources for health: overcoming the crisis. *Lancet.* 2004;364:1984–1990. https://doi.org/10.1016/S0140-6736(04)17482-5.
108. Meskó B, Hetényi G, Gyorffy Z. Will artificial intelligence solve the human resource crisis in healthcare? *BMC Health Serv Res.* 2018;18:1–4. https://doi.org/10.1186/S12913-018-3359-4/PEER-REVIEW.
109. Topol EJ. High-performance medicine: the convergence of human and artificial intelligence. *Nat Med.* 2019;25:44–56. https://doi.org/10.1038/s41591-018-0300-7.
110. Miller DD, Brown EW. Artificial intelligence in medical practice: the question to the answer? *Am J Med.* 2018;131:129–133. https://doi.org/10.1016/J.AMJMED.2017.10.035.
111. Datta S, Li Y, Ruppert MM, et al. Reinforcement learning in surgery. *Surgery.* 2021;170:329–332. https://doi.org/10.1016/J.SURG.2020.11.040.
112. Marc Overhage J, McCallie D. Physician time spent using the electronic health record during outpatient encounters a descriptive study. *Ann Intern Med.* 2020;172:169–174. https://doi.org/10.7326/M18-3684.
113. Bilkey GA, Burns BL, Coles EP, Mahede T, Baynam G, Nowak KJ. Optimizing precision medicine for public health. *Front Public Heal.* 2019;7:42. https://doi.org/10.3389/FPUBH.2019.00042/BIBTEX.
114. Ginsburg GS, Phillips KA. Precision medicine: from science to value. *Health Aff (Millwood).* 2018;37:694–701. https://doi.org/10.1377/hlthaff.2017.1624.

115. Kogan JN, Empey P, Kanter J, Keyser DJ, Shrank WH. Delivering on the value proposition of precision medicine: the view from healthcare payers. *Am J Manag Care*. 2018;24.
116. Barker RW. Is precision medicine the future of healthcare? *Per Med*. 2017;14:459–461. https://doi.org/10.2217/PME-2017-0060.
117. Agarwal A, Pritchard D, Gullett L, Amanti KG, Gustavsen G. A quantitative framework for measuring personalized medicine integration into US healthcare delivery organizations. *J Pers Med*. 2021;11:196. https://doi.org/10.3390/JPM11030196.
118. Manolio TA, Chisholm RL, Ozenberger B, et al. Implementing genomic medicine in the clinic: the future is here. *Genet Med*. 2013;154(15):258–267. https://doi.org/10.1038/gim.2012.157.
119. Pritchard DE, Moeckel F, Villa MS, Housman LT, McCarty CA, McLeod HL. Strategies for integrating personalized medicine into healthcare practice. *Per Med*. 2017;14:141–152. https://doi.org/10.2217/PME-2016-0064/SUPPL_FILE/SUPPL_MATERIAL2.DOCX.
120. Majumder B, Baraneedharan U, Thiyagarajan S, et al. Predicting clinical response to anticancer drugs using an ex vivo platform that captures tumour heterogeneity. *Nat Commun*. 2015;6:6169. https://doi.org/10.1038/NCOMMS7169.
121. Yang H, Wei Q, Zhong X, Yang H, Li B. Cancer driver gene discovery through an integrative genomics approach in a non-parametric Bayesian framework. *Bioinformatics*. 2017;33:483–490. https://doi.org/10.1093/BIOINFORMATICS/BTW662.
122. Yamankurt G, Berns EJ, Xue A, et al. Exploration of the nanomedicine-design space with high-throughput screening and machine learning. *Nat Biomed Eng*. 2019;34 (3):318–327. https://doi.org/10.1038/s41551-019-0351-1.
123. Alafeef M, Srivastava I, Pan D. Machine learning for precision breast cancer diagnosis and prediction of the nanoparticle cellular internalization. *ACS Sensors*. 2020;5: 1689–1698. https://doi.org/10.1021/ACSSENSORS.0C00329/SUPPL_FILE/SE0C00329_SI_001.PDF.
124. Liu L, Zhang L, Feng H, et al. Prediction of the blood-brain barrier (BBB) permeability of chemicals based on machine-learning and ensemble methods. *Chem Res Toxicol*. 2021;34:1456–1467. https://doi.org/10.1021/ACS.CHEMRESTOX.0C00343/SUPPL_FILE/TX0C00343_SI_001.XLSX.
125. Singh AV, Chandrasekar V, Janapareddy P, et al. Emerging application of nanorobotics and artificial intelligence to cross the BBB: advances in design, controlled maneuvering, and targeting of the barriers. *ACS Chem Neurosci*. 2021;12:1835–1853. https://doi.org/10.1021/ACSCHEMNEURO.1C00087.
126. Shityakov S, Roewer N, Broscheit JA, Förster C. In silico models for nanotoxicity evaluation and prediction at the blood-brain barrier level: a mini-review. *Comput Toxicol*. 2017;2:20–27. https://doi.org/10.1016/J.COMTOX.2017.02.003.
127. Kingston BR, Syed AM, Ngai J, Sindhwani S, Chan WCW. Assessing micrometastases as a target for nanoparticles using 3D microscopy and machine learning. *Proc Natl Acad Sci USA*. 2019;116:14937–14946. https://doi.org/10.1073/PNAS.1907646116/-/DCSUPPLEMENTAL.
128. Liu Z, Shi Y, Chen H, et al. Machine learning on properties of multiscale multisource hydroxyapatite nanoparticles datasets with different morphologies and sizes. *NPJ Comput Mater*. 2021;71(7):1–11. https://doi.org/10.1038/s41524-021-00618-1.
129. Priyadarshni N, Dutta S, Chanda N. Diethylenetriaminepentaacetic acid-functionalized gold nanoparticles for the detection of toxic chromium assisted by a machine-learning approach. *ACS Appl Nano Mater*. 2021;4:10713–10724. https://doi.org/10.1021/ACSANM.1C02171/SUPPL_FILE/AN1C02171_SI_001.PDF.
130. Shah S, Liu Y, Hu W, Gao J. Modeling particle shape-dependent dynamics in nanomedicine. *J Nanosci Nanotechnol*. 2011;11:919. https://doi.org/10.1166/JNN.2011.3536.

131. Pellegrino F, Isopescu R, Pellutiè L, et al. Machine learning approach for elucidating and predicting the role of synthesis parameters on the shape and size of TiO_2 nanoparticles. *Sci Reports*. 2020;101(10):1–11. https://doi.org/10.1038/s41598-020-75967-w.
132. Mekki-Berrada F, Ren Z, Huang T, et al. Two-step machine learning enables optimized nanoparticle synthesis. *NPJ Comput Mater*. 2021;71(7):1–10. https://doi.org/10.1038/s41524-021-00520-w.
133. Soltani M, Kashkooli FM, Souri M, et al. Enhancing clinical translation of cancer using nanoinformatics. *Cancers (Basel)*. 2021;13:2481. https://doi.org/10.3390/CANCERS13102481.
134. Stillman NR, Balaz I, Tsompanas MA, et al. Evolutionary computational platform for the automatic discovery of nanocarriers for cancer treatment. *NPJ Comput Mater*. 2021;71(7):1–12. https://doi.org/10.1038/s41524-021-00614-5.
135. Harrison PJ, Wieslander H, Sabirsh A, et al. Deep-learning models for lipid nanoparticle-based drug delivery. *Nanomedicine*. 2021;16:1097–1110. https://doi.org/10.2217/NNM-2020-0461/ASSET/IMAGES/LARGE/FIGURE7.JPEG.
136. Zapke SE, Willmann S, Grebe SO, Menke K, Thürmann PA, Schmiedl S. Comparing predictions of a PBPK model for cyclosporine with drug levels from therapeutic drug monitoring. *Front Pharmacol*. 2021;12:1134. https://doi.org/10.3389/FPHAR.2021.630904/BIBTEX.
137. Espié P, Tytgat D, Sargentini-Maier ML, Poggesi I, Watelet JB. Physiologically based pharmacokinetics (PBPK). *Drug Metab Rev*. 2009;41:391–407. https://doi.org/10.1080/10837450902891360.
138. Gao H, Wang W, Dong J, Ye Z, Ouyang D. An integrated computational methodology with data-driven machine learning, molecular modeling and PBPK modeling to accelerate solid dispersion formulation design. *Eur J Pharm Biopharm*. 2021;158:336–346. https://doi.org/10.1016/J.EJPB.2020.12.001.
139. Empey PE. Pharmacogenomics to achieve precision medicine. *Am J Health Syst Pharm*. 2016;73:1906–1907. https://doi.org/10.2146/AJHP160682.
140. Haq IU, Haq I, Xu B. Artificial intelligence in personalized cardiovascular medicine and cardiovascular imaging. *Cardiovasc Diagn Ther*. 2021;11:91123–91923. https://doi.org/10.21037/CDT.2020.03.09.
141. Panayides AS, Pattichis MS, Leandrou S, Pitris C, Constantinidou A, Pattichis CS. Radiogenomics for precision medicine with a big data analytics perspective. *IEEE J Biomed Heal Inform*. 2019;23:2063–2079. https://doi.org/10.1109/JBHI.2018.2879381.
142. Cunningham P. Dimension reduction. *Cogn Technol*. 2008;91–112. https://doi.org/10.1007/978-3-540-75171-7_4.
143. Wu PY, Cheng CW, Kaddi CD, Venugopalan J, Hoffman R, Wang MD. Omic and electronic health records big data analytics for precision medicine. *IEEE Trans Biomed Eng*. 2017;64:263. https://doi.org/10.1109/TBME.2016.2573285.
144. Ienca M, Ferretti A, Hurst S, Puhan M, Lovis C, Vayena E. Considerations for ethics review of big data health research: a scoping review. *PLoS One*. 2018;13. https://doi.org/10.1371/JOURNAL.PONE.0204937.
145. Pastorino R, De Vito C, Migliara G, et al. Benefits and challenges of Big Data in healthcare: an overview of the European initiatives. *Eur J Public Health*. 2019;29:23. https://doi.org/10.1093/EURPUB/CKZ168.
146. Isaacson RS, Ganzer CA, Hristov H, et al. The clinical practice of risk reduction for Alzheimer's disease: a precision medicine approach. *Alzheimers Dement*. 2018;14:1663. https://doi.org/10.1016/J.JALZ.2018.08.004.
147. Reitz C. Toward precision medicine in Alzheimer's disease. *Ann Transl Med*. 2016;4. https://doi.org/10.21037/ATM.2016.03.05.
148. Yamazaki Y, Zhao N, Caulfield TR, Liu CC, Bu G. Apolipoprotein E and Alzheimer disease: pathobiology and targetingstrategies. *Nat Rev Neurol*. 2019;15:501. https://doi.org/10.1038/S41582-019-0228-7.

149. O'Bryant SE, Zhang F, Johnson LA, et al. A precision medicine model for targeted NSAID therapy in Alzheimer's disease. *J Alzheimers Dis.* 2018;66:97–104. https://doi.org/10.3233/JAD-180619.
150. Aisen PS, Schafer KA, Grundman M, et al. Effects of rofecoxib or naproxen vs placebo on Alzheimer disease progression: a randomized controlled trial. *J Am Med Assoc.* 2003;289:2819–2826. https://doi.org/10.1001/JAMA.289.21.2819.
151. Hampel H, Caraci F, Cuello AC, et al. A path toward precision medicine for neuroinflammatory mechanisms in Alzheimer's disease. *Front Immunol.* 2020;11. https://doi.org/10.3389/FIMMU.2020.00456.
152. Toups K, Hathaway A, Gordon D, et al. Precision medicine approach to Alzheimer's disease: successful proof-of-concept trial. *medRxiv.* 2021. https://doi.org/10.1101/2021.05.10.21256982. 2021.05.10.21256982.
153. Ritchie K, Carrière I, Su L, et al. The midlife cognitive profiles of adults at high risk of late-onset Alzheimer's disease: The PREVENT study. *Alzheimers Dement.* 2017;13:1089–1097. https://doi.org/10.1016/J.JALZ.2017.02.008.
154. Kourtis LC, Regele OB, Wright JM, Jones GB. Digital biomarkers for Alzheimer's disease: the mobile/wearable devices opportunity. *NPJ Digit Med.* 2019;2. https://doi.org/10.1038/S41746-019-0084-2.
155. Schneider SA, Alcalay RN. Precision medicine in Parkinson's disease: emerging treatments for genetic Parkinson's disease. *J Neurol.* 2020;267:860. https://doi.org/10.1007/S00415-020-09705-7.
156. Akirov A. The role of precision medicine in Parkinson disease—Neurology advisor. *Neurol Advis.* 2020.
157. von Linstow CU, Gan-Or Z, Brundin P. Precision medicine in Parkinson's disease patients with LRRK2 and GBA risk variants—let's get even more personal. *Transl Neurodegener.* 2020;9. https://doi.org/10.1186/S40035-020-00218-X.
158. Stoddard-Bennett T, Reijo Pera R. Treatment of Parkinson's disease through personalized medicine and induced pluripotent stem cells. *Cells.* 2019;8:26. https://doi.org/10.3390/CELLS8010026.
159. Lipsmeier F, Taylor KI, Kilchenmann T, et al. Evaluation of smartphone-based testing to generate exploratory outcome measures in a phase 1 Parkinson's disease clinical trial. *Mov Disord.* 2018;33:1287–1297. https://doi.org/10.1002/MDS.27376.
160. Zhan A, Mohan S, Tarolli C, et al. Using smartphones and machine learning to quantify Parkinson disease severity the mobile Parkinson disease score. *JAMA Neurol.* 2018;75:876–880. https://doi.org/10.1001/jamaneurol.2018.0809.
161. Wurtman RJ. Personalized medicine strategies for managing patients with Parkinsonism and cognitive deficits. *Metab Clin Exp.* 2013;62:S27–S29. https://doi.org/10.1016/J.METABOL.2012.08.025.
162. Kay C, Skotte NH, Southwell AL, Hayden MR. Personalized gene silencing therapeutics for Huntington disease. *Clin Genet.* 2014;86:29–36. https://doi.org/10.1111/CGE.12385.
163. Andrzejewski KL, Dowling AV, Stamler D, et al. Wearable sensors in Huntington disease: a pilot study. *J Huntington Dis.* 2016;5:199–206. https://doi.org/10.3233/JHD-160197.
164. Libbrecht MW, Noble WS. Machine learning applications in genetics and genomics. *Nat Rev Genet.* 2015;16:321–332. https://doi.org/10.1038/NRG3920.
165. Zinman B, Wanner C, Lachin JM, et al. Empagliflozin, cardiovascular outcomes, and mortality in type 2 diabetes. *N Engl J Med.* 2015;373:17–18. https://doi.org/10.1056/NEJMOA1504720.
166. Jensen PB, Jensen LJ, Brunak S. Mining electronic health records: towards better research applications and clinical care. *Nat Rev Genet.* 2012;13:395–405. https://doi.org/10.1038/NRG3208.

167. Aronson SJ, Rehm HL. Building the foundation for genomics in precision medicine. *Nature*. 2015;5267573(526):336–342. https://doi.org/10.1038/nature15816.
168. Shen L, Bai J, Wang J, Shen B. The fourth scientific discovery paradigm for precision medicine and healthcare: challenges ahead. *Precis Clin Med*. 2021;4:80–84. https://doi.org/10.1093/PCMEDI/PBAB007.
169. Hartl D, de Luca V, Kostikova A, et al. Translational precision medicine: an industry perspective. *J Transl Med*. 2021;19. https://doi.org/10.1186/S12967-021-02910-6.
170. Agresti A. *Categor Data Anal*. 2003.
171. Cirillo D, Valencia A. Big data analytics for personalized medicine. *Curr Opin Biotechnol*. 2019;58:161–167. https://doi.org/10.1016/J.COPBIO.2019.03.004.
172. Papiez A, Marczyk M, Polanska J, Polanski A. Batch I: batch effect identification in high-throughput screening data using a dynamic programming algorithm. *Bioinformatics*. 2019;35:1885–1892. https://doi.org/10.1093/BIOINFORMATICS/BTY900.
173. Yamada R, Okada D, Wang J, Basak T, Koyama S. Interpretation of omics data analyses. *J Hum Genet*. 2021;66:93–102. https://doi.org/10.1038/S10038-020-0763-5.
174. Baker M. Biorepositories: building better biobanks. *Nature*. 2012;486:141–146. https://doi.org/10.1038/486141A.
175. Wilkinson MD, Dumontier M, Ij J, et al. The FAIR guiding principles for scientific data management and stewardship. *Sci Data*. 2016;3. https://doi.org/10.1038/SDATA.2016.18.
176. Torous J, Andersson G, Bertagnoli A, et al. Towards a consensus around standards for smartphone apps and digital mental health. *World Psychiatry*. 2019;18:97–98. https://doi.org/10.1002/WPS.20592.
177. Kamel Boulos MN, Brewer AC, Karimkhani C, Buller DB, Dellavalle RP. Mobile medical and health apps: state of the art, concerns, regulatory control and certification. *J Pub Health Inform*. 2014;5. https://doi.org/10.5210/OJPHI.V5I3.4814.
178. Ho D, Quake SR, McCabe ERB, et al. Enabling technologies for personalized and precision medicine. *Trends Biotechnol*. 2020;38:497. https://doi.org/10.1016/J.TIBTECH.2019.12.021.
179. Frank R, Hargreaves R. Clinical biomarkers in drug discovery and development. *Nat Rev Drug Discov*. 2003;2:566–580. https://doi.org/10.1038/NRD1130.
180. Silver Spring (MD). *Food and Drug Administration (US)*. Bethesda (MD): National Institutes of Health (US), BEST (Biomarkers, EndpointS, and other Tools) Resource, BEST (Biomarkers, EndpointS, Other Tools) Resour; 2016.
181. Park JJH, Siden E, Zoratti MJ, et al. Systematic review of basket trials, umbrella trials, and platform trials: a landscape analysis of master protocols. *Trials*. 2019;20. https://doi.org/10.1186/S13063-019-3664-1.
182. Tyson RJ, Park CC, Powell JR, et al. Precision dosing priority criteria: drug, disease, and patient population variables. *Front Pharmacol*. 2020;11. https://doi.org/10.3389/FPHAR.2020.00420.
183. Peck RW. Precision dosing: an industry perspective. *Clin Pharmacol Ther*. 2021;109:47–50. https://doi.org/10.1002/CPT.2064.
184. Buntz B. How precision drug-dosing supports individualized treatment. *Drug Discov Dev*. 2021.
185. Polasek TM, Shakib S, Rostami-Hodjegan A. Precision dosing in clinical medicine: present and future. *Expert Rev Clin Pharmacol*. 2018;11:743–746. https://doi.org/10.1080/17512433.2018.1501271.
186. Darwich AS, Ogungbenro K, Vinks AA, et al. Why has model-informed precision dosing not yet become common clinical reality? Lessons from the past and a roadmap for the future. *Clin Pharmacol Ther*. 2017;101:646–656. https://doi.org/10.1002/CPT.659.

187. Tutton R. Pharmacogenomic biomarkers in drug labels: what do they tell us? *Pharmacogenomics*. 2014;15:297–304. https://doi.org/10.2217/PGS.13.198.
188. Pacanowski MA, Zineh I. Pharmacogenomic strategies in drug safety. *Drug Discov Today Ther Strateg*. 2011. https://doi.org/10.1016/j.ddstr.2011.04.002.
189. Dolsten M, Søgaard M. Precision medicine: an approach to R&D for delivering superior medicines to patients. *Clin Transl Med*. 2012;1:7. https://doi.org/10.1186/2001-1326-1-7.
190. Romine PE, Harkins SK, Gray SW. Quality in the age of precision medicine: the clinician perspective. *J Oncol Pract*. 2016;12:839–843. https://doi.org/10.1200/JOP.2016.015321.
191. Pregelj L, Hwang TJ, Darrow JJ, et al. Precision medicines have faster approvals based on fewer and smaller trials than other medicines. *Health Aff*. 2018;37:724–731. https://doi.org/10.1377/HLTHAFF.2017.1580/ASSET/IMAGES/LARGE/FIGUREEX3.JPEG.
192. Dana G. *Precision Medicine Won't Change Healthcare, but Governance Could*. World Economic Forum; 2020.
193. Andreu-Perez J, Poon CCY, Merrifield RD, Wong STC, Yang GZ. Big data for health. *IEEE J Biomed Heal Inform*. 2015;19:1193–1208. https://doi.org/10.1109/JBHI.2015.2450362.
194. Cvach M. Monitor alarm fatigue: an integrative review. *Biomed Instrum Technol*. 2012;46:268–277. https://doi.org/10.2345/0899-8205-46.4.268.
195. Fiala C, Diamandis EP. The outcomes of scientific debates should be published: the arivale story. *J Appl Lab Med*. 2020;5:1070–1075. https://doi.org/10.1093/JALM/JFAA110.
196. Jiang F, Jiang Y, Zhi H, et al. Artificial intelligence in healthcare: past, present and future. *Stroke Vasc Neurol*. 2017;2:230–243. https://doi.org/10.1136/SVN-2017-000101.

CHAPTER FOUR

Precision medicine with multi-omics strategies, deep phenotyping, and predictive analysis

Zeeshan Ahmed[a,b,*]

[a]Institute for Health, Health Care Policy and Aging Research, Rutgers, The State University of New Jersey, New Brunswick, NJ, United States

[b]Department of Medicine, Rutgers Robert Wood Johnson Medical School, Rutgers Biomedical and Health Sciences, New Brunswick, NJ, United States

*Corresponding author: e-mail address: zahmed@ifh.rutgers.edu

Contents

1. Introduction 102
2. Multi-omics strategies & precision medicine 104
3. Deep phenotyping & precision medicine 107
4. Discussion & future recommendations 111
Acknowledgments 115
Competing interests 115
Author contribution 115
Biographical note 115
Funding information 116
References 116

Abstract

Precision medicine is driven by the paradigm shift of empowering clinicians to predict the most appropriate course of action for patients with complex diseases and to improve routine medical and public health practice. Understanding patients' multi-omics make-up in conjunction with the clinical data will lead to determining predisposition, diagnostic, prognostic and predictive biomarkers and to optimal paths providing personalized care for diverse and targeted chronic, acute, and infectious diseases. Precision medicine promotes integrating collective and individualized clinical data with patient-specific multi-omics data to develop therapeutic strategies and knowledge bases for predictive and personalized medicine in diverse populations. Artificial intelligence approaches and machine learning algorithms will add additional capabilities to precision medicine that will leverage and extend the information contained within the original data and facilitate modeling patient-specific multi-omics data against publicly available annotation data for better understanding disease mechanisms. This chapter

Progress in Molecular Biology and Translational Science, Volume 190
ISSN 1877-1173
https://doi.org/10.1016/bs.pmbts.2022.02.002

discusses emerging, significant, and recently reported multi-omics, deep phenotyping, and translational approaches to facilitate the implementation of precision medicine, as well as innovative, smart, and robust big-data platforms that are necessary to improve the quality and transition of healthcare by analyzing heterogeneous healthcare and multi-omics data.

1. Introduction

Various life threatening, long lasting, rare, emerging, and prevalent known chronic, acute, and infectious, and unknown diseases are the leading causes of deaths in the US[1] and even around the globe.[2–5] Factors that drive these diseases the most, include but are not limited to environmental, behavioral, metabolic, and genetic risks.[6] Availability of designer drugs, precise targeting of molecular fingerprints for tumor treatment and for selection of optimal drug therapy regimens, and predicting individual susceptibility to disease, diagnosis, and treatment of complex disorders are but a few of the many transformations expected in the decades to come.[7,8] The state-of-the-art is multi-omics to investigate the molecular mechanisms of disease. Various bench protocols and analytic approaches, involving genomics, epigenomics, transcriptomics, proteomics, and metabolomics data breakdown all are involved.[9,10] Due to the complex nature, progression, and heterogeneity in complex disorders, the scientific approach would be to perform analysis of individual patient profiles, giving rise to a new form of preventive and personalized medicine in healthcare.[11–17]

Precision medicine is one of the most significant developments in life and medical science, as it promotes intelligently integrative clinical (text and images) and multi-omics data analysis to provide new insights across diverse populations.[9–21] The potential benefits of precision medicine include, but not limited to, selecting treatments that are patient-specific; identifying the best strategies to diagnose and treat patients especially at risk for medical complications; analyzing clinical trials and patient records to identify follow-on indications; analyzing large volumes of clinical and multi-omics data into information that can be used to predict and prevent crisis; and analyzing patients' life style, medical history and genomic profile to perform proactive care. Precision medicine is developing quickly and identifying causes of end-stage disease and providing better personalized treatments through predictive analysis and deep phenotyping are being developed actively. Effectively implementing precision medicine approaches requires a secure information technology framework with novel taxonomies,[9,12,14,16] preferably integrated into medical institutions and linked to operational electronic health records (EHR), pathology and patient recruitment systems for consenting and sample collection (e.g., plasma, serum, nasopharyngeal swabs, tissue, blood, and their derivatives (DNA, RNA, stem cells, etc.)), and genomics and molecular labs/cores. To provide timely, reproducible, intelligent, integrative, and effective clinical, metabolomics, genomics, and other omics data analysis, it is important to link dynamic electronic databases to annotation data (e.g., diseases, genes, variants, biomarkers, drugs) and develop bioinformatics applications with user-friendly interfaces (e.g., desktop, web, mobile). The challenge is to efficiently standardize and integrate heterogenous data from disparate groups with non-aligned formats to provide decision-making information to healthcare providers without requiring massive computational resources.[22]

To effectively meet the goals of healthcare data analytics, significant efforts are required from the experts (e.g., administration, research investigators, physicians, surgeons, nurses, data and bench scientists, etc.) in multidisciplinary sciences (e.g., medical science, pathology, epidemiology, metabolomics, genomics, proteomics, bioinformatics, data science, precision medicine, etc.) located within one or multiple organizational units (e.g., hospitals, investigating dry and wet/bench laboratories, cores, security, information technology, etc.). One of the major unsolved challenges is to establish an efficient and secure workflow that can connect all internal and external organizational units, people, and systems to streamline transparent and reproducible data flow, quality inspection, integration, management, analysis, visualization, reporting, and sharing (Fig. 1). Despite current progress, there is still no stand-alone platform available to efficiently integrate clinical, metabolomics, and genomics data acquisition for practicing precision medicine.[23] Platforms like The Cancer Genome Atlas (TCGA),[24] database of Genotypes and Phenotypes (dbGaP),[25] Sequence Read Archive (SRA),[26] European Nucleotide Archive (ENA),[27] UK-Biobank,[28] "All of Us" program,[28] Trans-omics for precision medicine (TOPMed),[29] provide a great resource for genomics and other omics data but offer limited capacity for clinical information because they are not directly integrated with clinical health systems like Epic and Cerner, etc.[9] These platforms do not allow clinicians and researchers to expand from one to multiple diseases, incorporate clinical and multi-omics data into statistical diagnostic models to predict unsuspected high-risk factors,[30] understand complex and influencing relationships between genotypes and phenotypes, and place recommendations back into the medical records to improve outcomes.[31] The inability of disparate platforms to effectively integrate is largely due to the

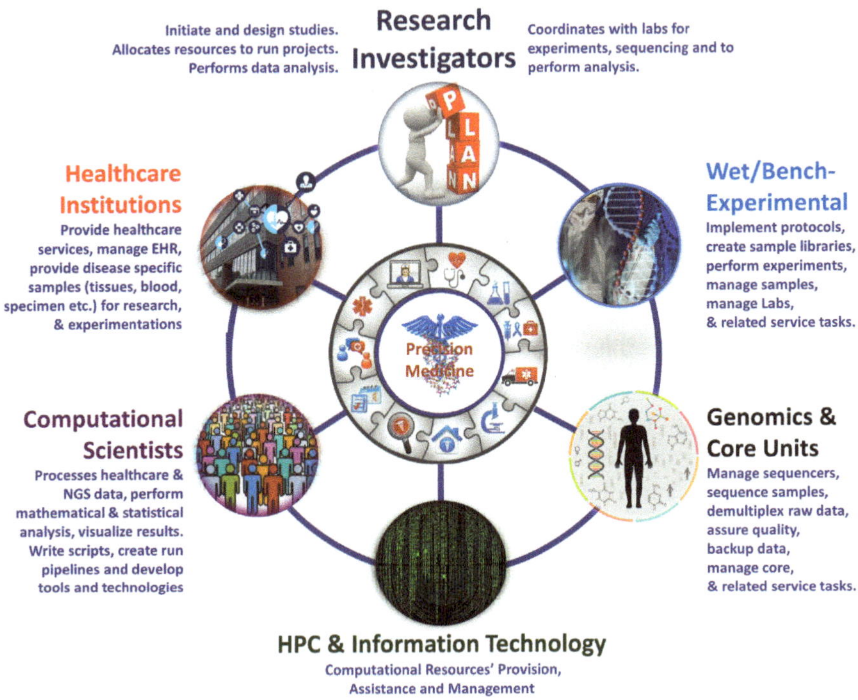

Fig. 1 Connecting multiple organizational units, people and systems to streamline healthcare data flow, integration, management, analysis and sharing.

high volume, limited focus, and heterogeneous nature of the data types they contain, which are acquired from variable sources, each with unique data structures.[32] We need innovative solutions for addressing the demands of users in clinical research settings with high data volume, large network, low resources, and acute care, where accurate diagnoses and actionable interpretations must be made rapidly and in a cost-effective manner to streamline patient treatment.[33]

2. Multi-omics strategies & precision medicine

In contrast to the classical single biomarker approaches, genomic signatures have the potential to provide more sensitive, specific, and robust predictions of disease-associated outcomes.[34] Over the last few decades, genomics has changed our views on conducting biomedical research, studying diseases, and understanding diversity in our society across the human species.[35] Although not yet widely accessible, there is a consistent rise in the numbers of individuals opting for elective genome and transcriptome sequencing.[36] Supporting precision medicine, several genomics applications have been developed and widely shared for gene expression and enrichment analysis,[37] and to investigate susceptibility in whole genome and exome sequenced data for variant discovery, annotation, prediction, and genotyping.[38] Furthermore, to understand the genetic basis of common diseases, numerous clinical, genomics, and clinical-genomics databases (e.g., Human Gene Mutation Database (HGMD), Catalogue Of Somatic Mutations In Cancer (COSMIC), Online Mendelian Inheritance in Man (OMIM), ClinVar, Ensembl, GeneCards, MalaCard, DISEASES, Disease Ontology, DiseaseEnhancer, DisGeNET, eDGAR, miR2Disease, DNetDB, GenCode, Novoseek, Swiss-Prot, LncRNADisease, Orphanet, and PROMIS-APP-SUITE (PAS), etc.) have been surfaced to support gene-disease data integration, analysis, and annotation.[35,39–46] Such, authentic information would then be used in real time to improve the care of current and future patients.

The robust identification of causal genes and disease mechanisms can provide the foundation for wide-ranging research on variable-complex disorders with the development of novel therapeutic approaches. Clinicians now have access to an increasing array of tests that allow them to determine which genetic variants exist in their patients. In the recent past, many genomics, transcriptomics, epigenomics and pharmacogenomics studies have been reported in which the authors have analyzed heterogenous next generation sequence (NGS) information to significantly improve the provision of precision medicine.[47–72] These include (I) understanding family history with genomic testing to find carriers for cancer syndromes[51]; (II) personalized treatments for the patients with cardiovascular and neurodegenerative conditions[52]; (III) genomic data model to support in clinical decision-making[53]; (IV) integrating targeted somatic and germline aberrations for supporting oncology assessment[54]; (V) RNA-seq driven expression and enrichment analysis to investigate CVD genes with associated phenotypes among high-risk heart failure patients[56]; (VI) RNA transcriptomic analysis of COVID-19 patients[57]; (VII) identifying disease-related biomarkers and pathways with single-cell sequencing[58]; (VIII) identification of biomarker for clinical diagnosis with epigenetic-guided prediction of tissue[63]; (IX) identify tumor-specific drug responsive markers using DNA methylation alterations[64]; (X) Identification of robust epigenetic biomarkers and clinical translation[65]; (XI) epigenetic map of maladaptive responses in patients with heart failure[66]; (XII) epigenetic precision medicine with DNA methylation defects and aberrant covalent histone modifications[67]; and, (XIII) reporting methodologies to assemble, regulate, and deliver epigenetic effectors to exploit the endurance of the therapeutic effect.[68]

Metabolites with significant correlations with clinical biochemical indexes can help improve the diagnostic capabilities of clinicians and suggest new metabolic targets for researchers attempting to understand the molecular mechanisms of weight gain and obesity.[9,10,73] Metabolic state based on presence or absence of a particular metabolite in absolute or relative concentration with an associated genetic signature will aid in characterizing disease phenotypes.[9,10] The analysis of metabolites and proteins involved in the cellular processes, and influenced by genetics, lifestyle, and environmental factors (the "exposome"), can provide a broader understanding of health, especially when used in conjunction with NGS.[74] Timely metabolomics data analysis has the potential to support precision medicine with the identification of biomarkers and quantification of a broad range of endogenous and exogenous metabolites, all of which may provide new insights into disease pathophysiology.[75–78] Metabolomics includes multiple data generation techniques[79] such as gas chromatography–mass spectrometry (GC–MS)[80–82] and liquid chromatography–mass spectrometry (LC–MS).[81] To analyze and interpret metabolomics data for precision medicine, various bioinformatics applications and annotation databases have been developed.[83–85] Metabolomics has progressed enormously in the last few decades, with improved methods for precision medicine,[86–95] narrated in Table 1. Furthermore, by harnessing the power of metabolomics, we can profile a patient's metabolome and correlate it with their BMI. A high BMI is a major risk factor for several chronic diseases, but the biology underlying these associations is not well understood. Generating a patient-specific metabolic profile associated with BMI can identify potential metabolic markers relevant to their disease.[10] Expected and reported results produced by these precision medicine approaches can guide researchers involved in biomarker discovery on how to make better use of available public knowledge and data sources. Despite their valuable scientific contributions, one of the common limitations among all these approaches is that most of the computational work accomplished has been performed manually, not automatically. For this reason, practicing precision medicine requires in-depth bioinformatics knowledge, biomedical informatics background, programming expertise, big data management and processing environment, high budget, and excessive timeline to reproduce and adopt results in clinical settings by integrating with EHR.

Table 1 Metabolomics driven approaches for precision medicine.

Number	Approach	Reference
1	Predicting drug response (e.g., Aspirin, Simvastatin, Acamprosate, and Antihypertensive drugs) and identifying biomarkers of disease susceptibility	86
2	Identifying effective cancer treatment therapies	87
3	Elucidating the pathophysiology, diagnosis, prognosis, and treatment of sepsis with metabolomics	88
4	Using metabolomics for the personalized care of trauma patients	89
5	Applying pharmacometabolomics for rare cancer, microbiological investigation, and therapy	90
6	Discovering dynamic biomarkers for disease progression and associated treatment response	91
7	Integrating metabolomics and proteomics data for biomarkers discovery and support personalized medicine in multiple sclerosis	92
8	Rapid and quantitative metabolite calculations with high-throughput metabolite array for precision medicine	93
9	Characterization and quantitation of global metabolites for precision cardiovascular medicine	94
10	Advanced precision medicine with integration of multifaceted multi-omics approaches with metabolomics and metabolic phenotyping	95

3. Deep phenotyping & precision medicine

Disease classification is routinely accomplished by healthcare units generating different streams of data, including imaging, pathology, genomics, and electrophysiology.[96] To get new insights into disease taxonomy, etiology, and pathogenesis, it is important to understand how diseases are related to each other. The clinical interpretation of the significance of specific gene variants can be unique to a patient. Therefore, it is critical to understand the genetic basis of common diseases, which genes predispose one to a particular medical condition, and how rare genetic variants contribute to diseases.

Deep phenotyping assesses genotype and phenotype associations among population data for various disease areas.[97] It has the potential to support the identification of potential variants by performing a global analysis of key genes targeting major components of the immune system, and investigating the role of immunity gene variants in influencing susceptibility in targeted disorders, as well as their functional implications and associations with other diseases. However, integrating genomic data with deep phenotyping to reveal cases where absence of genotype-phenotype associations likely resulted in uncertainty in patient care remains a problem. Furthermore, variability in interpretation for sequence variants is due, in part, to the lack of standard curated information to support clinical decision-making. Precision medicine has the potential to help doctors tailor therapy by integrating individuals' personal medical history with metabolomics and genetic makeup.[98]

The efficient use of EHR for research involves patients, researchers, clinicians, healthcare systems, and funders of clinical care and research.[33,99,100] Given that the primary purpose of the EHR is to facilitate patient care and improve clinical outcomes, integrating EHR with multi-omics and environmental data is of paramount importance in next-generation medicine development.[101] Precision medicine-supported, EHR-based healthcare data analytics involves database development, preparation, modeling, and creating the patient and knowledge databases. From the perspective of population-based studies, the integration of genomics data in the health system with longitudinal EHRs will enable the assessment of genetic risk in a wide range of diseases.[102–104] The next generation sequencing (NGS) data life cycle requires high-level expertise from various disciplines and is categorized into five major steps: (I) sample collection; (II) library preparation and sequencing; (III) demultiplexing raw data into structured file formats; (IV) QC metrics calculations using genomics data processing pipelines; and (V) data mining and downstream analysis (Fig. 2). Despite multiple recent successes, genomic sequencing has still not been fully automated and standardized in healthcare systems for the detection of genetic risks in healthy individuals.[105,106] In recent years, numerous efforts have been made to develop laboratory information management systems (LIMS) and genomics data processing systems, which aid in managing and building pipelines.[38,70,107–110] These are helpful in workflow creation and execution, but critical problems of integrating and processing sequence data from multiple sources still exists because of daunting complexity.[111] The traditional way is still mostly based on running a series of command-line applications, which require advanced programming skills.[112–116] It does not enable

Fig. 2 NGS data preparation, sequencing, and analysis cycle.

linking information generated at different stages of conducted experiments. Due to higher-throughput and lower-cost DNA sequencing technologies, healthcare and biological data are accumulating at larger scales and at exponential rates; therefore, it is essential to develop innovative bioinformatics applications to improve research quality and data sharing.

The genome, when studied together with the transcriptome and epigenome,[117] has differentiated human diseases[118] into multifactorial, mitochondrial,[119] chromosomal,[120] and monogenic[121] diseases. In the past, various efforts have been made in deciphering diseases to facilitate predictive diagnosis and thereby guide treatment factors.[122–139] However, inconsistencies in gene annotation and in specificity of disease classification terminologies add to complexity. Lack of an intelligent and integrative searchable system makes it difficult to elucidate underlying implications.[140] Most of the publicly available mutation databases are nascent and incomplete in accommodating standardized clinical needs for connecting genomic variants to phenotypes, and interpretation of structural and noncoding variants.[141,142] Thorough investigation of existing databases is required to assess the potential significance of even one sequence variant, but that is a cumbersome, time-consuming, and increasingly unfeasible process due to the absence of a standard centralized platform for connecting genes to their disease phenotype.[143]

Understanding how genetic variations and gene and protein expression relate to metabolites and then contribute to disease is an important avenue in advancing precision medicine. To understand how metabolite profiles link with external environment and associate with disease risk, we need to identify biomarkers for health assessment and customized therapy by combining consented individuals' functional measurement of observed metabolites with WGS and RNA-seq.[87,144–149] Linking genetic and metabolic profiles to phenotypes will facilitate the processes for identifying active genes that impact the functional health state. Correlating metabolites with clinical biochemical indexes will increase the diagnostic capabilities of clinicians and suggest new metabolic targets for researchers attempting to understand the molecular mechanisms of weight gain and obesity.[150] Multiple deep phenotyping approaches have been reported to support precision medicine, given in Table 2.[143,151–164] In future, we need deep digital phenotyping solution to integrate clinical, multi-omics and real-world digital (e.g., data generated using smart phones and watches, and other similar kinds of IoT devices) data with data.[165,166]

Table 2 Deep phenotyping approaches for precision medicine.

Number	Approach	Reference
1	Systematic deep phenotyping to deliver personalized care for oncology	151
2	Integrative approaches to cardiovascular disease prevention and treatment	152
3	Phenotypes, endotypes, genotypes, regiotypes, and theratypes for precision diagnosis and personalized treatment of allergic diseases	143
4	Deep phenotyping with genetic analysis of common disorders	153
5	Diagnosis of hereditary disorders with phenotype and genetic data integration	154
6	Investigating intensity and disability deep phenotyping, caused by headache attributed to COVID-19 patients	155
7	Identifying distinct immune signatures in multiple myeloma	156
8	Improve pregnancy outcomes by deciphering physiology with deep phenotyping	157

Table 2 Deep phenotyping approaches for precision medicine.—cont'd

Number	Approach	Reference
9	Understanding pathophysiology of heart failure with detailed phenotyping of pulsatile hemodynamics	158
10	Deep phenotyping to reveal unanticipated heterogeneity of clinical features in Xeroderma pigmentosum (XP)	159
11	Deep phenotyping of arterial hemodynamics in HFpEF to enhance our clinical, translational, and therapeutic approach to HFpEF	160
12	Deep phenotyping to identify patients with specific clinical neonatal and childhood features harboring the same de novo variant in KCNA2	161
13	Varying geno- and phenotypes to investigate the association between long-term complications and possible predictors of clinical outcome In classical galactosemia patients	162
14	Approaches for deep clinical phenotyping to identify biological markers of Parkinson's disease progression	163
15	Deep phenotyping and genetic characterization of Chromosome-5p minus syndrome to find statistically significant functional gender-based differences	164

4. Discussion & future recommendations

Precision medicine is a paradigm shift in the application and interpretation of informed medicine, moving from a deterministic conceptualization to a probabilistic interpretation of genetic risk. There is no single comprehensive solution that can link demographic information, personal lifestyle, medical history, recent visits to physicians and other health care providers, diagnoses performed, lab tests, longitudinal images, medications and procedures, samples taken for wet and dry lab experimentation, and treatment of incurable diseases. It has not yet been possible to track and prospectively follow the clinical progress and outcomes in patients and update clinical effects with time (e.g., having a critical predictor of future clinical events where a patient may show the disease months or years in the future). Moreover, there is no compact solution available for automatic data anonymization from identified to de-identified data sets for internal and global research objectives and to share and effectively draw comparisons with recently generated and existing data sets of high volume and complexity. Establishment of secure, interoperable, distributed, federated, product-line large-scaled digital health information infrastructure for research and analytics in healthcare can lead to significant improvements in provision of care and better health of populations at lower costs together with better work-life for clinicians and staff.

To enable the widespread adoption of precision medicine in clinical practice, the following integrative omics data challenges must be addressed: (I) multi-omics data aggregation, scalability, and integration into EHR; (II) handling inherent error rates in genome-wide data; (III) assessment of continuous and longitudinal data; (IV) classifying complex patterns of metabolite heterogeneity depending on gene expression; and (V) implementing artificial intelligence (AI)/machine learning (ML) to identify causes of the emergence of metabolic multimodality.[167–170] Furthermore, it is important to train physicians[171] and address ethical issues related to forensic identification, lost privacy, genetic discrimination, misuse of human genomics, and ownership and control of genetic information of participants.[172] To effectively implement data analytics, five big data management challenges must be overcome.[173–177] These are: (I) rapid growth in heterogeneous data and existence of multiple data standards, structures, types, and formats; (II) inadequacy of analyzable clinical data, and understanding algorithms for clinical data interpretation and drawing inference; (III) lack of effective open source tools to model biological interactions; (IV) integration of clinical and analytic systems and interdisciplinary field barriers; (V) and implementation of secure and reliable frameworks for healthcare data collection, simplification, management, analysis, and sharing.[178–180] Addressing multi-omics and healthcare data analytics challenges not only will solve many complex healthcare data-oriented problems of implementing secure networking, management, pervasive computing, advanced analytics, process modeling, data representation, integrity, privacy, reliability, and exchange, but it will also establish a cross-disciplinary collaborative research environment that will lead to new fundamental insights by analyzing original as well as aggregated healthcare data. The question is how to overcome these challenges by implementing a prototype healthcare information system embedded within a precision medicine initiative.[181]

Going forward, precision medicine requires implementation and application of novel and clinic-friendly platforms for enhanced patient care with intelligently integrative clinical, metabolomics and genomics data analysis, visualization, and interpretation (Fig. 3). We need a Findable, Accessible, Intelligent, and Reproducible (FAIR) approach to facilitate implementation

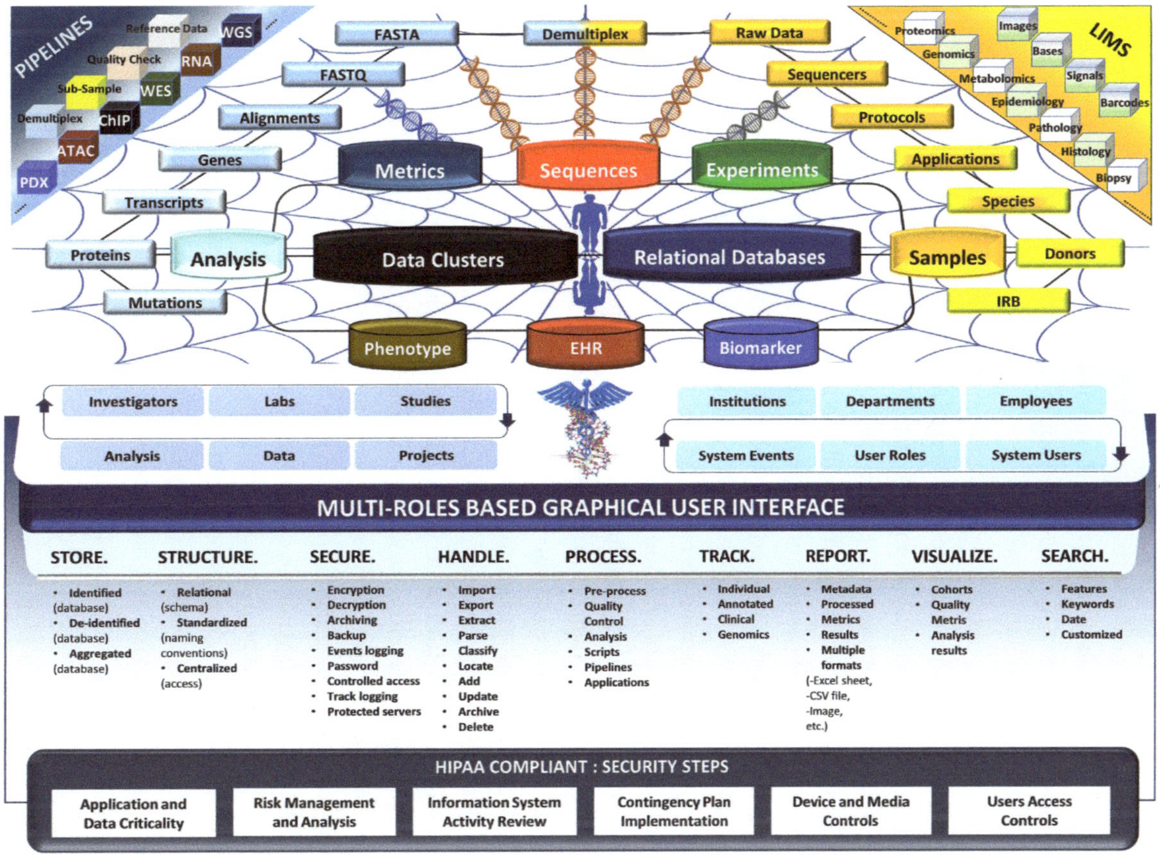

Fig. 3 Concept of new large-scale, user-friendly, interactive, cross-platform, encrypted, multi-role based, automated, customized, and centralized multi-database management framework.

of precision medicine and to accelerate diagnostic and preventive care delivery strategies beyond traditional symptom-driven, disease-causal medical practice. Its users will include geneticists, biologist, clinicians, and computational scientists. The overall concept is to help support and implement a new healthcare data analytic and research process that can connect people from different backgrounds and specialties with EHR and processed multi-omics data to facilitate decision making with analytical queries and informative outputs.

We need accurate and reproducible use of ML and AI algorithms in such platforms to predict life-threatening disease risk, starting from the most common to rare among the population data.[11,13] Applied AI and ML approaches will allow learning from a continuum of dataset sizes displaying heterogeneous levels of granularity. This will also lead to the identification of patterns revealing predictive and modifiable risk factors to support earlier diagnosis of targeted disorders and sequela. Novel and evidence-based studies are required to improve care quality for individuals of any gender, race, ethnicity, and age. The next steps would be to create an extended engagement with home health and long-term care industry stakeholders to identify community-based approaches to better support patients living with complex disorders. Additionally, the rich and valuable data resources developed around the globe need to be standardized and integrated to study the fundamentals of development and progression of human diseases.[182,183] This will enable exploitation of prior knowledge effectively and create interpretable biomarker models to understand wide variety of biological problems.

We must develop new large-scale, user-friendly, interactive, cross-platform, encrypted, multi-role based, automated, HIPAA-compliant, customized, and centralized multi-database management systems that will store (e.g., metadata in database and actual data in clusters and cloud), structure (e.g., relational schema, standardized naming conventions and data structure), and secure (e.g., encryption, decryption, archiving, backup, event logging, limited access and password management) clinical, genomics, and other kinds of omics, phenotypic and biomarker data in a manipulable framework (e.g., export, import, extract, parse, classify, locate, add/update, archive/delete) for tracking (individualized, clustered, and annotated), visualizing (two- and three-dimensional graphs), searching, reporting and sharing analyzed results in common formats (e.g., database, excel sheets, and CSV, text, images, PDF, html and xml files) (Fig. 3). These management systems should incorporate data analytics pipelines that can support investigating disease-causing variants among highly expressed genes to find the root causes of uncertainties in patient care, such as autosomal recessive carriers, heterozygotes with phenotypic manifestations, and genetic associations with metabolite changes. This will also help performing robust gene-based collapsing analysis[184] to identify functional variants associated with irregular metabolite levels, and/or monogenic and polygenic variants[35] that cause emergence of metabolic multimodality[99] and/or that affect BMI. It is not sufficient for diagnostic purposes to simply correlate a particular genomic variant with assertion of pathogenicity. The development of user-friendly graphical interfaces should be promoted to enable users from multidisciplinary backgrounds with little or no programming skills to perform complicated bioinformatics, biomedical informatics, and data science-oriented tasks but will also equip experienced bioinformatics professionals with more effective ways of data management, integration, and analysis.

Acknowledgments

We appreciate great support by the Rutgers Institute for Health, Health Care Policy and Aging Research (IFH); Department of Medicine, Rutgers Robert Wood Johnson Medical School (RWJMS); and Rutgers Biomedical and Health Sciences (RBHS), at the Rutgers, The State University of New Jersey. We thank members and collaborators of Ahmed Lab at the Rutgers (IFH, RWJMS, RBHS) for their support, participation, and contribution to this study.

Competing interests

The authors declare no competing financial interests.

Author contribution

Z.A. proposed the study and drafted the manuscript, and approved for the publication.

Biographical note

Z.A. is the Assistant Professor of Medicine—Tenure Track and Core Member at the Rutgers Institute for Health, Health Care Policy and Aging Research; and Department of Medicine—Division of General Internal Medicine, Rutgers Robert Wood Johnson Medical School, Rutgers Biomedical and Health Sciences, Rutgers University-New Brunswick, NJ. Z.A. is the Adjunct Assistant Professor at the Department of Genetics and Genome Sciences, and Pat and Jim Calhoun Cardiology Center, UConn School of Medicine, UConn Health, CT. Z.A. is the Academic Member of the Rutgers Microbiology and Molecular Genetics; Center for Cancer Health Equity, Rutgers Cancer Institute of New Jersey; Rutgers Human Genetics Institute of New Jersey, NJ.

Funding information

This work was supported by the Institute for Health, Health Care Policy and Aging Research (IFH), and Rutgers Robert Wood Johnson Medical School, Rutgers Biomedical and Health Sciences at the Rutgers, The State University of New Jersey.

References

1. Ahmad FB, Anderson RN. The leading causes of death in the US for 2020. *JAMA*. 2021;325(18):1829–1830. https://doi.org/10.1001/jama.2021.5469.
2. GBD 2019 Diseases and Injuries Collaborators. Global burden of 369 diseases and injuries in 204 countries and territories, 1990-2019: a systematic analysis for the global burden of disease study 2019. *Lancet (London, England)*. 2020;396(10258): 1204–1222. https://doi.org/10.1016/S0140-6736(20)30925-9.
3. Murray CJ, Lopez AD. Global mortality, disability, and the contribution of risk factors: global burden of disease study. *Lancet (London, England)*. 1997;349(9063):1436–1442. https://doi.org/10.1016/S0140-6736(96)07495-8.
4. Murray CJ, Lopez AD. Mortality by cause for eight regions of the world: global burden of disease study. *Lancet (London, England)*. 1997;349(9061):1269–1276. https://doi.org/10.1016/S0140-6736(96)07493-4.
5. Walsh JA. Disease problems in the third world. *Ann N Y Acad Sci*. 1989;569: 1–16. https://doi.org/10.1111/j.1749-6632.1989.tb27354.x.
6. GBD 2017 Risk Factor Collaborators. Global, regional, and national comparative risk assessment of 84 behavioural, environmental and occupational, and metabolic risks or clusters of risks for 195 countries and territories, 1990-2017: a systematic analysis for the global burden of disease study 2017. *Lancet (London, England)*. 2018;392(10159): 1923–1994. https://doi.org/10.1016/S0140-6736(18)32225-6.
7. Ahmed Z. Intelligent health system for the investigation of consenting COVID-19 patients and precision medicine. *Pers Med*. 2021;18(6):573–582. https://doi.org/10.2217/pme-2021-0068.
8. Groft SC, Posada de la Paz M. Preparing for the future of rare diseases. *Adv Exp Med Biol*. 2017;1031:641–648. https://doi.org/10.1007/978-3-319-67144-4_34.
9. Ahmed Z. Practicing precision medicine with intelligently integrative clinical and multi-omics data analysis. *Hum Genomics*. 2020;14(1):35. https://doi.org/10.1186/s40246-020-00287-z.
10. Ahmed Z, Zeeshan S, Foran D, Kleinman L, Wondisford F, Dong X. Integrative clinical, genomics and metabolomics data analysis for mainstream precision medicine to investigate COVID-19. *BMJ Innov*. 2021;7(1):6. https://doi.org/10.1136/bmjinnov-2020-000444.
11. Ahmed Z, Mohamed K, Zeeshan S, Dong X. Artificial intelligence with multi-functional machine learning platform development for better healthcare and precision medicine. *Database (Oxford)*. 2020;2020, baaa010. https://doi.org/10.1093/database/baaa010.
12. Ginsburg GS, Phillips KA. Precision medicine: from science to value. *Health Aff*. 2018;37(5):694–701. https://doi.org/10.1377/hlthaff.2017.1624.
13. Gameiro GR, Sinkunas V, Liguori GR, Auler-Júnior J. Precision medicine: changing the way we think about healthcare. *Clinics (Sao Paulo)*. 2018;73, e723. https://doi.org/10.6061/clinics/2017/e723.
14. König IR, Fuchs O, Hansen G, von Mutius E, Kopp MV. What is precision medicine? *Eur Respir J*. 2017;50(4):1700391. https://doi.org/10.1183/13993003.00391-2017.

15. Elemento O. The future of precision medicine: towards a more predictive personalized medicine. *Emerg Topics Life Sci.* 2020;4(2):175–177. https://doi.org/10.1042/ETLS20190197.
16. Faulkner E, Holtorf AP, Walton S, et al. Being precise about precision medicine: what should value frameworks incorporate to address precision medicine? A report of the personalized precision medicine special interest group. *Value Health.* 2020;23(5):529–539. https://doi.org/10.1016/j.jval.2019.11.010.
17. Yeh VM, Bergner EM, Bruce MA, et al. Can precision medicine actually help people like me? African American and Hispanic perspectives on the benefits and barriers of precision medicine. *Ethn Dis.* 2020;30(suppl 1):149–158. https://doi.org/10.18865/ed.30.S1.149.
18. Pinho J. *Precision Medicine.* vol. 15(1). Sao Paulo, Brazil: Einstein; 2017:VII–X. https://doi.org/10.1590/S1679-45082017ED4016.
19. Bilkey GA, Burns BL, Coles EP, Mahede T, Baynam G, Nowak KJ. Optimizing precision medicine for public health. *Front Public Health.* 2019;7:42. https://doi.org/10.3389/fpubh.2019.00042.
20. McAlister FA, Laupacis A, Armstrong PW. Finding the right balance between precision medicine and personalized care. *CMAJ.* 2017;189(33):E1065–E1068. https://doi.org/10.1503/cmaj.170107.
21. Goetz LH, Schork NJ. Personalized medicine: motivation, challenges, and progress. *Fertil Steril.* 2018;109(6):952–963. https://doi.org/10.1016/j.fertnstert.2018.05.006.
22. Lazaridis KN, et al. Implementing individualized medicine into the medical practice. *Am J Med Genet C Semin Med Genet.* 2014;166C:15–23.
23. Sboner A, Elemento O. A primer on precision medicine informatics. *Brief Bioinform.* 2016;17:145–153.
24. Tomczak K, Czerwińska P, Wiznerowicz M. The Cancer genome atlas (TCGA): an immeasurable source of knowledge. *Contemp Oncol.* 2015;19(1A):A68–A77.
25. Mailman MD, et al. The NCBI dbGaP database of genotypes and phenotypes. *Nat Genet.* 2007;39(10):1181–1186.
26. Leinonen R, et al. The sequence read archive. *Nucleic Acids Res.* 2011;39(Database issue):D19–D21.
27. Leinonen R, et al. The European nucleotide archive. *Nucleic Acids Res.* 2011;39 (Database issue):D28–D31.
28. Sudlow C, et al. UK biobank: an open access resource for identifying the causes of a wide range of complex diseases of middle and old age. *PLoS Med.* 2015;12(3):e1001779.
29. All of Us Research Program Investigators, et al. The "all of us" research program. *N Engl J Med.* 2019;381(7):668–676.
30. Prokopenko D, et al. Whole-genome sequencing in severe chronic obstructive pulmonary disease. *Am J Respir Cell Mol Biol.* 2018;59(5):614–622.
31. Esteva A, et al. A guide to deep learning in healthcare. *Nat Med.* 2019;25 (24–29):30617335.
32. Sonawane AR, Weiss ST, Glass K, Sharma A. Network medicine in the age of biomedical big data. *Front Genet.* 2019;10:294.
33. Ahmed Z, Kim M, Liang BT. MAV-clic: management, analysis, and visualization of clinical data. *JAMIA Open.* 2018;2(1):23–28. https://doi.org/10.1093/jamiaopen/ooy052.
34. Glaab E. Using prior knowledge from cellular pathways and molecular networks for diagnostic specimen classification. *Brief Bioinform.* 2016;17(3):440–452.
35. Zeeshan S, Xiong R, Liang BT, Ahmed Z. 100 Years of evolving gene-disease complexities and scientific debutants. *Brief Bioinform.* 2020;21(3):885–905. https://doi.org/10.1093/bib/bbz038.

36. Hou YC, et al. Precision medicine integrating whole-genome sequencing, comprehensive metabolomics, and advanced imaging. *Proc Natl Acad Sci U S A*. 2020;117 (6):3053–3062. 31980526.
37. Ahmed Z, Renart EG, Zeeshan S, Dong X. Advancing clinical genomics and precision medicine with GVViZ: FAIR bioinformatics platform for variable gene-disease annotation, visualization, and expression analysis. *Hum Genomics*. 2021;15(1):37. https://doi.org/10.1186/s40246-021-00336-1.
38. Ahmed Z, Renart EG, Zeeshan S. Genomics pipelines to investigate susceptibility in whole genome and exome sequenced data for variant discovery, annotation, prediction and genotyping. *PeerJ*. 2021;9, e11724. https://doi.org/10.7717/peerj.11724.
39. Ahmed Z, Zeeshan S, Mendhe D, Dong X. Human gene and disease associations for clinical-genomics and precision medicine research. *Clin Transl Med*. 2020;10(1): 297–318. https://doi.org/10.1002/ctm2.28.
40. Ahmed Z, Zeeshan S, Xiong R, Liang BT. Debutant iOS app and gene-disease complexities in clinical genomics and precision medicine. *Clin Transl Med*. 2019;8(1): 26. https://doi.org/10.1186/s40169-019-0243-8.
41. Yang H, Wang K. Genomic variant annotation and prioritization with ANNOVAR and wANNOVAR. *Nat Protoc*. 2015;10(10):1556–1566. https://doi.org/10.1038/nprot.2015.105.
42. Yu Y, Wang Y, Xia Z, et al. PreMedKB: an integrated precision medicine knowledgebase for interpreting relationships between diseases, genes, variants and drugs. *Nucleic Acids Res*. 2019;47(D1):D1090–D1101. https://doi.org/10.1093/nar/gky1042.
43. Li YH, Yu CY, Li XX, et al. Therapeutic target database update 2018: enriched resource for facilitating bench-to-clinic research of targeted therapeutics. *Nucleic Acids Res*. 2018;46(D1):D1121–D1127. https://doi.org/10.1093/nar/gkx1076.
44. Barbarino JM, Whirl-Carrillo M, Altman RB, Klein TE. PharmGKB: a worldwide resource for pharmacogenomic information. Wiley interdisciplinary reviews. *Syst Biol Med*. 2018;10(4), e1417. https://doi.org/10.1002/wsbm.1417.
45. Stenson PD, Mort M, Ball EV, et al. The human gene mutation database: towards a comprehensive repository of inherited mutation data for medical research, genetic diagnosis and next-generation sequencing studies. *Hum Genet*. 2017;136(6):665–677. https://doi.org/10.1007/s00439-017-1779-6.
46. Griffith M, Spies NC, Krysiak K, et al. CIViC is a community knowledgebase for expert crowdsourcing the clinical interpretation of variants in cancer. *Nat Genet*. 2017;49(2):170–174. https://doi.org/10.1038/ng.3774.
47. Brittain HK, Scott R, Thomas E. The rise of the genome and personalised medicine. *Clin Med (Lond)*. 2017;17(6):545–551. https://doi.org/10.7861/clinmedicine.17-6-545.
48. Roth SC. What is genomic medicine? *J Med Libr Assoc*. 2019;107(3):442–448. https://doi.org/10.5195/jmla.2019.604.
49. Alzu'bi AA, Zhou L, Watzlaf V. Genetic variations and precision medicine. *Perspect Health Inf Manag*. 2019;16(Spring):1a.
50. Roden DM, Tyndale RF. Genomic medicine, precision medicine, personalized medicine: what's in a name? *Clin Pharmacol Ther*. 2013;94(2):169–172. https://doi.org/10.1038/clpt.2013.101.
51. Bylstra Y, Lim WK, Kam S, et al. Family history assessment significantly enhances delivery of precision medicine in the genomics era. *Genome Med*. 2021;13 (1):3. https://doi.org/10.1186/s13073-020-00819-1.
52. Strianese O, Rizzo F, Ciccarelli M, et al. Precision and personalized medicine: how genomic approach improves the Management of Cardiovascular and Neurodegenerative Disease. *Genes*. 2020;11(7):747. https://doi.org/10.3390/genes11070747.

53. Kim HJ, Kim HJ, Park Y, Lee WS, Lim Y, Kim JH. Clinical genome data model (cGDM) provides interactive clinical decision support for precision medicine. *Sci Rep.* 2020;10(1):1414. https://doi.org/10.1038/s41598-020-58088-2.
54. Yap TA, Johnson A, Meric-Bernstam F. Precision medicine in oncology-toward the integrated targeting of somatic and germline genomic aberrations. *JAMA Oncol.* 2021;7 (4):507–509. https://doi.org/10.1001/jamaoncol.2020.7988.
55. Roos A, Byron SA. Genomics-enabled precision medicine for Cancer. *Cancer Treat Res.* 2019;178:137–169. https://doi.org/10.1007/978-3-030-16391-4_5.
56. Ahmed Z, Zeeshan S, Liang BT. RNA-seq driven expression and enrichment analysis to investigate CVD genes with associated phenotypes among high-risk heart failure patients. *Hum Genomics.* 2021;15(1):67. https://doi.org/10.1186/s40246-021-00367-8.
57. Prokop JW, Hartog NL, Chesla D, et al. High-density blood transcriptomics reveals precision immune signatures of SARS-CoV-2 infection in hospitalized individuals. *Front Immunol.* 2021;12, e11724. https://doi.org/10.3389/fimmu.2021.694243.
58. Kim J, Park J. Single-cell transcriptomics: a novel precision medicine technique in nephrology. *Korean J Intern Med.* 2021;36(3):479–490. https://doi.org/10.3904/kjim.2020.415.
59. Tsimberidou AM, Fountzilas E, Bleris L, Kurzrock R. Transcriptomics and solid tumors: the next frontier in precision cancer medicine. *Semin Cancer Biol.* 2020. https://doi.org/10.1016/j.semcancer.2020.09.007. S1044-579X(20)30196-6.
60. Khodadadian A, Darzi S, Haghi-Daredeh S, et al. Genomics and transcriptomics: the powerful technologies in precision medicine. *Int J General Med.* 2020;13: 627–640. https://doi.org/10.2147/IJGM.S249970.
61. Kronfol MM, Dozmorov MG, Huang R, Slattum PW, McClay JL. The role of epigenomics in personalized medicine. *Expert Rev Precis Med Drug Dev.* 2017;2(1): 33–45. https://doi.org/10.1080/23808993.2017.1284557.
62. Rasool M, Malik A, Naseer MI, et al. The role of epigenetics in personalized medicine: challenges and opportunities. *BMC Med Genomics.* 2015;8(suppl 1):S5. https://doi.org/10.1186/1755-8794-8-S1-S5.
63. Moran S, Martinez-Cardús A, Boussios S, Esteller M. Precision medicine based on epigenomics: the paradigm of carcinoma of unknown primary. *Nat Rev Clin Oncol.* 2017;14(11):682–694. https://doi.org/10.1038/nrclinonc.2017.97.
64. Yan W, Herman JG, Guo M. Epigenome-based personalized medicine in human cancer. *Epigenomics.* 2016;8(1):119–133. https://doi.org/10.2217/epi.15.84.
65. Xu Y, Doonan SR, Ordog T, Bailey RC. Translational opportunities for microfluidic technologies to enable precision epigenomics. *Anal Chem.* 2020;92(12):7989–7997. https://doi.org/10.1021/acs.analchem.0c01288.
66. Napoli C, Benincasa G, Donatelli F, Ambrosio G. Precision medicine in distinct heart failure phenotypes: focus on clinical epigenetics. *Am Heart J.* 2020;224:113–128. https://doi.org/10.1016/j.ahj.2020.03.007.
67. Werner RJ, Kelly AD, Issa JJ. Epigenetics and precision oncology. *Cancer J.* 2017;23 (5):262–269. https://doi.org/10.1097/PPO.0000000000000281.
68. Sgro A, Blancafort P. Epigenome engineering: new technologies for precision medicine. *Nucleic Acids Res.* 2020;48(22):12453–12482. https://doi.org/10.1093/nar/gkaa1000.
69. Weinshilboum RM, Wang L. Pharmacogenomics: precision medicine and drug response. *Mayo Clin Proc.* 2017;92(11):1711–1722. https://doi.org/10.1016/j.mayocp.2017.09.001.
70. Ahmed Z, Ucar D. I-ATAC: interactive pipeline for the management and pre-processing of ATAC-seq samples. *PeerJ.* 2017;5, e4040. https://doi.org/10.7717/peerj.4040.

71. Singh DB. The impact of pharmacogenomics in personalized medicine. *Adv Biochem Eng Biotechnol.* 2020;171:369–394. https://doi.org/10.1007/10_2019_110.
72. Nassar SF, Raddassi K, Ubhi B, Doktorski J, Abulaban A. Precision medicine: steps along the road to combat human cancer. *Cell.* 2020;9(9):2056. https://doi.org/10.3390/cells9092056.
73. Dandekar T, Fieselmann A, Majeed S, Ahmed Z. Software applications toward quantitative metabolic flux analysis and modeling. *Brief Bioinform.* 2014;15(1):91–107. https://doi.org/10.1093/bib/bbs065.
74. Bekri S. The role of metabolomics in precision medicine. *Exp Rev Precis Med Drug Dev.* 2016;1(6):517–532.
75. Beger RD, Dunn W, Schmidt MA, et al. Metabolomics enables precision medicine: "a white paper, community perspective". *Metabolomics.* 2016;12(10):149. https://doi.org/10.1007/s11306-016-1094-6.
76. Cacciatore S, Loda M. Innovation in metabolomics to improve personalized healthcare. *Ann N Y Acad Sci.* 2015;1346(1):57–62. https://doi.org/10.1111/nyas.12775.
77. Everett JR. Pharmacometabonomics in humans: a new tool for personalized medicine. *Pharmacogenomics.* 2015;16(7):737–754. https://doi.org/10.2217/pgs.15.20.
78. Jacob M, Lopata AL, Dasouki M, Abdel Rahman AM. Metabolomics toward personalized medicine. *Mass Spectrom Rev.* 2019;38(3):221–238. https://doi.org/10.1002/mas.21548.
79. Azad RK, Shulaev V. Metabolomics technology and bioinformatics for precision medicine. *Brief Bioinform.* 2019;20(6):1957–1971. https://doi.org/10.1093/bib/bbx170.
80. Ahmed Z, Zeeshan S, Huber C, et al. 'Isotopo' a database application for facile analysis and management of mass isotopomer data. *Database (Oxford).* 2014;2014:bau077. https://doi.org/10.1093/database/bau077.
81. Ahmed Z, Zeeshan S, Huber C, et al. Software LS-MIDA for efficient mass isotopomer distribution analysis in metabolic modelling. *BMC Bioinf.* 2013;14:218. https://doi.org/10.1186/1471-2105-14-218.
82. Ahmed Z, Mayr M, Zeeshan S, Dandekar T, Mueller MJ, Fekete A. Lipid-pro: a computational lipid identification solution for untargeted lipidomics on data-independent acquisition tandem mass spectrometry platforms. *Bioinformatics (Oxford, England).* 2015;31(7):1150–1153. https://doi.org/10.1093/bioinformatics/btu796.
83. Schmidt JC, Dougherty BV, Beger RD, Jones DP, Schmidt MA, Mattes WB. Metabolomics as a truly translational tool for precision medicine. *Int J Toxicol.* 2021;40(5):413–426. https://doi.org/10.1177/10915818211039436.
84. Ahmed Z, Zeeshan S, Dandekar T. Developing sustainable software solutions for bioinformatics by the "Butterfly" paradigm. *F1000Research.* 2014;3:71. https://doi.org/10.12688/f1000research.3681.2.
85. Wishart DS. Emerging applications of metabolomics in drug discovery and precision medicine. *Nat Rev Drug Discov.* 2016;15(7):473–484. https://doi.org/10.1038/nrd.2016.32.
86. Li B, He X, Jia W, Li H. Novel applications of metabolomics in personalized medicine: a Mini-review. *Molecules (Basel, Switzerland).* 2017;22(7):1173. https://doi.org/10.3390/molecules22071173.
87. Puchades-Carrasco L, Pineda-Lucena A. Metabolomics applications in precision medicine: an oncological perspective. *Curr Top Med Chem.* 2017;17(24):2740–2751. https://doi.org/10.2174/1568026617666170707120034.
88. Eckerle M, Ambroggio L, Puskarich MA, et al. Metabolomics as a driver in advancing precision medicine in Sepsis. *Pharmacotherapy.* 2017;37(9):1023–1032. https://doi.org/10.1002/phar.1974.
89. Jayaraman SP, Anand RJ, DeAntonio JH, et al. Metabolomics and precision medicine in trauma: the state of the field. *Shock (Augusta, Ga).* 2018;50(1):5–13. https://doi.org/10.1097/SHK.0000000000001093.

90. Cristoni S, Bernardi LR, Malvandi AM, et al. A case of personalized and precision medicine: pharmacometabolomic applications to rare cancer, microbiological investigation, and therapy. *Rapid Commun Mass Spectrom.* 2021;35(2), e8976. https://doi.org/10.1002/rcm.8976.
91. Kohler I, Hankemeier T, van der Graaf PH, Knibbe C, van Hasselt J. Integrating clinical metabolomics-based biomarker discovery and clinical pharmacology to enable precision medicine. *Eur J Pharm Sci.* 2017;109S:S15–S21. https://doi.org/10.1016/j.ejps.2017.05.018.
92. Del Boccio P, Rossi C, di Ioia M, Cicalini I, Sacchetta P, Pieragostino D. Integration of metabolomics and proteomics in multiple sclerosis: from biomarkers discovery to personalized medicine. *Proteomics Clin Appl.* 2016;10(4):470–484. https://doi.org/10.1002/prca.201500083.
93. Xie G, Wang L, Chen T, et al. A metabolite Array technology for precision medicine. *Anal Chem.* 2021;93(14):5709–5717. https://doi.org/10.1021/acs.analchem.0c04686.
94. Lam SM, Wang Y, Li B, Du J, Shui G. Metabolomics through the lens of precision cardiovascular medicine. *J Genet Genomics.* 2017;44(3):127–138. https://doi.org/10.1016/j.jgg.2017.02.004.
95. Long NP, Nghi TD, Kang YP, et al. Toward a standardized strategy of clinical metabolomics for the advancement of precision medicine. *Metabolites.* 2020;10(2):51. https://doi.org/10.3390/metabo10020051.
96. Eisen MB, Spellman PT, Brown PO, Botstein D. Cluster analysis and display of genome-wide expression patterns. *Proc Natl Acad Sci U S A.* 1998;95(25):14863–14868.
97. Delude CM. Deep phenotyping: the details of disease. *Nature.* 2015;527(7576):S14–S15. https://doi.org/10.1038/527S14a.
98. Vogenberg FR, Isaacson Barash C, Pursel M. Personalized medicine: Part 1: evolution and development into theranostics. *P & T.* 2010;35(10):560–576.
99. Bourgeois FT, et al. Development of the precision link biobank at Boston Children's Hospital: challenges and opportunities. *J Pers Med.* 2017;7(4):21.
100. Ahmed Z, Liang BT. Systematically dealing practical issues associated to healthcare data analytics. In: Arai K, Bhatia R, eds. *Advances in Information and Communication. Lecture Notes in Networks and Systems.* Springer Nature; 2019:69.
101. Frey LJ, Lenert L, Lopez-Campos G. EHR big data deep phenotyping. Contribution of the IMIA genomic medicine working group. *Yearb Med Inform.* 2014;9(1):206–211. https://doi.org/10.15265/IY-2014-0006.
102. Reed D, Raina K, Agarwal R. Nutraceuticals in prostate cancer therapeutic strategies and their neo-adjuvant use in diverse populations. *NPJ Precis Oncol.* 2018;2:15.
103. Dewey FE, et al. Distribution and clinical impact of functional variants in 50,726 whole-exome sequences from the DiscovEHR study. *Science.* 2016;354:aaf6814.
104. Van Hout CV, et al. Exome sequencing and characterization of 49,960 individuals in the UK Biobank. *Nature.* 2020;586:749–756. https://doi.org/10.1038/s41586-020-2853-0.
105. Evans JP, Powell BC, Berg JS. Finding the rare pathogenic variants in a human genome. *JAMA.* 2017;317:1904–1905.
106. Murray MF. The path to routine genomic screening in health care. *Ann Intern Med.* 2018;2018(169):407–408.
107. Ahmed Z, Renart EG, Mishra D, Zeeshan S. JWES: a new pipeline for whole genome/exome sequence data processing, management, and gene-variant discovery, annotation, prediction, and genotyping. *FEBS Open Bio.* 2021;11(9):2441–2452. https://doi.org/10.1002/2211-5463.13261.
108. Qu K, et al. Integrative genomics analysis by interoperation of bioinformatics tools in GenomeSpace. *Nat Methods.* 2016;13(3):245–247.

109. Scholtalbers J. Galaxy LIMS for next-generation sequencing. *Bioinformatics.* 2013;29:1233–1234.
110. Ahmed Z, et al. MAV-Seq: an interactive platform for the management, analysis, and visualization of sequence data. *Hum Genomics.* 2016;10(12):19–20.
111. Chaussabel D, Ueno H, Banchereau J, Quinn C. Data management: it starts at the bench. *Nat Immunol.* 2016;10:1225–1227.
112. Sittig DF, Murphy DR, Smith MW, Russo E, Wright A, Singh H. Graphical display of diagnostic test results in electronic health records: a comparison of 8 systems. *J Am Med Inform Assoc.* 2015;22(4):900–904.
113. Mamykina L, Vawdrey DK, Hripcsak G. How do residents spend their shift time? A time and motion study with a particular focus on the use of computers. *Acad Med.* 2016;91(6):827–832.
114. Oxentenko AS, West CP, Popkave C, Weinberger SE, Kolars JC. Time spent on clinical documentation: a survey of internal medicine residents and program directors. *Arch Intern Med.* 2010;170:377–380.
115. Arndt BG, et al. Tethered to the EHR: primary care physician workload assessment using EHR event log data and time-motion observations. *Ann Fam Med.* 2017;15 (5):419–426.
116. Howe JL, Adams KT, Hettinger AZ, Ratwani RM. Electronic health record usability issues and potential contribution to patient harm. *JAMA.* 2018;319:1276–1278.
117. Abbott A. Project set to map marks on genome. *Nature.* 2010;463:596–597.
118. Frazer KA. Decoding the human genome. *Genome Res.* 2012;22:1599–1601.
119. Falk MJ, Sondheimer N. Mitochondrial genetic diseases. *Curr Opin Pediatr.* 2010;22:711–716.
120. Lobo I, Zhaurova K. Birth defects: causes and statistics. *Nat Educ.* 2008;1:18.
121. Chial H. Mendelian genetics: patterns of inheritance and single-gene disorders. *Nat Educ.* 2008;1:63.
122. Lage K, et al. A human phenome-interactome network of protein complexes implicated in genetic disorders. *Nat Biotechnol.* 2007;25:309–316.
123. Kohler S, et al. The human phenotype ontology project: linking molecular biology and disease through phenotype data. *Nucleic Acids Res.* 2014;42:D966–D974.
124. Zhou X, et al. Human symptoms-disease network. *Nat Commun.* 2014;5:4212.
125. Blair DR, et al. A nondegenerate code of deleterious variants in Mendelian loci contributes to complex disease risk. *Cell.* 2013;155:70–80.
126. Jensen AB, et al. Temporal disease trajectories condensed from population-wide registry data covering 6.2 million patients. *Nat Commun.* 2014;5(4022).
127. Davis DA, Chawla NV. Exploring and exploiting disease interactions from multi-relational gene and phenotype networks. *PLoS One.* 2011;6: e22670.
128. Hidalgo CA, et al. A dynamic network approach for the study of human phenotypes. *PLoS Comput Biol.* 2009;5: e1000353.
129. Wahl B, et al. Artificial intelligence (AI) and global health: how can AI contribute to health in resource-poor settings? *BMJ Glob Health.* 2018;3(4): e000798.
130. Guo J, Li B. The application of medical artificial intelligence Technology in Rural Areas of developing countries. *Health Equity.* 2018;2(1):174–181.
131. Jones LD, et al. Artificial intelligence, machine learning and the evolution of healthcare: a bright future or cause for concern? *Bone Joint Res.* 2018;7(3):223–225.
132. Goh KI, et al. The human disease network. *Proc Natl Acad Sci U S A.* 2007; 104:8685–8690.
133. Liu YI, et al. The "etiome": identification and clustering of human disease etiological factors. *BMC Bioinf.* 2009;10(suppl 2):S14.
134. Hamaneh MB, Yu YK. DeCoaD: determining correlations among diseases using protein interaction networks. *BMC Res Notes.* 2015;8:226.

135. Li Y, Agarwal P. A pathway-based view of human diseases and disease relationships. *PLoS One*. 2009;4: e4346.
136. Lu M, et al. An analysis of human microRNA and disease associations. *PLoS One*. 2008;3: e3420.
137. Duran-Frigola M, et al. A chemo-centric view of human health and disease. *Nat Commun*. 2014;5:5676.
138. Ma W, et al. An analysis of human microbe-disease associations. *Brief Bioinform*. 2017;18 (1):85–97. https://doi.org/10.1093/bib/bbw005 [electronic].
139. Joyner MJ, Paneth N. Seven questions for personalized medicine. *JAMA*. 2015; 314:999–1000.
140. Luo Y, Ahmad FS, Shah SJ. Tensor factorization for precision medicine in heart failure with preserved ejection fraction. *J Cardiovasc Transl Res*. 2017;10:305–312.
141. Evanko D. The power of a crowd. *Nat Methods*. 2014;11:31.
142. Hayden EC. Geneticists push for global data-sharing. *Nature*. 2013;498:16–17.
143. Agache I, Akdis CA. Precision medicine and phenotypes, endotypes, genotypes, regiotypes, and theratypes of allergic diseases. *J Clin Invest*. 2019;129(4):1493–1503. https://doi.org/10.1172/JCI124611.
144. Clish CB. Metabolomics: an emerging but powerful tool for precision medicine. *Cold Spring Harb Mol Case Stud*. 2015;1: a000588.
145. Gamazon ER, Skol AD, Perera MA. The limits of genome-wide methods for pharmacogenomic testing. *Pharmacogenet Genomics*. 2012;22:261–272.
146. Kaddurah-Daouk R, Weinshilboum R. Pharmacometabolomics research network. Metabolomic signatures for drug response phenotypes: pharmacometabolomics enables precision medicine. *Clin Pharmacol Ther*. 2015;98:71–75.
147. Fiehn O. Metabolomics—the link between genotypes and phenotypes. *Plant Mol Biol*. 2002;48:155–171.
148. Holmes E, Wilson ID, Nicholson JK. Metabolic phenotyping in health and disease. *Cell*. 2008;134:714–717.
149. Wishart DS. Is cancer a genetic disease or a metabolic disease? *Exp Biol Med*. 2015;2:478–479.
150. Mancinelli L, Cronin M, Sadée W. Pharmacogenomics: the promise of personalized medicine. *AAPS PharmSci*. 2000;2(1):E4.
151. Yurkovich JT, Tian Q, Price ND, Hood L. A systems approach to clinical oncology uses deep phenotyping to deliver personalized care. *Nat Rev Clin Oncol*. 2020;17 (3):183–194. https://doi.org/10.1038/s41571-019-0273-6.
152. Leopold JA, Loscalzo J. Emerging role of precision medicine in cardiovascular disease. *Circ Res*. 2018;122(9):1302–1315. https://doi.org/10.1161/CIRCRESAHA.117.310782.
153. Gershon ES, Pearlson G, Keshavan MS, et al. Genetic analysis of deep phenotyping projects in common disorders. *Schizophr Res*. 2018;195:51–57. https://doi.org/10.1016/j.schres.2017.09.031.
154. Son JH, Xie G, Yuan C, et al. Deep phenotyping on electronic health records facilitates genetic diagnosis by clinical exomes. *Am J Hum Genet*. 2018;103(1):58–73. https://doi.org/10.1016/j.ajhg.2018.05.010.
155. Planchuelo-Gómez Á, Trigo J, de Luis-García R, Guerrero ÁL, Porta-Etessam J, García-Azorín D. Deep phenotyping of headache in hospitalized COVID-19 patients via principal component analysis. *Front Neurol*. 2020;11, e8976. https://doi.org/10.3389/fneur.2020.583870.
156. Papadimitriou K, Tsakirakis N, Malandrakis P, et al. Deep phenotyping reveals distinct immune signatures correlating with prognostication, treatment responses, and MRD status in multiple myeloma. *Cancer*. 2020;12(11):3245. https://doi.org/10.3390/cancers12113245.

157. Paquette AG, Hood L, Price ND, Sadovsky Y. Deep phenotyping during pregnancy for predictive and preventive medicine. *Sci Transl Med.* 2020;12(527):eaay1059. https://doi.org/10.1126/scitranslmed.aay1059.
158. Chirinos JA. Deep phenotyping of systemic arterial hemodynamics in HFpEF (part 1): physiologic and technical considerations. *J Cardiovasc Transl Res.* 2017;10(3): 245–259. https://doi.org/10.1007/s12265-017-9735-3.
159. Fassihi H, Sethi M, Fawcett H, et al. Deep phenotyping of 89 xeroderma pigmentosum patients reveals unexpected heterogeneity dependent on the precise molecular defect. *Proc Natl Acad Sci U S A.* 2016;113(9):E1236–E1245. https://doi.org/10.1073/pnas.1519444113.
160. Chirinos JA. Deep phenotyping of systemic arterial hemodynamics in HFpEF (part 2): clinical and therapeutic considerations. *J Cardiovasc Transl Res.* 2017;10(3):261–274. https://doi.org/10.1007/s12265-017-9736-2.
161. Hully M, Lo Barco T, Kaminska A, et al. Deep phenotyping unstructured data mining in an extensive pediatric database to unravel a common KCNA2 variant in neurodevelopmental syndromes. *Genet Med.* 2021;23:968–971. https://doi.org/10.1038/s41436-020-01039-z.
162. Welsink-Karssies MM, Ferdinandusse S, Geurtsen GJ, et al. Deep phenotyping classical galactosemia: clinical outcomes and biochemical markers. *Brain Commun.* 2020;2(1), fcaa006. https://doi.org/10.1093/braincomms/fcaa006.
163. Dorsey ER, Omberg L, Waddell E, et al. Deep phenotyping of Parkinson's disease. *J Parkinsons Dis.* 2020;10(3):855–873. https://doi.org/10.3233/JPD-202006.
164. Nevado J, Bel-Fenellós C, Sandoval-Talamantes AK, et al. Deep phenotyping and genetic characterization of a cohort of 70 individuals with 5p minus syndrome. *Front Genet.* 2021;12, 645595. https://doi.org/10.3389/fgene.2021.645595.
165. Fagherazzi G. Deep digital phenotyping and digital twins for precision health: time to dig deeper. *J Med Internet Res.* 2020;22(3), e16770. https://doi.org/10.2196/16770.
166. Hartl D, de Luca V, Kostikova A, et al. Translational precision medicine: an industry perspective. *J Transl Med.* 2021;19(1):245. https://doi.org/10.1186/s12967-021-02910-6.
167. Beam AL, Kohane IS. Big data and machine learning in health care. *JAMA.* 2018;319:1317–1318.
168. Obermeyer Z, Lee TH. Lost in thought - the limits of the human mind and the future of medicine. *N Engl J Med.* 2017;377:1209–1211.
169. Bates DW, et al. Big data in health care: using analytics to identify and manage high-risk and high-cost patients. *Health Aff (Millwood).* 2014;33:1123–1131.
170. Krumsiek J, et al. Gaussian graphical modeling reconstructs pathway reactions from high-throughput metabolomics data. *BMC Syst Biol.* 2011;5:21.
171. McGrath SP, Peabody Jr AE, Walton D, Walton N. Legal challenges in precision medicine: what duties arising from genetic and genomic testing does a physician owe to patients? *Front Med.* 2021;8, 663014. https://doi.org/10.3389/fmed.2021.663014.
172. Overby CL, Tarczy-Hornoch P. Personalized medicine: challenges and opportunities for translational bioinformatics. *Pers Med.* 2013;10(5):453–462. https://doi.org/10.2217/pme.13.30.
173. Kim MO, Coiera E, Magrabi F. Problems with health information technology and their effects on care delivery and patient outcomes: a systematic review. *J Am Med Inform Assoc.* 2017;7(24):246–260.
174. Sligo J, Gauld R, Roberts V, Villa L. A literature review for large-scale health information system project planning, implementation and evaluation. *Int J Med Inf.* 2017;97:86–97.
175. Marx V. Biology: the big challenges of big data. *Nature.* 2013;498(255–60):23765498.

176. Raghupathi W, Raghupathi V. Big data analytics in healthcare: promise and potential. *Health Inf Sci Syst.* 2014;2:3.
177. Alyass A, Turcotte M, Meyre D. From big data analysis to personalized medicine for all: challenges and opportunities. *BMC Med Genomics.* 2015;8:33.
178. Wang X, Williams C, Liu ZH, Croghan J. Big data management challenges in health research—a literature review. *Brief Bioinform.* 2019;20(1):156–167, bbx086. https://doi.org/10.1093/bib/bbx086.
179. Duffy DJ. Problems, challenges and promises: perspectives on precision medicine. *Brief Bioinform.* 2016;17(3):494–504.
180. Frey LJ, Bernstam EV, Denny JC. Precision medicine informatics. *J Am Med Inform Assoc.* 2016;23(4):668–670.
181. Seymour CW, Gomez H, Chang CH, et al. Precision medicine for all? Challenges and opportunities for a precision medicine approach to critical illness. *Crit Care.* 2017;21(1):257. https://doi.org/10.1186/s13054-017-1836-5.
182. Thier SO. The future of disease prevention. *J Gen Intern Med.* 1990;5(5 suppl): S136–S137. https://doi.org/10.1007/BF02600860.
183. Thier SO. Health care reform: who will lead? *Ann Intern Med.* 1991;115(1): 54–58. https://doi.org/10.7326/0003-4819-115-1-54.
184. Tonn MK, et al. Stochastic modelling reveals mechanisms of metabolic heterogeneity. *Commun Biol.* 2019;2:108.

CHAPTER FIVE

Progress in molecular biology and translational science addressing the needs of nano-rare patients

Stanley T. Crooke*
n-Lorem Foundation, Carlsbad, CA, United States
*Corresponding author: e-mail address: stan.crooke@nlorem.org

Contents

1. Introduction 128
2. Parsing patient populations 129
3. Mutation focused drug discovery 131
4. Prerequisites to respond to the needs of nano-rare patients 131
   4.1 Genotype, phenotype, clinical investigator 131
   4.2 A drug discovery platform capable of rapidly creating personalized experimental treatments 132
   4.3 Rapidity 132
   4.4 Cost-effectiveness 132
   4.5 Meeting any of the criteria summarized above is essentially impossible for most drug discovery technologies 133
5. Creating an industrialized approach to the non-profit treatment of nano-rare patients 135
   5.1 Access to appropriately characterized patients 135
   5.2 Establishing quality systems 136
   5.3 Access to treatment 136
   5.4 Treatment goals and methods of assessment 137
   5.5 Baseline (natural history) characterization and evaluation of ASO treatment effects 138
   5.6 Analysis of aggregate performance and sharing experience with all stakeholders 139
   5.7 The optimal ASO for each patient 139
   5.8 Demand for treatment 140
   5.9 Creating a broad network of collaborators 140
6. The future 141
   6.1 Broadening treatment opportunities 141
   6.2 Facilitating earlier diagnosis 142

Progress in Molecular Biology and Translational Science, Volume 190
ISSN 1877-1173
https://doi.org/10.1016/bs.pmbts.2022.04.002

6.3 Reducing isolation 143
7. Remaining challenges 144
References 145

Abstract

The healthcare systems in the developed economies were established primarily to address the more prevalent diseases and have been customized to support the provision of therapeutics to rare patients. However, with the ever-broader implementation of genomic sequencing, it is clear that there are significantly more disease-causing mutations in the human genome than realized and that many mutations are much rarer than current definitions of rare disease populations. Given this, I propose parsing patient populations and defining patient populations more precisely. Nano-rare patients are defined as patients having disease-causing mutations that are unique to a single patient or having a known worldwide prevalence of less than 30. These patient populations present unique challenges to healthcare systems that demand the development of novel models for delivery of therapeutics and novel, more efficient drug discovery technologies, such as antisense technology. The challenges presented by nano-rare patients, a novel non-profit model as a means of providing experimental treatments rather than the traditional commercial model, and progress in establishing a non-profit solution are discussed.

1. Introduction

Quite appropriately, the healthcare systems in place in developed economies were established to enhance the diagnosis and treatment of diseases with substantial prevalence.[1,2] Thanks to the efforts of many patient advocacy groups and other concerned groups, those systems have been customized to contend with diseases of much smaller prevalence. Initially, the Orphan Drug Act in the United States defined rare diseases as having a prevalence of less than 200,000 patients in the United States and provided incentives to stimulate the commercial development of drugs to treat rare diseases.[3,4] Though the specific definitions of rare diseases vary slightly and the incentives differ a bit, similar approaches have been adopted in the European Union and other regulatory jurisdictions.[5–11] These legislative initiatives have resulted in significant advances in the treatment of rare diseases, played a significant role in the maturation of the Biotechnology industry and have been criticized as contributing to the rising costs of drugs. This is turn has put significant added pressure on financing healthcare systems and the development of semi-quantitative approaches for assessing the net economic value of drugs for rare diseases, in particular, and for drugs, in general.[12–14] In fact, the total economic burden of rare diseases in the U.S. has been estimated to be nearly $1 trillion with drugs for rare diseases costing about $48 billion annually.[15] Moreover, as more human genomes are sequenced, more disease-causing mutations will be identified, increasing the numbers of rare diseases, patients afflicted, costs attributed to these populations and costly drug therapies.[1,2,5,6]

2. Parsing patient populations

As more human genomes have been sequenced, ever rarer patient populations have been identified, resulting in new terms that remain poorly defined. The term "ultra-rare" has been in vogue and used to describe patient populations with a worldwide prevalence of as many as 12,000 patients.[5,6] I propose to subdivide patient populations into common, rare, ultra-rare, micro-rare and nano-rare. Nano-rare patient populations are defined as patients with a mutation expressed in 1–30 patients worldwide. Micro-rare patients have a mutation expressed in 31–300 patients worldwide. Why sub-divide patient populations in this way? Though there is certainly overlap, patients in each subgroup differ, present different types of problems and may require different solutions and even different business models.

As is well known, the more prevalent diseases present significant challenges because, unless they are infectious diseases, these diseases are multifactorial. Most have a genetic trait(s) that increases the probability of exhibiting the diseases, but environmental factors, lifestyle and many unknown factors contribute to the expression of the disease. Indeed, approximately 50% of deaths in developed economies are considered secondary to chronic systemic inflammation (CSI), which is known to be secondary to numerous genetic, lifestyle and environmental factors.[16] Thus, discriminating causal factors from secondary consequences and compensatory efforts to maintain homeostasis to identify potential attractive molecular targets for drug therapy is quite challenging. Adding to the challenges are current regulatory requirements for commercial approval of many drugs designed to treat many of the common diseases still demand that benefit be demonstrated in a broad population of patients with the same signs and symptoms, but who may have a variety of different molecular causes and secondary or different types of responses to maintain homeostasis.

In contrast, as the patient populations become rarer, identifying the molecular cause and therefore a molecular target becomes much simpler as extremely rare patient populations are almost always the result of a mutation in a single gene. Interestingly, as more patients undergo genomic sequencing, it is now understood that even for more prevalent genetically caused diseases, there is often a relatively common mutation and a number of even rarer mutations in the same gene that result in the syndrome and variations in the presentation of patients with the disease.[5,6] However, the rarer the patient population, the more difficult it is to economically justify development of a commercially focused drug discovery program. In fact, in the nano-rare patient population (1–30), commercial development is literally or logistically impossible given current regulatory requirements. Many nano-rare patients have mutations unique to them. In the instance in which a single patient is known to have a specific mutation, proving safety and efficacy in a manner that meets current regulatory requirements is not possible. Further, since most nano-rare mutations identified to date are de novo, patients are not localized by founder effect and they are identified at widely varying stages in the progression of their disease,[1,2] rendering conducting a controlled clinical trial practically or logistically impossible. Moreover, to generate a reasonable return on investment, a company would need to charge affected patients unprecedented annual costs. In principle, "curing" a disease with a single administration of a gene replacement may be possible, but to date definitive evidence showing life-long effects of replacement genes is lacking. However, even in the case in which a single administration of a gene might "cure" a disease, the economics are similarly challenging. This is because the total return on investment is determined by the total number of patients treated and the cost of a unit of treatment. Consequently, to make commercial development of a single dose treatment economically feasible for a population, let's say of 30, the cost to those patients would still have to be extremely high. Obviously, in the vast majority of instances, the term "cure" is inappropriate to use for drug therapy because continued therapeutic intervention is required. Consequently, a non-commercial model must be considered for the nano-rare patient.

For patients in the micro-rare category, though they typically have a genetic etiology, it is usually difficult to justify a multimillion-dollar commitment to drug development when the development time and risks of failure are unclear. Given that the discount for risk is so large at the initiation of development for drugs for micro-rare patients, a hybrid model may be needed in which investments are made to demonstrate value and reduce risk by an organization with high risk tolerance and a very small infrastructure, then the de-risked product candidate could be licensed to a commercial company. For such hybrid solutions, the term venture philanthropy has been coined.

3. Mutation focused drug discovery

Traditional definitions of disease are millennia old. For example, congestive heart failure (CHF) was described in multiple ancient civilizations and the modern understanding of CHF began with Harvey in the 17th century[17] and psoriasis was first described by ancient Romans.[18] Irrespective of when and who first established traditional definitions of diseases, all definitions are simply descriptions of the manifestations of molecular pathologic events and efforts to maintain homeostasis that have progressed to such a degree that a patient is symptomatic. Though the traditional disease definitions have been of value historically, as they codified what was understood and provided foci for research to better understand the molecular pathology and the discovery of diagnostic and therapeutic agents, once a disease is identified as genetic and the causal mutation identified as the therapeutic target, the name of the disease is irrelevant.

4. Prerequisites to respond to the needs of nano-rare patients

4.1 Genotype, phenotype, clinical investigator

The first step in the journey of a patient with any disease is to achieve a diagnosis. The extreme rarity of nano-rare diseases, and their unique presentations, means the physician does not have the information necessary to make a differential diagnosis. This issue leads to multiple referrals to various specialists and the failure to achieve a diagnosis. In fact, the vast majority of nano-rare patients are never diagnosed, and for the fortunate few who are referred to a tertiary care center capable of characterizing the genotype and phenotype of the patient, the average time from symptom onset is >8 years.[19] These data and much of what is known about undiagnosed diseases derive from work conducted by the Undiagnosed Disease Network (UDN), a consortium of tertiary care centers that, over the past 5–10 years, has had success in genotyping and phenotyping previously undiagnosed patients, determining the function of the mutant gene, and characterizing the nature of the mutations.[20] Though the UDN has made important strides in addressing the needs of undiagnosed diseases, even with all the collective expertise in that consortium, only about one third of patients accepted have achieved a diagnosis.[20]

4.2 A drug discovery platform capable of rapidly creating personalized experimental treatments

To treat nano-rare patients with personalized genetically-based experimental medicine, unique drug discovery technologies with the following attributes must be available.

4.3 Rapidity

Given the fact that the small fraction of nano-rare patients who are genetically diagnosed typically have invested multiple years to achieve a diagnosis during which their diseases progress, all are desperate, and many are in urgent need of therapeutic intervention, the drug discovery platform must be capable of rapidly moving from an understanding of the disease-causing mutation to a therapeutic agent. Ideally, the technology would support a mutation-directed discovery program. Given the severity of disease and the desperation of nano-rare patients, the total time from acceptance to treatment must be minimized and ideally less than 18 months. This is essentially impossible for most drug discovery platforms.

4.4 Cost-effectiveness

Numerous factors contribute to the cost of creating a drug and, to a large extent, the factors that drive cost are similar whether the drug is to be administered to a single or many patients. The discovery of the experimental treatment must be a relatively small contributor to cost and time. The key advantage in nano-rare patients is that the cause, a well characterized mutation, is known. For diseases for which the molecular pathology is unknown or for multifactorial diseases, years and many scores of millions of dollars must be invested in an often futile search to identify a relevant target for drug discovery efforts. Ideally, of course, a drug discovery technology that directly moves from the mutation to the patient is necessary. In development, many costs are relatives fixed. For example, the cost of conducting toxicology studies is relatively fixed, so the key variable in this cost center is the number and duration of species and toxicology studies that must be conducted. Manufacturing the drug is another potentially significant driver of cost and is greatly affected by the potency and duration of effect of the drug. Ideally, potent drugs with prolonged durations of effect should be coupled to low manufacturing costs. Further, the drug should be stable enough to assure that one manufacturing campaign could supply drug to the patient for 10 years or more. Finally, preparing the formulated drug is also a relatively fixed and significant cost. For drugs administered parenterally, the formulated drug must be filled sterilely in vials.

4.5 Meeting any of the criteria summarized above is essentially impossible for most drug discovery technologies

4.5.1 *Versatility*

There are multiple types of mutations that cause disease, ranging from full loss of gene function, null mutations, to toxic gain-of-function mutations. Further, mutations are not limited to the coding sequence and may alter pre-mRNA processing, the stability of the RNA or protein or even affect translational regulatory elements in mRNAs that are known to down-regulate the translation of approximately 50% of human proteins.[21,22] Therefore, though no drug discovery technology is capable of addressing all nano-rare patients, sufficient versatility to support treating a meaningful number of nano-rare patients is important.

4.5.2 *Predictability*

Critical questions that must be addressed for all patients to be treated with any experimental therapeutic are what the initial dose level should be, by what route of administration should the agent be administered and how frequently should the agent be administered. Additionally, it would be particularly valuable to have at least a general sense of the potential types of side effects.

4.5.3 *Scalability*

Though each patient is unique, it estimated that the total nano-rare patient population is many millions. Moreover, as more human genomes are sequenced, more mutations, more diseases and more patients will be identified. Additionally, typically when a treatment opportunity is shown to be viable, patients who previously were undiagnosed or are diagnosed, but have chosen not seek treatment, emerge, expanding the known prevalence of patients with a specific mutation. Thus, ideally, an approach for nano-rare patients should be readily scalable.[1,2,5,6]

Clearly, the attributes discussed above present a formidable challenge and thus it is highly unlikely that any single drug discovery platform could meet the needs of all nano-rare patients. Nevertheless, one of most important therapeutic dicta is that "we treat the patients we can treat" today while investing in research that in the long term will enhance and broaden the treatments for all patients.

Antisense oligonucleotide (ASO) technology is capable of meeting the needs of many nano-rare patients today. Once a causal mutation is identified, ASO technology is a direct path to treatment. In fact, it is ideally suited to be used in a mutation-focused effort.[23,24] Relative to other drug discovery platforms, the discovery of the optimal ASO for an RNA target is very rapid and inexpensive, particularly if the organization involved has coupled massively parallel screening of millions of ASOs, efficient automation and ever more sophisticated machine learning, with decades of advances in understanding the molecular mechanisms driving the observed phenotypes.[1,2] Relatively recent advances have added new post RNA binding mechanisms that enhance the versatility of the technology. ASOs can be administered by any route of administration, including oral, and are administered either monthly or quarterly or less frequently after initiation of ASO therapy. These drugs distribute broadly after systemic delivery and broadly in the CNS after intrathecal administration. Given the potency and duration of effects of modern ASOs, the total amount of an ASO needed to treat a patient for life is very modest. Importantly within a chemical class and there are several chemical classes that are very well understood and have been studied in thousands of patients–each ASO behaves similarly. This is critical as it facilitates informed decisions about the dose level, the route of administration and the frequency of dosing needed to achieve a therapeutic effect. Further, for several of the more commonly used and studied chemical classes, databases that integrate all safety observations in non-human primates and all controlled clinical trials have been published and are available to regulatory agencies.[23–27] The availability of these data is important because they reveal potential types of side effects. Finally, ASO technology is readily scalable and could, over time, meet the needs of many thousands of nano-rare patients.[24,28–32]

ASO technology does have limitations, the most significant of which is that it cannot be used to address null mutations. Only some form of gene replacement therapy can be used in the case of null mutations. After systemic administration, the commonly used ASOs distribute unevenly with relatively limited distribution at attractive therapeutic doses to skeletal muscles, the heart, some visceral organs such as the pancreas and some malignancies.[30] Because high doses are required to achieve therapeutic concentrations in those tissues, today, for nano-rare patients, the technology is appropriate to consider after systemic administration for targets in the liver and kidney, after intravitreal administration for targets in the eye, after intrathecal administration for CNS targets and after aerosol administration for some targets in the lung.[1,2]

4.5.4 A supportive regulatory environment

Not only are nano-rare patients unique and present unique needs and challenges to the healthcare system, the concept of a non-profit foundation, creating a personalized experimental treatment one patient at a time and providing the experimental treatment for free for life, was a concept never considered in any of the legislative bases for the FDA or any of the guidances issued by the FDA. However, the FDA has responded promptly to the needs of nano-rare patients and the opportunity to treat some of these patients with ASOs. General draft administrative guidance was issued in 2021 and in 2022 draft preclinical, manufacturing, and clinical guidance documents were issued. The preclinical guidance supports initiating clinical trials after completing a three-month rodent safety study and appropriate in vitro or in vivo proof of pharmacology, the manufacturing guidance provides directions for manufacturing, stability testing, and controls and the clinical guidance describes general approaches to clinical administration of ASOs in nano-rare patients. Importantly the guidances are limited to patients with life threatening or severely debilitating diseases and are limited to ASOs of well-studied chemical classes.[1,2]

5. Creating an industrialized approach to the non-profit treatment of nano-rare patients

Though it has been shown that it is possible to take advantage of antisense technology to discover and develop a personalized ASO treatment, an industrialized approach is required if the needs of many patients are to be addressed. By industrialized, I mean that a systematic approach that assures that each step in the process from making the complex benefit/risk decision to treat a patient with an experimental ASO, to the evaluation of the performance of such ASO in each patient, must be conducted in a manner that assures that each step is taken with the highest quality possible. This is particularly true for the process that identifies the optimal personalized ASO. Moreover, an industrialized approach is required to assure that the process can be scaled to meet the demand, which is projected to many thousands of patients in the coming years. These approaches have been developed by a single non-profit foundation, n-Lorem, but are broadly applicable.[1,2]

5.1 Access to appropriately characterized patients

Clearly, access to genotypically and phenotypically characterized patients, clinical investigators, and institutions capable of conducting investigator-initiated experimental treatment and evaluation of nano-rare patients is critical. Ideally, the genomic characterization includes the identification and proof of the causative gene and mutation. Additionally, the function of the gene must be known to understand the potential effects of altering the normal allele, and the mutation must be clearly categorized, e.g., missense, nonsense, null, loss-of-function, toxic gain-of-function. Phenotypic characterization must be as thorough, and is as critical, as the genotype. *What is the age and gender of the patent, at what age did the patient become symptomatic, what organs are affected, what are the primary and secondary manifestations of the mutation, have developmental delays resulted from the mutation and what domains are affected* are a few of the questions that must be clearly understood?

5.2 Establishing quality systems

The FDA guidelines for ASOs of well understood chemical classes to treat nano-rare patients with experimental ASOs support initiating clinical administration with minimal preclinical data. Therefore, once the mutation is fully understood and demonstrated to be causative, the gene is functionalized and an ASO shown to be pharmacologically effective in vitro and, in some cases, if a rodent model exists and is relevant (typically mutation knock-in mice), a single 3-month rodent toxicity study is sufficient to initiate exposure of a patient to the personalized ASO. While the guidance facilitates rapidly responding to the needs of these desperate patients, the limited preclinical data increase risk and place an ever-greater premium on assuring the maximum possible quality at each step. The process can be divided into the steps shown in (Fig. 1).

5.3 Access to treatment

To initiate the process to determine if a patient is appropriate for experimental ASO treatment, the patient or parent and a research physician must complete the application available on the n-Lorem website. As the application requires significant genotypic and phenotypic information, a physician investigator must, typically, participate in completing the application. Once received and blinded as to patient identity, a preliminary assessment of whether the prevalence of the mutation, the organ(s) affected, the nature of the mutation and the functions of the normal gene is conducted. If there is insufficient information, queries of the applicants may be needed or, in some cases, additional work may be done by n-Lorem. Once sufficient information is available, n-Lorem staff present a detailed summary of the case and a

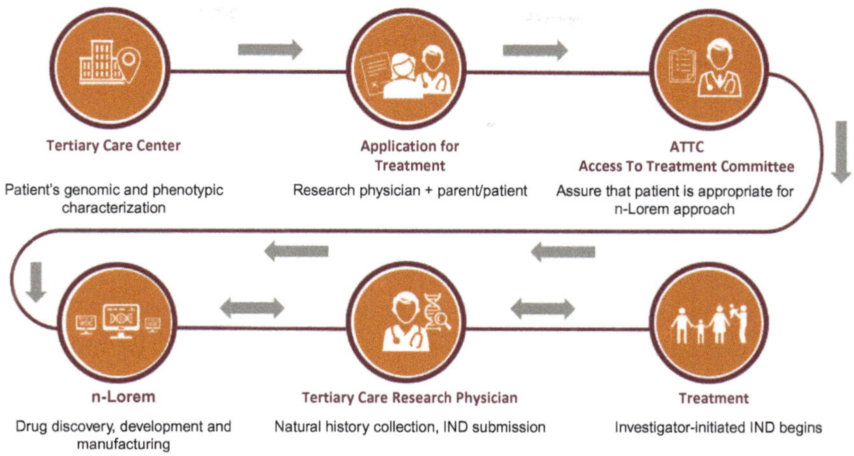

Fig. 1 n-Lorem process from application to treatment.

recommendation to the Access To Treatment Committee (ATTC). This committee is composed of geneticists, organ and disease experts, ASO and clinical trial experts as well as bioethicists and provides a recommendation to the n-Lorem leadership team that makes the final decision. The participation of the ATTC in reaching a decision is critical because nano-rare patients are very complex and the risk/benefit decisions quite challenging.

5.4 Treatment goals and methods of assessment

Though it is axiomatic that before exposure of a patient to any therapeutic, a clear set of treatment goals must be defined, it is even more important in the nano-rare patient because these individuals are typically quite ill, the manifestations of the disease complex and the potential risk of side effects must be mitigated by the opportunity to provide a meaningful benefit. Consequently, primary, secondary, and exploratory treatment goals are agreed with the responsible physician, then the specific clinical measurements are defined. Given that most patients to be treated are unique, it is not possible to develop novel biomarkers of disease that can be used to assess the effect of an experimental agent. This is especially true because a crossover design is not practical because the patients are quite ill and current ASOs have very long durations of action that would require months on and off treatment to compare effects. Therefore, validated clinical assessments are used such as proteinuria, creatinine and calculated glomerular filtration rate for patients with renal disease. Developmental delays are common for mutations expressed in infancy. Because it is unclear to what extent developmental delays can be recovered with effective treatment of the cause of the disease, often development delays are selected as exploratory treatment goals.

5.5 Baseline (natural history) characterization and evaluation of ASO treatment effects

Given there is no regulatory path to commercial approval and n-Lorem certainly does not plan to commercialize experimental ASOs for nano-rare patients, the clinical guidance under which n-Lorem operates do not provide specific clinical trial requirements. Nevertheless, it is imperative that we learn as much from each patient and the aggregate experience as possible. Two critical questions that we must try to answer as rigorously as possible is does a personalized experiment ASO appear beneficial and did the ASO cause any side effects? Though, single patient clinical trials have been conducted and several designs have been employed, all studies involve some form of cross-over design comparing an experimental treatment observations to placebo or an approved agent.[3,33–35] Unfortunately, such approaches are not feasible to evaluate the performance of experimental ASOs in nano-rare patients because nano-rare patients are typically in urgent need of treatment and modern ASOs have relatively long durations of effect (one to several months).

For a patient population comprising a single patient, or even 30 patients dispersed around the world and at varying stages of disease, the natural history of the disease is unknown and the delays in diagnosis caused by the extreme rarity of these patients mean that nearly all of the patients applying for treatment are severely ill and progressing rapidly. Thus, an individualized approach to the evaluation of the effect of experimental treatment is required. However, because the discovery and development of an experimental ASO at n-Lorem takes 12–18 months, it is possible collect data during those 12–18 months focused on the planned treatment goals using the clinical measurements to be employed during the first year of experimental treatment. These data then constitute a proximal baseline or natural history for the patient to be compared to the same parameters while the experimental ASO is administered. Importantly, because n-Lorem cannot pay clinical sites to conduct these observational studies, we encourage investigators to perform any assessments they may find of interest, but n-Lorem asks only for the results of clinical tests performed as consistent with optimal medical care. This approach extends to the management of the "clinical trials." Compared to traditional trials conducted to gain commercial approval, monitoring of investigators and sites is limited and to a large extent remote. On the other hand, given the demand we have already experienced, it seems likely that in a few years we will be conducting 100–125 individual patient baseline and 1 year "clinical trials" simultaneously. Consequently, the partnership we have established with Parexel, a clinical contract research organization experienced in conducting studies in very small patient populations is one the important partnerships effected.

5.6 Analysis of aggregate performance and sharing experience with all stakeholders

In addition to assessing the impact of experimental ASOs in individual patients, it is important to assess the overall performance of an approach and the portfolio of experimental ASOs. Here again, a novel solution is required if the analysis is to extend beyond simple parameters such as the number of applications received per year, the number accepted and the number treated. While the goal is to assess ASO performance in individual patients with as much granularity as possible, the diversity of patients, organs involved, and disease manifestations mean that estimating the value of the approach relative to the overall patient population treated must be relatively crude. For example, in what fraction of patients did the investigator report significant improvement in a primary or secondary treatment goal, or what fraction of patients experienced an adverse event?

Investigators should be encouraged to publish case reports describing their experience with experimental ASOs in individual patients. In addition, the entity should openly publish the results of analyses of aggregate performance yearly in peer-reviewed journals and disseminate results at scientific meetings, colloquia, etc.

5.7 The optimal ASO for each patient

As described above, ASO technology is ideally positioned to rapidly provide experimental treatments to many nano-rare patients. However, all ASOs are not created equal. There are several chemical classes with different properties, multiple designs to exploit different post-RNA binding mechanisms and each target RNA is different. Since RNA in cells is highly structured, extensively protein bound and undergoing various processing and degradation events, it is essential, even for the best understood chemical classes, to screen numerous potential ASO binding sites in each RNA. Experience has shown that the more sites screened the better the ASO and a minimum number of sites to screen is ~500.[32] Fortunately, n-Lorem accesses fully automated screening systems, sophisticated algorithms derived from more than 30 years of research and millions of ASOs screened, as well as in depth knowledge of the molecular mechanisms driving the screening observations.[29,30] Additionally safety databases are available that integrate all the safety observations made in non-human primate and controlled clinical studies for all the chemical classes used to treat nano-rare patients.[23,25–27] Thus, any organization that intends to discover experimental ASOs, even for a single patient, would be wise to attempt to access similar experience or collaborate with an organization like n-Lorem to assure that the optimal ASO is identified.

5.8 Demand for treatment

Considering the experience at n-Lorem as exemplary of the challenges encountered in establishing a broad industrialized approach to the treatment of nano-rare patients, demand for treatment is likely to be substantial and grow exponentially. At n-Lorem only, nearly 150 new applications have been received in two years. More than 50 patients have been accepted for experimental ASO treatment. The reasons for rejection of patients include mutations that cannot be addressed by ASO treatment, involvement in organs that are beyond the current scope of n-Lorem or a prevalence in excess of 30 patients worldwide. Others have been rejected or deferred if insufficient information was available about the mutation, the gene function, or the evidence that the causality of the mutation was insufficient. In many cases that have been deferred, n-Lorem scientists and collaborators have performed the necessary experiments to gain approval. The bulk of cases have been neurological diseases, but renal, ophthalmic, and hepatic applications are increasing. The overall demand continues to increase and since n-Lorem has limited its activities to the United States, as n-Lorem expands to other countries, the demand will continue to increase.

5.9 Creating a broad network of collaborators

No single, non-profit foundation is likely to be able to meet the demand for treatment. Indeed, the term "it takes a village" seems quite apropos and the response to the opportunity to meet the needs of nano-rare patients using antisense technology has been quite gratifying (Fig. 2). Particularly impressive and important has been the response in the biotechnology and pharmaceutical industry. Members of essentially every reagent, equipment and service provider to the industry are contributing with price reductions, and

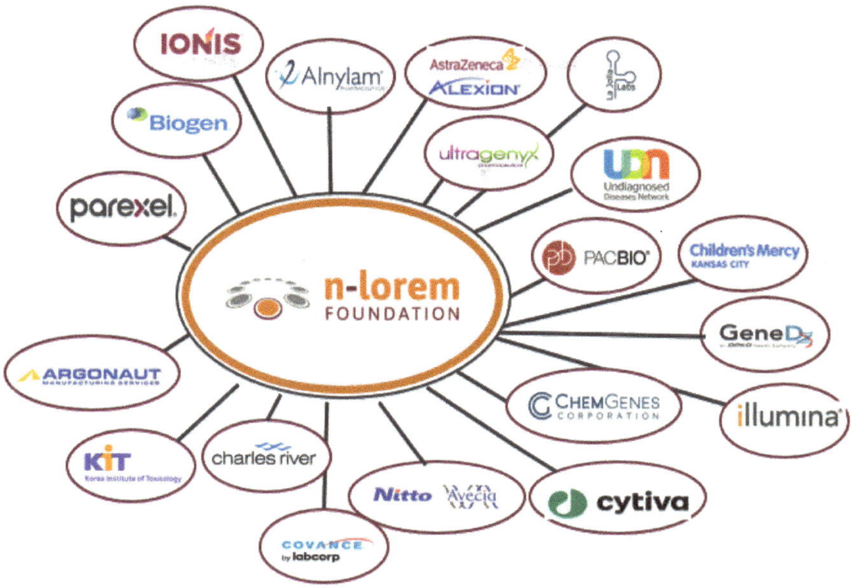

Fig. 2 The current n-Lorem network of donors and collaborators.

in-kind or cash contributions. Several biotechnology and pharmaceutical companies have joined Ionis and Biogen in supporting and collaborating with n-Lorem. The price discounts and in-kind contributions from vendors to the industry have supported a ~40% reduction in the cost per patient while facilitating more rapid development of experimental ASOs by providing "slots" in their busy schedules to assure that each step is coordinated and efficiently coupled. For example, it is critical to have manufacturing and sterile vial fill slots available when the results of toxicology studies are available. A failure to coordinate these activities could cause multi-month delays that would not be easily tolerated by patients. Because of the efficiency of the technology, automation, potency and duration of action in organs selected and the routes of delivery that are appropriate as well as the very limited preclinical testing required by the special guidance, it is anticipated that at steady state the average cost per patient will be $0.6–0.8 Million.

6. The future

6.1 Broadening treatment opportunities

Though the progress made by the UDN, other personalized medicine efforts, n-Lorem, and other efforts to provide experimental treatments, the nano-rare patient population remains largely undiscovered and the fortunate few who achieve a diagnosis are isolated and unlikely to benefit from effective treatment. The immediate role of an organization like n-Lorem should be helping to meet one the most important principle in therapeutics: treat the patients today that are amenable to existing approaches while investing in research to enhance and broaden treatments in the future. Certainly, an important goal for n-Lorem is to expand into countries other than the U.S. and continue to broaden awareness of our efforts to increase the number of patients treated. It certainly is gratifying that n-Lorem is progressing toward treating an ever-growing number of patients, but about two-thirds of the patients who apply for treatment are not appropriate candidates for experimental ASO treatment. In the future, as other technologies mature, we look forward to new approaches that can expand the number of patients who can be treated. It would be particularly important to add gene therapy as an option for patients with null mutations and the NIH has announced an initiative in gene therapy for patients with very rare mutations. However, it is not clear how rapidly gene therapy can be "industrialized" sufficiently to scale to meet the demand. Perhaps more advances in delivery and significant reductions in cost will be required before gene therapy can be broadly applied to the nano-rare patient population. As other forms of mutation directed genetic medicines advance, perhaps they can be added to the arsenal of technologies that can address the needs of nano-rare patients.

6.2 Facilitating earlier diagnosis

An urgent need and a prerequisite to actually gaining insight into the nature and scale of the health issues caused by extremely rare mutations in humans is to define the incidence and prevalence of these disease-causing mutations and to diagnose the afflicted patients much earlier than is possible today. The extreme rarity of these mutations means that while highlighting the existence of this patient population and introducing the topic in continuing medical education programs and in medical school curriculum may help, a more systematic population-wide approach is needed. Fortunately, the technology to answer these questions exists and is genomic sequencing. Though the costs of genomic sequencing are likely to decline, even at today's costs, introducing genomic sequencing with appropriate bioethical safeguards to standard newborn evaluation protocols is practical and could yield enormous dividends in basic knowledge. An example would be *how the frequency of de novo mutations varies as a function of ethnicity, environment, and age of parents.*

What fraction of nano-rare mutations causes disease, how many of the diseases are severe, and what fraction are expressed during infancy are other important issues? If genomics were coupled to systematic long-term follow up, new insights into the natural histories of mutation-caused diseases, including nano-rare diseases, could be accumulated.

Beyond the advances in basic knowledge, an even more important and immediate benefit would be to achieve more rapid diagnosis of nano-rare patients. For the first time, there would be information about the incidence and prevalence of nano-rare diseases and these patients could be considered for treatment before the disease progresses, and for diseases identified in infancy, early enough to prevent, or limit developmental delays. Clearly, for adoption of genomic sequencing as a part of routine newborn screening, meaningful policy and financial issues must be discussed and, in the U.S., insurers convinced that such approaches could save healthcare dollars. In some countries, for example the United Kingdom, efforts to introduce newborn genomic screening are in progress. Irrespective of the type of healthcare delivery system, the potential of a diagnosis resulting in potential treatment enhances the economic benefit and heightens interest in achieving diagnosis. Consequently, n-Lorem hopes to play a role in encouraging the introduction of genomic screening of newborns, by being one opportunity to couple diagnosis to treatment.

6.3 Reducing isolation

An issue that may not be obvious but is a significant factor in the plight of nano-rare patients is their isolation. Quite obviously, when a mutation is expressed in a single patient, there is no community, no patient advocacy group, or any organization that the patient and family can turn to be better informed about the mutation and the disease, to provide emotional support or help the patient be referred to appropriate physicians to achieve a diagnosis and be directed toward treatments if available. In responses, initiating a patient empowerment program that will focus exclusively on nano-rare patients is important. The programs should include a podcast series that is both educational and provides a site for concerns of all stakeholders to be heard, a web site that will provide help to patients seeking a diagnosis or a physician capable of caring for the patient and annual meetings of patients and families. Once again, as a demonstration that this approach is possible, n-Lorem plans to launch a program such as described above this year.

7. Remaining challenges

As we have discussed, many challenges to treating nano-rare patients exist. The challenges faced by n-Lorem provide a good example. Though n-Lorem is off to a solid start, we have barely left the starting line. Certainly, the most immediate challenge is to respond to the demand and achieve the goal of moving from approval of a patient for treatment to the initiation of dosing with 12–18 months. Additionally, it is important that awareness of the opportunity for treatment provided by n-Lorem be steadily broadened, and that n-Lorem expand operations beyond the U.S. While some nano-rare patients have been treated with experimental ASOs[36], the most important next step is to treat many more patients and show that the systems we have created work. We look forward to more treatment successes and reporting regularly on the performance of experimental ASOs in nano-rare patients.

Though we are pleased with the progress to date, an important goal is to substantially increase the efficiency of the efforts to genotype, characterize the type of mutation and the functions of the gene and documenting the phenotype of nano-rare patients in a fashion that facilitates a decision about the appropriate therapeutic approach. Today, applications are frequently submitted with inadequate sequencing, inadequate characterization of the mutation or functionalization of the gene. We then need to work with our genomic partners to perform added sequencing and to characterize the mutation and gene function in our laboratory. This results in delays and additional cost. To address these issues, we are working with key investigators and institutions to create what we call "partners in excellence" who will work seamlessly with n-Lorem to develop the information needed to make a treatment decision in most efficient and cost-effective fashion possible.

It is also important that n-Lorem establish more holistic solutions that provide support to nano-rare patients. Initiating the n-Lorem nano-rare patient podcast series is a first step, but more services are needed. In particular, nano-rare patients need help in identifying appropriate personalized medicine centers and clinical investigators. We hope to create a network of systems that any patient or parent can access to reduce the journey to diagnosis. Every day matters for nano-rare patients, but we consider these approaches interim solutions to be implemented while we and others work toward the implementation of new-born genomic screening.

Finally, n-Lorem must robustly answer this question: is a charitable model scalable and sustainable? Once again, the response to n-Lorem and the progress to date are encouraging. However, numerous new partners and donors must be added before the sustainability of the n-Lorem non-profit is proven sustainable.

References

1. Crooke ST. Meeting the needs of patients with ultrarare diseases. *Trends Mol Med.* 2022;28(2):87–96.
2. Crooke ST. *n-Lorem Foundation Notes Latest in Guidance from FDA on Developing ASO Drug Products for Patients With Nano-Rare Diseases. Guidance Issued by FDA Provides a Supportive Environment to Develop Experimental ASO Medicines for Patients With Nano-Rare Disease*; 2022. https://www.businesswire.com/news/home/20220106005046/en/n-Lorem-Foundation-Notes-Latest-in-Guidance-From-FDA-on-Developing-ASO-Drug-Products-for-Patients-With-Nano-rare-Diseases.
3. Kravitz RL, et al. A case for n-of-1 trials—reply. *JAMA Int Med.* 2019;179(3):453.
4. Mikami K. Orphans in the market: the history of orphan drug policy. *Soc His Med.* 2017;32(3):609–630.
5. Crooke ST. Addressing the needs of patients with ultra-rare mutations one patient at a time: the n-lorem approach. *Nucleic Acid Ther.* 2021.
6. Crooke ST. A call to arms against ultra-rare diseases. *Nat Biotechnol.* 2021.
7. Andel C, et al. The economics of healthcare quality and medical errors. *J Healthcare Fin.* 2012;39(1):39–50.
8. Nguengang Wakap S, et al. Estimating cumulative point prevalence of rare diseases: analysis of the Orphanet database. *Eur J Hum Genet.* 2020;28(2):165–173.
9. Jo A, et al. Prevalence and practice for rare diseases in primary care: a national cross-sectional study in the USA. *BMJ Open.* 2019;9:1–4.
10. Copley-Merriman K. Rare diseases: addressing the challenges in diagnosis, drug approval, and patient access. *Value Health.* 2018;21(5):491–492.
11. Richter T, et al. Rare disease terminology and definitions—a systematic global review: report of the ISPOR rare disease special interest group. *Value Health.* 2015;18(6): 906–914.
12. Raghupathi V, Raghupathi W. Healthcare expenditure and economic performance: insights from the United States data. *Front Public Health.* 2020;8:156.
13. Scheunemann LP, White DB. The ethics and reality of rationing in medicine. *Chest.* 2011;140(6):1625–1632.
14. De Vivo DC, et al. Nusinersen initiated in infants during the presymptomatic stage of spinal muscular atrophy: interim efficacy and safety results from the Phase 2 NURTURE study. *Neuromusc Disord.* 2019;29(11):842–856.
15. Group, T. E. F. a. L. *Economic Burden of Rare Diseases in the U.S. Approached $1 Trillion in 2019, Surpassing Cost Estimates for Many Chronic Diseases*; 2021.
16. Furman D, et al. Chronic inflammation in the etiology of disease across the life span. *Nat Med.* 2019;25(12):1822–1832.
17. Hajar R. Congestive heart failure: a history. *Heart Views.* 2019;20(3):129–132.
18. Benedek TG. Psoriasis and psoriatic arthropathy, historical aspects: part I. *J Clin Rheumatol.* 2013;19(4):193–198.
19. Spillmann RC, et al. A window into living with an undiagnosed disease: illness narratives from the Undiagnosed Diseases Network. *Orphanet J Rare Dis.* 2017;12(1):71.
20. Walley NM, et al. Characteristics of undiagnosed diseases network applicants: implications for referring providers. *BMC Health Serv Res.* 2018;18(1):652.

21. Liang XH, et al. Antisense oligonucleotides targeting translation inhibitory elements in 5' UTRs can selectively increase protein levels. *Nucleic Acids Res.* 2017;45(16): 9528–9546.
22. Liang XH, et al. Translation efficiency of mRNAs is increased by antisense oligonucleotides targeting upstream open reading frames. *Nat Biotechnol.* 2016;34(8):875–880.
23. Crooke ST, et al. The effects of 2'-O-methoxyethyl oligonucleotides on renal function in humans. *Nucleic Acid Ther.* 2018;28(1):10–22.
24. Crooke ST, et al. RNA-targeted therapeutics. *Cell Metab.* 2018;27(4):714–739.
25. Crooke ST, et al. Integrated safety assessment of 2'-O-methoxyethyl chimeric antisense oligonucleotides in nonhuman primates and healthy human volunteers. *Mol Ther.* 2016;24(10):1771–1782.
26. Crooke ST, et al. The effects of 2'-O-methoxyethyl containing antisense oligonucleotides on platelets in human clinical trials. *Nucleic Acid Ther.* 2017;27(3):121–129.
27. Crooke ST, et al. Integrated assessment of the clinical performance of GalNAc(3)-conjugated 2'-O-methoxyethyl chimeric antisense oligonucleotides: I. Human volunteer experience. *Nucleic Acid Ther.* 2019;29(1):16–32.
28. Crooke ST, et al. Phosphorothioate modified oligonucleotide-protein interactions. *Nucleic Acids Res.* 2020;48(10):5235–5253.
29. Crooke ST, et al. The interaction of phosphorothioate-containing RNA targeted drugs with proteins is a critical determinant of the therapeutic effects of these agents. *J Am Chem Soc.* 2020;142(35):14754–14771.
30. Crooke ST, et al. Antisense drug discovery and development technology considered in a pharmacological context. *Biochem Pharmacol.* 2020;114196.
31. Crooke ST, et al. Antisense technology: an overview and prospectus. *Nat Rev Drug Discov.* 2021.
32. Crooke ST, et al. Antisense technology: a review. *J Biol Chem.* 2021;296:100416.
33. Margolis A, Giuliano C. Making the switch: from case studies to N-of-1 trials. *Epilep Behav Rep.* 2019;12, 100336.
34. Lillie EO, et al. The n-of-1 clinical trial: the ultimate strategy for individualizing medicine? *Pers Med.* 2011;8(2):161–173.
35. Duan N, et al. Single-patient (n-of-1) trials: a pragmatic clinical decision methodology for patient-centered comparative effectiveness research. *J Clin Epidemiol.* 2013;66 (8 Suppl):S21–S28.
36. Korobeynikov VA, et al. Antisense oligonucleotide silencing of FUS expression as a therapeutic approach in amyotrophic lateral sclerosis. *Nat Med.* 2022;28(1):104–116.

CHAPTER SIX

Precision medicine in epilepsy

Ryan James McGinn, Erica Leah Von Stein,
Jacqueline Elizabeth Summers Stromberg, and Yi Li*
Department of Department of Neurology & Neurological Sciences, Stanford University School of Medicine, Stanford, CA, United States
*Corresponding author: e-mail address: lyi@stanford.edu

Contents

1. Introduction 148
2. Precision therapy in genetic epilepsy syndromes 148
   2.1 Current genetic testing tools and their yield in epilepsy 148
   2.2 Precision medicine approach in genetic epilepsy syndromes 151
   2.3 New avenues of personalized treatment in genetic epilepsy 171
3. Precision therapy with personalized pharmacogenomics 173
4. Precision therapy with neuromodulation 174
   4.1 Responsive neurostimulation (RNS system) 174
   4.2 Deep brain stimulation (DBS) 175
5. Conclusion 177
Funding 177
References 177

Abstract

Epilepsy is a common neurological disorder affecting over 50 million people worldwide. Although more than 20 new anti-seizure medications have been developed in the past three decades, about one-third of patients still experience recurrent seizures. The heterogeneity of epilepsy types and the highly variable inter-individual response to the therapies makes disease management challenging. A precision medicine approach of treatment for epilepsy patients that considers individual patient characteristics has long been advocated. With advances in DNA sequencing technology, it has been estimated than >30% of all epilepsy syndromes have genetic components. Here we summarize recent approaches in genetic testing and current personalized treatments that are already applicable in genetic epilepsy syndromes. We also discuss current precision therapy utilized in personalized pharmacogenomics, as well as tailored treatment with neuromodulation for epilepsy patients. In the near future, we envision precision medicine in epilepsy will be applicable on a large scale involving not only targeted therapies, but also diagnosis, prognosis, and comorbidity management.

Progress in Molecular Biology and Translational Science, Volume 190
ISSN 1877-1173
https://doi.org/10.1016/bs.pmbts.2022.04.001

1. Introduction

Epilepsy is a common neurological disorder affecting approximately 1% of the population and over 50 million people worldwide. Although more than 20 new anti-seizure medications have been approved by the FDA in the past three decades, about one-third of patients still experience recurrent seizures. The goal of epilepsy treatment is to achieve seizure remission while not causing adverse reactions and to improve the overall quality of life for the patients. Precision medicine refers to a tailored approach of medical treatment with consideration of individual characteristics of each patient.[1] It, in many regards, has been present for many years in the past with the localization of epileptic foci and surgical treatments for epilepsy being performed in a highly selective and individual manner. As our understanding of the molecular pathogenesis of epilepsy and human genomics improve, so too has our ability to appreciate the causal nature of specific mutations in various epilepsies, and also the nature of one's genetic makeup in conferring risk of drug intolerance and adverse effects. At the same time, as our appreciation of the organization of the brain and its functional and structural connections has improved, so too has our ability to modulate these networks to improve seizure control, which so far has taken the shape of neurostimulation. Here we explore these relatively new areas of precision medicine in various epilepsies of interest and point out some areas of emerging therapeutic advances.

2. Precision therapy in genetic epilepsy syndromes

Genetic factors have been shown to contributes substantially to, or cause, epilepsy.[2,3] With the advance of next generation sequencing and tremendous drops in its cost, the development of genetics-driven precision medicine has been advocated in epilepsy patients. In this section, we will discuss the current genetic testing modalities that are utilized in clinical settings, as well as genetic epilepsy syndromes that are associated with targeted treatment regimen.

2.1 Current genetic testing tools and their yield in epilepsy

2.1.1 Microarray

Chromosomal microarrays are a type of cytogenetic testing modality that allows detection of aneuploidy as well as genetic copy number variants

(CNV). The use of single nucleotide polymorphism microarrays can detect uniparental disomy as well as consanguinity.[4] An advantage of chromosomal microarrays is they provide unbiased analysis of both targeted and untargeted loci by evaluating all chromosomes. The use of genomic microarrays for the detection of CNV is particularly important in children with epilepsy and comorbid developmental delay, autism spectrum disorder, or other congenital abnormalities. One study evaluating 1097 patients with epilepsy and comorbid neurodevelopmental disorder found pathogenic autosomal CNV in 10.9% of patients.[5] Although the diagnostic yield and cost-effectiveness of chromosomal microarrays are not as high as genetic panel testing or whole exome sequencing, it has been deemed the first line of testing in children with epilepsy and neurodevelopmental or other congenital comorbidities.[6]

2.1.2 Targeted gene panel testing

Genetic panels are another genetic testing modality utilized in the diagnosis of genetic epilepsy syndromes. Single gene and gene panel testing have traditionally involved the Sanger method of sequencing each gene's nucleotide sequence. Now, most genetic testing is completed using next generation sequencing technology, which allows for high-throughput parallel sequencing. There are many commercially available genetic epilepsy panels that include hundreds of targeted genes known to be associated with epilepsy. Compared to microarrays, gene panels can detect deletions and other mutations *within* genes. However, the human genome is comprised of exons (protein-encoding regions), introns (intragenic non-coding regions), and non-coding intergenic regions between genes, but targeted gene panel tests do not account for non-coding regions. Another common limitation of gene panels is that for some gene variants, the significance and pathogenicity is unknown.[7] In addition, gene panels may require frequent modification due to research implicating the involvement of new genes in epilepsy. The diagnostic yield of epilepsy gene panels varies depending on the specific panel as well as the population screened, but studies have determined the yield to be between 10% and 50%.[8] One study has shown that in an adult epilepsy cohort, targeted panel tests had a diagnostic yield of 23% for those with comorbid intellectual disability.[9] When comparing the diagnostic yield and cost-effectiveness among different tests, including whole exome sequencing, genetic panels, and chromosomal microarrays, genetic panels were determined to be the most cost effective.[6]

2.1.3 Whole-exome sequencing (WES) and whole-genome sequencing (WGS)

Only 1–2% of the human genome encodes proteins, but approximately 85% of mutations associated with disease-related traits occur in these protein-coding regions.[10] Whole exome sequencing (WES) targets the entirety of exons in the genome, termed the exome; while whole genome sequencing (WGS) targets the entirety of the human genome: introns, exons, and intergenic regions.[10] Yang et al.[11] performed an observational study ($n = 2000$) employing WES in primarily pediatric patients with suspected genetic clinical diagnoses, with a yield of approximately 25% positive molecular diagnoses. Similar diagnostic rates of WES have been demonstrated in literature focusing on the pediatric population.[12,13] On the other hand, WGS, although currently more expensive than WES,[14] has a higher yield of diagnosis (40–60%); which is nearly double that reported across literature for WES.[15] Diagnostic advantages of WGS include additional analysis of non-coding regions and mitochondrial DNA, as well as more robust yield with certain structural variants including breakpoints in repetitive regions and copy-neutral variants (inversions and translocations).[15,16]

2.1.4 Multiomic data in research

Single alterations of the genome such as deletions, duplications and translocations are of great importance in human disease in general, but the importance of a higher level of organization also been acknowledged. For example, epigenetics, the alteration of gene expression without a change in DNA sequence has been an active field of study and may involve methylation of DNA, histone modification, and non-coding regulatory segments of the genome. While there are some well-known syndromes of epigenetic origin in epilepsy, such as Prader-Willi and Angelman syndromes, a broader, more systematic search of epigenetic factors is underway and comprises the fields of epigenomics. Several pathways of interest include those involved in development, ion channel function, cell-cell interactions, notch signaling pathways, and axonal structure.[17,18] Furthermore, proteomics, the complex interactions between proteins post-translationally, has also been investigated in epilepsy. A great number of proteins appear to be overexpressed or underexpressed and this may have a role in the generation or maintenance of hyperexcitability, or may even represent a biological response to count seizure generation. The proposed mechanisms reported are many, including mitochondrial function, reactive oxygen species, neuronal plasticity and axonal organization, cytoskeletal organization, metabolic pathways, vesicle release, and neurotransmission. One of the limitations of the field is the necessity for directly affected tissue, for which the underlying pathology, duration of epilepsy, spatial/functional relationship to the seizure focus and other factors may play an important role. This limits comparisons among patients. Despite this, it is certainly possible that future work will elicit targets for rational drug design, an understanding of mechanisms and nature of drug resistance, and an ability to predict overall prognosis from pathology.[19]

Lastly, larger scale functional and structural connectivity within the human brain, a field known as connectomics, is increasingly being recognized as having a partial genetic basis, and a field of genetic connectomics is emerging.[20] Recent data suggest that structural connectivity may be at least in part explained via genetic factors, and perhaps particularly so for the so-called rich-club interconnectivity between major hubs of the brain. Genetic connectomics has already found application in fields such as dementia,[21] as well as autism and schizophrenia.[20] The application to epilepsy is likely to follow suit, and may find promise in predicting surgical outcome, selecting stimulation sites for those with responsive neurostimulation or deep brain stimulation, or even selection of medication in a personalized manner.

2.2 Precision medicine approach in genetic epilepsy syndromes

More than 1000 inherited syndromes are known in which epilepsy is a prominent clinical feature, and the number is growing steadily. With a growing list of causative genes encoding ion channels, proteins and metabolic pathways, precision medicine will enable rational drug discovery, supplementation of missing enzymes, and narrowing of anti-seizure medication choice, among other benefits. In this section, we will explore some of the genetic epilepsies for which a known gene has clear therapeutic ramifications (Table 1). While much of the work done to date may better be described as personalized medicine, as it entails the tailoring of known treatment to a nonspecific defect in a causative gene, we review such tailoring here, in part as a view towards the future in which individual mutations may themselves be a target for truly precise interventions.

2.2.1 ALDH7A1/PNPO/PROSC

The pyridoxine-dependent epilepsies consist of a series of genetic epilepsies that are profoundly pyridoxine-responsive. Seizures may include clonic, myoclonic and tonic seizures, which typically start in infancy, often within

Table 1 Summary of genetic epilepsy syndromes with available precision medication approaches.

Gene variant	Encoded protein	Disorder	Specific therapy regimen
ALDH7A1/ PNPO/ PROSC	alpha-AASA dehydrogenase/ Pyridoxamine 5′-phosphate oxidase/ Proline Synthetase	Pyridoxine Dependent Epilepsy	Pyridoxine ± Pyridoxal 5′-phosphate ± riboflavin
ATP7A	P-type copper transporting ATPase 1	Menkes disease	Copper histidine
FOLR1	Folate receptor alpha	Cerebral folate deficiency syndrome	Folinic acid
GAMT	Guanidinoacetate Methyltransferase enzyme	Cerebral Creatine Deficiency Syndrome 2	Creatine supplementation, Ornithine supplementation, arginine-restricted diet
AGAT	Arginine:Glycine amidinotransferase enzyme	Cerebral Creatine Deficiency Syndrome 3	Creatine supplementation
SLC6A8	Creatine Transporter 1 protein (CT1)	Cerebral Creatine Deficiency Syndrome 1	Creatine supplementation
GRIN2A,B,D	gluN2A,B,D subunits of the NMDA receptor	Various, including CSWS, Landau Klefner Syndrome, Atypical Benign Partial Epilepsy of Childhood	Memantine, Dextromethorphan, Ketamine, Magnesium
KCNA2	Kv1.2 voltage-gated potassium channel	Various, including epileptic encephalopathy, progressive myoclonic epilepsy	4-Aminopyridine
KCNQ2	Voltage-gated potassium channel	Early Onset Epileptic Encephalopathy (EOEE), Benign familial seizures of early infancy	Potassium channel openers (i.e., Retigabine)
KCNT1	Potassium channel subfamily T, member 1	Various, including Sleep-related hypermotor epilepsy, Early onset epileptic encephalopathy, epilepsy of infancy with migrating focal seizures	Quinidine
MOSC1	Molybdenum Cofactor (MOCO) Sulfurase C-Terminal Domain Containing Protein 1	Molybdenum cofactor deficiency type A	Cyclic Pyranopterin Monophosphate (cPMP)
POLG1	Mitochondrial DNA polymerase gamma	Mitochondrial disorders, includes Alpers-Huttenlocher Syndrome (AHS)	Vatiquinone (EPI-743) inhibits ferroptosis

PRRT2	Proline-rich transmembrane protein 2	Paroxysmal kinesigenic dystonia, Benign familial infantile epilepsy	Carbamazepine
SCN1A	$Na_v1.1$ voltage gated sodium channel	Dravet Syndrome, GEFS+, Migrating partial seizures of infancy	Stiripentol, Cannabidiol, Fenfluramin
SCN2A	$Na_v1.2$ voltage-gated sodium channel	Various, including infantile epileptic encephalopathy, benign infantile encephalopathy, Ohtahara syndrome, Lennox-Gastaut syndrome	Lidocaine, Mexiletine
SCN8A	$Na_v1.6$ voltage-gated sodium channel	SCN8A-related epilepsy with encephalopathy	Oxcarbazepine, Carbamazepine, Phenytoin
SLC2A1	Glucose transporter protein type 1 (GLUT1)	GLUT1 deficiency syndrome	Ketogenic diet
SLC6A1	GABA transporter protein type 1 (GAT1)	Various, including Doose syndrome	Valproic acid, Tiagabine, Vigabatrin, Ketogenic diet
SLC19A3	Thiamine monophosphate transporter	Thiamine-biotin responsive basal ganglia disease	High-dose thiamine and biotin
SLC35A2	Golgi-localized Uridine diphosphate (UDP)-galactose transporter	Glycosylation disorder can present as EOEE	ACTH therapy, galactose supplementation
STXBP1	Syntaxin binding protein-1	Various, including Ohtara syndrome, early myoclonic epilepsy, West syndrome, Dravet syndrome	4-Phenylbutarate
TPP1	Tripeptidyl Peptidase 1 (TPP1) lysosomal exopeptidase proenzyme	Neuronal Ceroid Lipofuscinosis Type 2	Cerliponase Alfa
TRPM6	Transient Receptor Potential Melastatin 6 (TRPM6) transepithelial magnesium transport channel	Hypomagnesemia with Secondary Hypocalcemia	Magnesium supplementation
TSC1, TSC2	Hammartin, Tuberin	Tuberous Sclerosis Complex	mTOR inhibitors (Rapamycin, Everolimus), Vibagatrin (inhibits GABA transaminase), Ketogenic diet

hours of birth, but later onset may occur up to 2 years of age. They may be accompanied by a number of other clinical presentations including dystonia, exaggerated startle, irritability, respiratory distress, hypothermia, hepatomegaly, shock, acidosis, hypocalcemia, hypomagnesemia, hypothyroidism and diabetes insipidus. The EEG may be normal but may also show burst suppression or hypsarrhythmia. MRI findings, if abnormal, include cortical dysplasia and hydrocephalus.[22] The treatment regimen consists of 50–100 mg of pyridoxine intravenously, followed by 5–10 mg/kg/day. Patients are further placed on a lysine restricted diet and may also receive competitive inhibitors of lysine transport.[23,24] Responsiveness depends on age of presentation, with quick response in infantile forms (typically within an hour), and up to a week within those of later onset.[22] Trials are ongoing regarding nutritional supplementation to support a lysine-restricted diet (clinicaltrials.gov identifier: NCT04672226). Nevertheless, even with an excellent seizure response to pyridoxine, about 75% of patients will have cognitive impairment.[25,26]

The genetic basis for pyridoxine dependent epilepsy is a mutation in the gene ALDH7A1 that is typically autosomal recessive in inheritance. The majority of the mutations are missense mutations (49%), but splice site-mutations (21%), insertion-deletion mutations, premature stop codons (7%), CNVs (6%), and single nucleotide variants (SNVs) at the boundary that still result in splicing errors (2%) may also occur.[24] ALDH7A1 is a gene involved in the breakdown of L-alpha-aminoadipic semialdehyde (alpha-AASA) via alpha-AASA dehydrogenase, an enzyme responsible for lysine oxidation. It is located on chromosome 5 at 5q23.2. This enzymatic deficiency leads to an over-abundance of alpha-AASA and L-delta-piperideine 6-carboxylate (P6C), which is the cyclic equivalent of alpha-AASA. P6C is then thought to sequester a vitamin B6 cofactor, pyridoxil 5′-phosphate (PLP).[22] Alpha-AASA can be tested in the urine in a standard urinary organic acid screen. Other forms of epilepsy responsive to pyridoxine exist; however, aside from supplementation, there are as yet no further therapeutic advances.

Pyridoxamine 5′-phosphate oxidase deficiency involves mutations in the PNPO gene, which is involved in oxidation of pyridoxine 5′-phosphate (PNP) into pyridoxal 5′-phosphate (PLP), using flavin mononucleotides (FMN) as cofactors. This results in a lack of the active form of vitamin B6 (i.e., PLP). This is also autosomal recessive, and has an associated phenotype of neonatal epilepsy, often within hours of birth, at times associated with decreased growth, hypochromic microcytic anemia, demyelinating neuropathy, hyperkinetic movement disorders, eye deviation, microcephaly, autism, ADHD, and irritability.[27] These are at times responsive to pyridoxine, but may need administration of PLP,[28] or even a combination of pyridoxine, PLP and riboflavin.[27] Those with responsiveness to PLP are dependent on it and may need increasing dosages for sick days. Possible negative side effects include abnormal liver function and cirrhosis.[29] Unfortunately, about 56% of those with seizure control continued to have developmental delay and intellectual disability.[27] Mutation in the proline synthetase co-transcribed homolog (PROSC) gene have also been described, and similarly inherited in an autosomal recessive manner. The protein appears to be an intracellular binding protein that regulates free PLP levels. Treatment with PLP was associated with improved seizure control, above and beyond pyridoxine alone. Most required other anti-seizure medications to control seizures.[30]

2.2.2 ATP7A

The gene ATP7A encodes for the P-type copper transporting ATPase 1 and is responsible for the transport of copper across the blood-brain barrier, as well as copper storage and release in polarized cells. It is inherited in an x-linked recessive manner.[31] One of the clinical presentations of ATP7A mutations is Menkes disease, an infantile onset disorder in which patients have failure to thrive, seizures, progressive neurological deterioration and developmental regression, connective tissue abnormalities and a classic kinked and coarse-hair known as *pilli torti* as well as hypopigmented streaks of the hair. Seizure types include focal seizures, infantile spasms, myoclonic seizures and generalized tonic-clonic seizures. EEG often shows focal or multifocal polyspikes, and in some, hypsarrhythmia or continuous spike and wave activity during slow sleep.[32] MRI may show widespread atrophy and ventriculomegaly, as well as tortuous blood vessels. Serum copper and ceruloplasmin are low, as are copper levels in the liver and brain, although copper levels are increased in fibroblasts and several other organ tissues. Other less severe phenotypes involving ATP7A mutations exist, namely, occipital horn syndrome, and ATP7A-related motor neuron disease; but these do not exhibit epilepsy and will not be discussed further here.[31]

Death generally occurs by the age of 2–3 years old, unless treatment is initiated within weeks of birth—a difficult task as neonates are generally asymptomatic at this age. Treatment of Menkes disease involves copper histidine supplementation.[33] Treatment is particularly effective if some residual ATP7A activity is still present, which is not the case for all mutations.

Preclinical data on gene therapy via adenovirus transport of the ATP7A gene and concomitant administration of copper histidine has shown prolonged survival and improved motor outcomes in mice,[34] although clinical trials in humans are not yet forthcoming.

2.2.3 *FOLR1*

FOLR1 gene encodes for the folate receptor alpha, for which loss of function mutation is associated with cerebral folate deficiency syndrome. This is an uncommon cause of epilepsy in childhood and 18 patients have been described to date.[35] Clinically, patients typically have myoclonic seizures, ataxia, and psychomotor decline beginning in the second or third year of life.[35] In advanced cases, progressive neurological and motor (e.g., spastic-dystonic) decline, cortical blindness, and dysphagia may also be present.[35] EEG may show multifocal epileptiform discharges.[35] Imaging may show, in more advanced cases, diffuse white matter T2 hyperintensity, cerebellar atrophy, and low choline and myoinositol in the white matter on magnetic resonance spectroscopy (MRS).[35] Folinic acid supplementation, when provided early can improve and at times reverse symptoms,[36] but little to no improvement may be seen in cases that are already advanced.[35]

2.2.4 *GAMT/AGAT/SLC6A8 (creatinine deficiency syndromes)*

Creatine is converted into creatine phosphate, which can buffer the concentration of ATP in cells with high energy demand.[38,39] Mutations in proteins involved in the synthesis and transportation of creatine constitute the Cerebral Creatine Deficiency Syndromes. The three genes implicated in these disorders are AGAT (arginine:glycine amidinotransferase), GAMT (guanidinoacetate methyltransferase), and SLC6A8 (Creatine Transporter 1). AGAT catalyzes the conversion of arginine and glycine into guanidinoacetate (GAA) and ornithine, in the kidney. GAMT catalyzes the conversion of GAA and S-adenosyl-L-methionine into creatine and S-adenosyl-L homocysteine in the liver. SLC6A8 transports creatine into tissues with high creatine kinase activity, especially the brain and muscle.[37] Diagnosis of the Cerebral Creatine Deficiency Syndromes (CCDS) is completed with both genetic testing as well as phenotypic testing to confirm biochemical creatine deficiency. Such phenotypic testing is often performed using MR spectroscopy of the brain, which demonstrates decreased or absent creatine in all three cerebral creatine deficiency syndromes.[38] Serum and urine creatine and guanidinoacetate levels can also distinguish between the various cerebral creatine deficiency syndromes.

The AGAT gene is located on chromosome 15q15.3 and encodes the arginine:glycine amidinotransferase enzyme; autosomal recessive inheritance of mutations in this gene causes cerebral creatine deficiency syndrome 3 (CCDS3). Clinically, CCDS3 presents with global developmental delay and myopathy.[39] Febrile seizures in infancy can sometimes occur, though epilepsy is rare and is not necessarily considered a hallmark feature of CCDS3.[40] The targeted treatment of CCDS3 is oral creatine supplementation, thereby bypassing the deficient synthetic pathway. Early initiation of treatment has been shown to result in typical development and suppression of phenotypic gene expression.[39,41] If left untreated, CCDS3 can manifest irreversible brain injury within the first year of life.[39]

The GAMT gene is located on chromosome 19.13.3 and encodes the guanidinoacetate methyltransferase enzyme; mutations in this gene causes cerebral creatine deficiency syndrome 2 (CCDS2). It presents with typical development in the first year of life, followed by regression or arrest of development and onset of seizures.[42] Severe epilepsy is seen in nearly all patients, and patients often have multiple seizure types—commonly febrile, generalized tonic clonic, and myoclonic astatic seizures.[40] Patients with CCDS2 can also develop movement disorders including ballismus, chorea, and dystonia.[40] Of all the cerebral creatine deficiency syndromes, CCDS2 is considered the most severe. The targeted treatment for CCDS2 is aimed at both increasing creatine and decreasing GAA. Therefore, patients should be given oral creatine supplementation, which both supplements creatine directly and negatively inhibits AGAT to decrease production of GAA.[37] Patients should also be put on a low-arginine diet and ornithine supplementation, to decrease substrate for AGAT and negatively inhibit AGAT, respectively.[43,44] The supplementation of sodium benzoate—which conjugates with glycine and therefore decreases the amount of glycine that can be converted by AGAT to GAA—has also demonstrated positive effect.[45] While epilepsy and movement disorders respond well to treatment, language and cognitive functions were less affected by therapy.[46,47] Although the precise best treatment regimen remains to be determined, creatine supplementation, ornithine supplementation, and low-arginine diet are the mainstays of treatment, and should be initiated as early as possible once the diagnosis is made.[45]

The SLC6A8/CT1 gene is located on chromosome Xq28 and encodes the creatine transporter 1 protein, which transports creatine into tissues with high creatine kinase activity, including the brain. CCDS1 is the result of X-linked mutation in the SLC6A8 gene, which results in significantly decreased creatine in brain cells, despite normal creatine synthetic function. Clinically, patients with CCDS1 present with moderate to severe intellectual disability, language impairment, behavioral problems, epilepsy, movement disorders, and gastrointestinal problems.[48] Seizures are often infrequent and easily treatable, and refractory epilepsy is rare. There is currently no efficacious targeted treatment for CCDS1, but creatine supplementation is generally recommended.[38] Development of a targeted treatment for CCDS1 has proven challenging, as the intrinsic impairment is in transport of creatine into the brain, rather than due to an enzymatic deficiency. One potential treatment is cyclocreatine, a creatine analog that has been demonstrated to improve cognition in a mouse model of CRT deficiency.[49] Another treatment approach under development involves the use of nanocapsules or nanoemulsions for drug delivery across the blood brain barrier.[50]

2.2.5 *GRIN2A,B,D*

The GRIN2A, B and D genes encode for the proteins GluN2A, B and D, which are isoforms of one of two subunits, along with gluN1 subunits, that comprise the N-methyl-D aspartate glutamatergic receptors (NMDARs). The phenotypes of mutations are highly varied. For example, GRIN2A mutations may cause childhood epilepsy with centrotemporal spikes, Continuous Spike Waves during Slow Wave Sleep (CSWS), Laundau-Kleffner syndrome and Atypical benign partial epilepsy of childhood.[51–53] Valproic acid, stiripentol, and clobazam may be of greater benefit overall as may steroids, particularly in cases of CSWS.[52] In patients with GRIN2A mutations, truncation mutations may be more responsive to valproic acid, clobazam and stiripentol, while duplications may better respond to steroids, topiramate or phenobarbital, although sample sizes for comparison remain small.[52] Of particular interest, is the emerging evidence of the NMDAR blocker memantine,[54] which may have only partial responsiveness to some mutations, such as one specific missense mutation of the GRIN2D gene.[51] Of great interest is the reporting of parallel in vitro testing of channel properties under various treatments in *Xenopus laevus* oocytes prior to clinical application of treatment in human patients. This specific GRIN2D gene *in vitro* showed partial responsiveness to dextromethorphan and ketamine as well as memantine, and a case report showed excellent response to combined magnesium and ketamine administration, of interest as the NMDAR requires release of magnesium block by depolarization prior to channel opening.[51] Trials are ongoing for the use of memantine in epileptic encephalopathy in general (clinicaltrials.gov identifier: NCT03779672), the true efficacy of this treatment remains to be seen, and may be specific to gain of function mutations, rather than loss of function mutations.[52] Although these treatments are not precise in the sense that they do not target individual mutations, they hint that an understanding of mutation is critical to a personalized approach, and may underly rational drug design in the future.

2.2.6 *KCNT1*

The KCNT1 gene encodes for a potassium channel that is voltage-dependant and sodium activated. Gain of function mutations in this gene encode for an incredible variety of genetic epilepsies, including sleep-related hypermotor epilepsy (SHE), previously known as autosomal dominant nocturnal frontal lobe epilepsy (ADNFLE), early onset epileptic encephalopathy (EOEE), and epilepsy of infancy with migrating focal seizures (EIMFS). Seizure types may include hypermotor seizures, bilateral tonic-clonic seizures, focal to bilateral tonic clonic seizures, hemiclonic seizures, focal impaired aware seizures, epileptic spasms and myoclonic seizures. The age of onset varies widely also, with SHE occurring in children and adults, EOEE occurring in infants and children and EIMFS occurring in neonates and infants.[55–57]

Treatment with quinidine, an antimalarial agent and also an anti-arrhythmic sodium and potassium channel blocker, has been found to be effective in controlling seizures in some patients. However, the results are variable, with some observational data suggesting 50% reduction and seizure freedom rates may be as low as 20% and 5%, respectively.[55] Age has been described as a possible variable in predicting response,[56] but may not be a consistent finding.[55] Likewise, the effect may be more significant in EIMFS and lennox-gastaut syndrome phenotypes,[57] but this is also not consistent across studies.[55] The dosage of quinidine that may be effective in epilepsy is unknown, but dosing in cardiac disease has been used as a point of reference, with increasing dose titration to 15–60 mg/kg/day. The QTc must be monitored through this titration.[57]

2.2.7 *KCNA2*

KCNA2 encodes an alpha subunit of the Kv1.2 voltage-gated potassium channel. It exists mostly as a *de novo* mutation; however, milder phenotypes with autosomal dominant inheritance have been described,[58] and both loss and gain of function mutations may have a phenotype including epilepsy.[59,60] KCNA2 has been described with variable phenotypes which may include epileptic encephalopathy or progressive myoclonic epilepsy.[59] Electrographic findings included a photoparoxysmal response at low frequency (1 Hz). Somatosensory evoked potentials, while not giant, were increased in N20-P25 amplitude.[59]

Treatment with 4-aminopyridine with dosing of 0.15–2.6 mg/kg/day has been shown in case series to produce seizure freedom in several patients. Furthermore, significant reduction in generalized tonic-clonic seizures were noted, despite those patients not being seizure free. Exceptions to treatment response were patients with solely generalized tonic-clonic seizures prior to 4-aminopyridine administration – these patients did not benefit from a seizure response to the drug. Response was also noted in motor, gait, cognitive and speech domains in many patients.[61]

2.2.8 KCNQ2

The gene KCNQ2 encodes voltage-gated potassium cation channels which are ubiquitous in the brain and are critical to the generation and transmission of action potentials.[62,63] There are two main epilepsy syndromes associated with mutations in KCNQ2: benign familial seizures of early infancy and EOEE.[62] While a diagnosis of the former portends an excellent prognosis, most patients with EOEE have a clinical course characterized by medically refractory epilepsy, intellectual disability, and encephalopathy.[62]

For evidence-based treatment of KCNQ2 mutation related epilepsy, antiepileptic agents with sodium channel blocking mechanism seemed to have efficacy in both the benign and EOEE groups.[62] Potassium channel openers are a promising class of medications being explored as personalized therapy in patients with KCNQ2 loss-of-function mutations. In particular, retigabine is an ASM which functions to open the Kv7 subfamily voltage-gated potassium channels.[64] The safety profile and efficacy of Retigabine were evaluated in Phase II and III trials (RESTORE 1 and RESTORE 2) as adjunct therapy in patients who were at baseline taking at least one (but less than four) additional ASMs.[64] Retigabine was ultimately withdrawn from the market due to occasional hepatotoxicity and reversible blue tissue discoloration, sequelae of drug metabolic oxidation producing electrophilic quinones.[65] However, Bock et al.[65] have identified sulfide analogues that function effectively as potassium channel openers, but lack the risk of quinone formation, thus theoretically avoiding the negative side-effect profile of Retigabine.

2.2.9 MOSC1/MOSC2/GPHN

Molybdenum cofactor deficiencies (MoCDs) are a group of autosomal recessive disorders characterized by dysfunctional metabolism of sulfur-containing amino acids.[66,67] Approximately 2/3 of MoCD cases are attributed to mutations in the MOSC1 gene, classified as MoCD type A, which results in defective synthesis of cyclic pyranopterin monophosphate (cPMP).[68] The second most common pathologic mutation is in the gene MOSC2, classified as MoCD type B which results in accumulation of cPMP that cannot be further utilized.[68] MoCD types A and B, and the less common Type C (due to mutations in the GPHN gene) have indistinguishable clinical phenotypes, typically presenting within several days of life with rapidly progressive postnatal encephalopathy and refractory epilepsy.[68] cPMP supplementation is the first efficacious and safe targeted intervention in patients with MoCD type A; however, there is no apparent benefit in patients with MoCD type B.[77] Good long-term developmental outcomes were observed when cPMP supplementation was administered early in the course of MoCD type A, prior to developing significant encephalopathy.[67,68]

2.2.10 POLG1

POLG1 gene (mitochondrial DNA polymerase gamma) is located on chromosome band 15q26, and functions in the repair and replication of mitochondrial DNA.[69,70] The most common polymorphism is POLG1 c.3428A > G variant (p.E1143G) in exon 21, with an estimated occurrence of 4.1% in the general European population and 2.9% in the North American population.[71–75] Phenotypically, POLG mutations present from infancy to late adulthood, with a continuum of overlapping features which characterize at least six major syndromes.[76] In particular, Alpers-Huttenlocher syndrome (AHS) presents in childhood with progressive and severe encephalopathy, in addition to refractory epilepsy and hepatic dysfunction.[76]

Prospective genetic testing of POLG may be helpful prior to initiating therapy with a common anti-seizure medication: valproic acid, which has the uncommon but potentially lethal side effect of idiosyncratic liver injury.[77] Heterozygous genetic variation in POLG has been shown to be significantly correlated with liver injury due to valproic acid toxicity (OR = 24).[77] In addition, a[72,73] novel targeted therapeutic agent Vatiquinone (EPI-743) is under clinical evaluation in patients with pediatric mitochondrial diseases associated with epilepsy, including POLG1 mutation. Vatiquinone potentially prevents ferroptosis—a type of iron and lipid dependent regulated cell death.[78] There is a current recruiting Phase 2/3 clinical trial (clinicaltrials.gov identifier: NCT04378075) to evaluate the safety and efficacy of Vatiquinone in patients with mitochondrial disease and refractory epilepsy.

2.2.11 *PRRT2*

PRRT2 is a gene located on chromosome 16p11.2, encoding proline-rich transmembrane protein 2. It is expressed highly in the cerebral cortex, basal ganglia, and cerebellum.[79,80] Within neurons it localizes primarily to the axonal region, and specifically in glutamatergic synapses associated with GRIN1A, a member of the AMPA receptor family.[81] It may also be related to vesicle release via interaction with the synaptic vesicle docking protein SNAP-25, a component of the SNARE complex.[82] PRRT2 mutation has been identified as the cause of a broad array of phenotypes including epilepsy, movement and headache disorders. Classically, it has been associated with paroxysmal kinesigenic dyskinesia (PKD), but has also been associated with benign familial infantile epilepsy (BFIE), and a disorder characterized by both phenotypes in the same patient known as paroxysmal kinesigenic dyskinesia with infantile convulsions.[83] The shared paroxysmal nature of symptoms in PRRT2 mutation syndrome suggest a common underlying pathophysiology.

BFIE is a disorder in which clusters of non-febrile, focal unaware or generalized tonic-clonic seizures occurring begin between the ages of 3 and 12 months of age. There may also be brief episodes of behavioral arrest, head and/or eye deviation, hypertonia, cyanosis, or limb jerks. Parieto-occipital sharps and aftercoming slow waves constitute ictal activity on the EEG and may secondarily generalize. Genetic testing is useful, as it indicates a more benign course, and treatment responsivity to antiseizure medications such as valproate, carbamazepine, and phenobarbital. It is also possible for these patients to present later in life with PKD.

PKD is an entity consisting of short-lived (generally <1 min) and frequent (oncer per 2 weeks to hundreds per day) episodes of hyperkinetic movements such as chorea or dystonia that are brought on by movements such as rising from a chair, ambulating, or startle.[84] There is some controversy over whether these movements constitute seizure or movement disorder; however, they tend to be exquisitely sensitive to carbamazepine, often needing only small doses in the range of 50–200 mg/day.[83] This carbamazepine responsiveness may be particularly common in patients with PKD who have a mutation in the PRRT2 gene compared to those with a similar phenotype but no known mutation.[85]

2.2.12 SCN1A

SCN1A encodes for the α-subunit of a voltage gated sodium channel ($Na_v1.1$), located on the long arm of chromosome 2. $Na_v1.1$ channel consists of both alpha and beta subunits, and six transmembrane regions.[86] It is voltage sensitive such that they undergo a conformational change when the transmembrane voltage reaches a threshold level during depolarization, and subsequently opens to allow influx of sodium at the initiation of the action potential. In this way, it is critical for network transmission and excitability in the central nervous system; its involvement in epilepsy is hence unsurprising.[87]

Several phenotypes are associated with the SCN1A mutation with varying disease severities, including Dravet syndrome, genetic epilepsy with febrile seizures plus (GEFS+), migrating partial seizures of infancy (MPSI).[86,88] These are thought to exist on a spectrum, with greatest mutational severity in Dravet syndrome and least mutation severity in typical febrile seizures.[87] Dravet syndrome is associated with largely *de novo* mutations,[89] while the GEFS+ phenotype is inherited in an autosomal dominant fashion with incomplete penetrance.[90] Dravet syndrome is typically characterized by normal development initially, typically up to the first year, followed by myoclonic, focal and atypical absence seizures accompanied by developmental plateau or regression between the ages of 1–4 years old. Complex febrile seizures may occur earlier, often around 6 months of age, and may include hemi-clonic seizures that may shift sides between seizures.[88] MPSI differs in that multiple focal seizure types are present in early infancy, becoming more frequent over time and are generally refractory to antiseizure medications (ASMs), and there tends to be severe developmental impairment from an early age.[91] GEFS+ is a syndrome present with great variability among family members and between patients, and may include both typical and complex febrile seizures, as well as generalized and focal epilepsy.[88] 40–50% of the mutations leading to Dravet syndrome are missense, with another 40% or so being truncation mutations.[88] Nearly all of the mutations leading to GEFS+ are missense in nature.[88] It may be that mutations in or near the pore forming region of the protein are more likely to cause a Dravet phenotype.[89] These mutations have been linked to different phenotypes, which is important for counseling in clinical practice. For example, more rapid cognitive decline is seen in Dravet patients with truncation mutations regardless of age of seizure onset, when compared to those with missense mutation.[92] The importance of testing also lies in the therapeutic options and limitations. Worsening of symptoms has been documented with sodium channel blockers such as carbemazepine and related medications, and lamotrigine.[93] Stiripentol is a medication that has been shown to have particular efficacy, when combined with other agents, in patients with Dravet syndrome.[94] Its efficacy was slightly better in Dravet patients with SCN1A missense mutations than those with truncation mutations.[95] Dravet syndrome is also one of the few epileptic syndromes for which cannabidiol has evidence of efficacy.[96] Fenfluramine is another medication recently approved by FDA in 2020 to control seizures in children from 2 years of age and older with Dravet syndrome.[97] Consensus guidelines suggest that first-line agents should include clobazam and valproic acid, while second line agents should include stiripentol and topiramate as well as the ketogenic diet, with clonazepam, levetiracetam, and zonisamide forming third line options.[98] There are also open-label studies currently enrolling for the use of antisense oligonucleotides whose purpose is to increase wild-type expression of the SCN1A protein, helping to restore normal sodium channel function (clinicaltrials.gov identifier: NCT04740476). Preclinical data suggests that this treatment may be effective in reducing both seizure incidence and mortality due to sudden unexplained death in epilepsy (SUDEP).[99]

2.2.13 SCN2A

SCN2A is a gene that encodes for the fast-inactivating $Na_V1.2$ voltage-gated sodium channel, a channel responsible for the initiation of action potentials in a number of neural subtypes.[100] The phenotype of SCN2A mutation includes autism, intellectual disability, and infantile epilepsy. There are two primary forms of epilepsy, an infantile epileptic encephalopathy which occurs as *de novo* mutations; and a benign infantile encephalopathy, which can be inherited in an autosomal dominant fashion and seizures may resolve by 2 years of age. There are even more rare subtypes which include epileptic encephalopathy with migrating partial seizures,[101] generalized epilepsy with epileptic encephalopathy with choreoathetoid movements, childhood onset epileptic encephalopathy, and benign infantile seizures with late-onset episodic ataxia.[100]

Infantile epileptic encephalopathy may begin with rhythmic movements in utero; however, the overall phenotype appears to be guided by age of onset. Onset before an age of 3 months is typically accompanied by a phenotype of Ohtahara syndrome, that is to say, tonic seizures and burst-suppression on EEG. Onset after 3 months of age is accompanied by a phenotype of West syndrome, accompanied by the classic infantile spasms and hypsarrhythmia, as well as later evolution to multiple seizure types and variable developmental delay, and hence, Lennox-Gastaut syndrome.[102] Other associated features may include hypotonia, microcephaly, cortical visual impairment and hyperkinetic movement disorders.[100]

A dramatic reduction in seizure frequency is seen with treatment of IV lidocaine at 5 mg/kg/min followed by mexiletine at 6 mg/kg/day with multiple drug failure otherwise having been reported.[103] Unfortunately, these children did have significant developmental delay despite the treatment. Both lidocaine and mexiletine are associated with life-threatening arrhythmias and must be discussed with due care with families prior to their clinical use.[103]

2.2.14 SCN8A

SCN8A encodes for the $Na_v1.6$ voltage gated sodium channel. Gain of function mutations lead to an epileptic encephalopathy that will be discussed here, while loss of function mutations may lead to autism, developmental delay, and ataxia.[104] The epileptic phenotype forms a spectrum in terms of age of onset, which may range from 1 day to 18 months; and seizure types which include epileptic spasms, febrile seizures, myoclonic seizures, tonic seizures, bilateral convulsive seizures, and tonic-clonic seizures, with later additional atonic seizures and absence seizures possible. Status epilepticus has been described as both non-convulsive and convulsive, and SUDEP may also be present. The degree of developmental delay is also variable. Additional clinical features can include hypotonia and hyperkinetic movement disorders such as dystonia, and choreoathetosis, ataxia and hyperreflexia. Many patients loose eye contact without a clear diagnosis of autism, although autism and ADHD may be present.[105] Hypoplasia of the corpus callosum and cerebral atrophy have been noted on MRI, although normal imaging has also been described. EEG similarly could be normal or contain focal or multifocal epileptiform discharges. Despite this, many patients develop at least moderate to severe background slowing on EEG.[105]

Regarding treatment, it has been shown that some patients could achieve seizure-free with sodium-channel blockers such as oxcarbazepine, carbamazepine, and phenytoin.[105] Currently, phase II trials in human are being conducted to complement preclinical data[106] on a molecule known as NBI-92352, which is a sodium-channel inhibitor specifically designed to inhibit Na_v 1.6 channels (clinicaltrials.gov identifier:NCT04873869).

2.2.15 *SLC2A1*

SLC2A1 is a gene encoding the glucose transporter type 1 (GLUT1) protein on chromosome 15q13.3 responsible for transporting glucose across the blood-brain barrier. GLUT1 Syndrome is most frequently inherited in an autosomal dominant manner, but may rarely be autosomal recessive. The classical presentation is in infancy, where patients present with seizures, impaired neurodevelopment and growth, and a variety of movement disorders. There is a typical progression, beginning with seizures and abnormal movements of the eyes and head, followed by developmental delays, ataxia, dystonia and other movement disorders over time. Multiple seizure and epileptic syndromes have been described, including early-onset absence epilepsy, and Doose syndrome (myoclonic-atonic seizures); focal seizures are overall less common.[107] There are also presentations in adulthood which may include only movement disorders such as dystonia, but can include medically refractory seizures of adult onset as well. SLC2A1 mutations have also been associated with the entity of paroxysmal exercise-induced dyskinesia in which patients experience movement disorders such as dystonia brought on by extended periods of exercise or ambulation.[108]

Early diagnosis is important and can be achieved through a comparison of glucose in the cerebrospinal fluid (CSF) and blood. The CSF ratio of CSF: blood is classically <0.45, and CSF glucose concentration <60 mg/dL.[109] Treatment involves primarily the ketogenic diet, which has shown efficacy in treating seizures,[110] and improving overall cognitive outcome over time, which generally ranges from mild to moderate cognitive impairment.[111] In adult patients, particularly if therapy is initiated in adulthood, the modified Atkins diet may also be considered.[112]

2.2.16 *SLC6A1*

The solute carrier family 6 member 1 (SLC6A1) gene encodes for the GABA transporter protein type 1 (GAT1) protein. The GAT1 protein is a presynaptic reuptake transporter of GABA in both neurons and glia. Its clinical phenotype includes Doose syndrome, early onset epilepsy with myoclonic atonic seizures, generalized epilepsy, focal epilepsy, autism spectrum disorder, intellectual disability, and even schizophrenia. Valproic acid, lamotrigine, and ethosuximide appears to be the most effective antiseizure medications.[113] However, cognitive outcome does not appear to be improved even in those patients with improved seizure control. Given the GABA receptor modulatory properties of tiagabine and vigabatrin, these may also have some merit as well.[114] Ketogenic diet has also been described as effective, at least for patients with myoclonic-atonic seizures.[115] Investigation is currently ongoing as to gene therapy involving antisense oligonucleotide or viral-mediated gene therapy.[114]

2.2.17 *SLC19A3*

SLC19A3 encodes thiamine monophosphate transporter, which functions involves the key steps in thiamine metabolism. Mutations in the disease cause an entity known as thiamin-biotin responsive basal ganglia disease. Clinically, patients encounter encephalopathy and neurological deterioration after metabolic challenges such as infection. Epileptic spasms, dystonia, parkinsonism, external ophthalmoplegia, and quadriplegia may be present. Interestingly, two patients have been described with a Wernicke-like encephalopathy and status epilepticus in the second decade of life, although the vast majority present in infancy.[116] MRI may show T2 hyperintensity, and cytotoxic edema, primarily affected the basal ganglia and especially the striatum, as well as thalami, cerebral cortex, and brainstem at times involved.[117] The condition may be fatal; however, prevention of mortality and further attacks has been reported with treatment of high-dose thiamine and biotin both at 10–15 mg/kg/day.[118]

2.2.18 *SLC35A2*

SLC35A2 gene is located at Xp11.23 and encodes the Golgi-localized Uridine diphosphate (UDP)-galactose transporter, with mutations resulting in galactose-deficient glycoproteins.[119] Although a multi-system disease, SLC35A2 phenotypically has predominant neurological features including early-onset epileptic encephalopathy.[120] Inheritance is X-linked dominant, thus affecting more females than males.[120] Adrenocorticotropic hormone (ACTH) supplementation is an evidence-based treatment for Epilepsy in this population.[119,121,122] Furthermore, there is evidence suggesting clinical improvement with dietary galactose supplementation, which may partially overcome the Golgi UDP-galactose deficiency and improve galactosylation.[120,122]

2.2.19 *STXBP1*

Syntaxin binding protein-1 (STXBP1) is a protein involved in the soluble N-ethylmaleimide-sensitive factor attachment protein receptors (SNARE) proteins—a complex involved in synaptic vesicle fusion and neurotransmitter release. Mutations in this gene span a large variety of infantile and childhood epileptic encephalopathies including Ohtahara syndrome, early myoclonic epilepsy, West syndrome, Dravet syndrome, Rett syndrome,[123] as well as broader syndromes such as autism and/or intellectual disability and epilepsy, and intellectual disability alone. Patients may also have movement disorders such as ataxia, hyperkinetic movement disorders such as dystonia and tremor, hypotonia, or spasticity. Seizure types included infantile spasms, focal seizures and tonic seizures. EEG showed multifocal epileptiform abnormalities, burst suppression, or hypsarrhythmia.[124]

Of the standard anti-seizure medications, valproic acid, levetiracetam and vigabatrin[124] as well as cannabidiol may be the most efficacious.[123] There is significant preclinical data supporting the use of 4-phenylbutarate, which has already been approved by FDA for treating urea cycle disorders. It has a role in membrane stabilization, and increase levels of STXBP1 in preclinical models.[123] The first clinical trial is currently underway (clinicaltrials.gov identifier: NCT04937062).

2.2.20 *TPP1*

The Tripeptidyl Peptidase 1 (TPP1) gene is located on chromosome 11p15 and encodes a lysosomal exopeptidase proenzyme, which once activated, degrades polypeptides by cleaving tripeptides in the cell lysosome.[125] Neuronal Ceroid Lipofuscinosis Type 2 (CLN2) is a genetic disorder that results from autosomal recessive inheritance of TPP1 gene mutations, causing TPP1 deficiency and consequently impaired lysosomal protein degradation. There have been at least 131 identified mutations in the TPP1 gene that are associated with CLN2.[126] However, two specific TPP1 variant alleles (c.509-1 G>C and c.622C>T) have been found to account for 50% of the disease-associated alleles.[126] The precise pathogenic mechanism remains uncertain, but like other lysosomal storage disorders, CLN2 results in oxidative stress and mitochondrial network fragmentation.[127] These cellular effects result in selective neuronal cell loss, particularly in the cerebellum and cerebral cortex, as well as retinal cell loss.[128] The typical clinical presentation involves seizures and language delay followed by rapid motor and cognitive delay, as well as visual blindness by the age of 7 years old.[129] The onset is typically between the age of 2–4 years old, followed by mortality in the first or second decades of a life.[129] Some patients with less common mutational variants have demonstrated a later symptom onset.[126] The natural history of the disease results in death in 7–8 years following initial symptom onset.[129] Diagnosis of CLN2 is done through both genetic testing and TPP1 enzyme activity assays, and expert opinion recommends completion of both diagnostic tests.[130]

In 2017, the FDA approved cerliponase alfa, a recombinant proenzyme form of human TPP1, for the treatment of CLN2. An open-label study evaluating the use of intraventricular infusion of cerliponase alfa every 2 weeks in children ages 3-16 with CLN2 demonstrated less decline in language and motor function compared to historical controls.[131] Of note, although one of this study's outcome measures (the CLN2 total score) factored in seizures, seizure frequency was not directly measured.[131] In addition to enzyme replacement therapy, there is ongoing research into targeted gene therapy. One study in mice has demonstrated efficacy of adeno-associated virus delivery of a CNL2 gene cassette with increased TPP1 levels.[132] In addition, a small trial of intraparenchymal administration of this vector in children with CLN2 demonstrated a slower rated of motor and language decline, but further research into the safety and efficacy of such therapy is required before more widespread use.[133]

2.2.21 TRPM6

The Transient Receptor Potential Melastatin 6 (TRPM6) gene is located on chromosome 9q22 and encodes a transepithelial magnesium transport channel expressed in the intestinal epithelium and in the distal convoluted tubule of the kidney.[134,135] Hypomagnesemia with secondary hypocalcemia (HSH) is a genetic disorder that results from autosomal recessive inheritance of TRPM6 gene mutations, causing impaired enteral absorption and renal reabsorption of magnesium.[135,136] More than 39 distinct TRPM6 mutations have been identified in patients with HSH, all of which result in loss of protein function.[137–139] Homozygous TRPM6 mutations confer impaired magnesium absorption, which results in low serum magnesium levels. Hypomagnesemia results in impaired synthesis and secretion of parathyroid hormone, which results in hypocalcemia.[140] Magnesium plays an important role in neuronal membrane stabilization, by interacting with NMDA glutamate receptors and inhibiting voltage-gated calcium channels, both of which result in membrane depolarization when activated.[141,142] Hypomagnesemia, therefore, cause seizures by leading to increased activation of NMDA receptors and increased excitatory neurotransmission.

Clinically, patients with HSH present with generalized seizures and neuromuscular tetany, usually within the first month of life, and are found to have severe hypomagnesemia (with serum magnesium levels often near 0.2 mmol/L) and hypocalcemia.[143] Diagnosis is initially made clinically, based on presentation and electrolyte levels, and confirmed by genetic sequencing. Targeted therapy for HSH is magnesium supplementation—first with intravenous supplementation and then long-term oral supplementation.[143] Early diagnosis and treatment are essential, as it has been shown to resolve hypocalcemia, mostly correct hypomagnesemia, stop seizures, and result in normal neurodevelopment.[136,143] Intellectual disability, failure to thrive, and even death can result from treatment non-compliance.[144] A long-term outcome study reported on five patients who had no serious complications after 40 years, on magnesium supplementation.[145]

2.2.22 TSC1 and TSC2

Tuberous Sclerosis Complex (TSC) is a neurocutaneous syndrome affecting 1 in 6000 births.[146] Mode of inheritance is autosomal dominant. The TSC1-TSC2 complex normally inhibits the mammalian Target of Rapamycin complex 1 (mTORC1) signaling pathway. TSC mutation follows the two-hit Knudsen tumor suppressor gene hypothesis, manifesting after inactivation of both alleles of either TSC1 or TSC2 genes.[147] Inactivation of TSC through loss of function mutations results in hyperactivation of mTORC1, with downstream increase in cell growth and protein translation, in addition to decrease in autophagy.[147] Eighty-five percent of TSC presentations have pathologic mutations in either the gene TSC1 which encodes the protein hamartin or TSC2 which encodes the protein tuberin.[146] TSC1 is located on chromosome 9q34 and is mutated in approximately 20% of clinical diagnoses.[147] TSC2 is located on chromosome 16q13 and is mutated in approximately 70% of clinical diagnoses.[147] TSC1 mutations are less common and present with less severe clinical phenotypes compared to TSC2. Approximately one out of three cases are familial, with the remainder being sporadic (2–3% *de novo* germline); somatic mosaicism has also been reported.[147,148]

TSC phenotype includes benign tumors in various organs, specifically within the central nervous system: subependymal nodules, subependymal giant cell astrocytomas (SEGAs), cortical tubers, and retinal hamartomas. Epilepsy is seen in 80–90% of patients.[146] Most cases present with infantile spasms and typically acquire other seizure types.[148] Greater than 90% of patients have cortical tubers - focal cerebral cortical hamartomatous glioneuronal lesions typically at the grey-white matter junction, believed to form before birth as part of corticogenesis.[146,148] Higher tuber burden is associated with younger epilepsy onset and lower cognition index[149] although epilepsy and tubers do not always correlate.[146,148,150]

TSC-related epilepsy is refractory in 30–40% of cases.[148] Attempted management of refractory epilepsy in TSC may include trial of anti-seizure medications, steroids, resection of tubers, vagus nerve stimulation, and ketogenic diet.[146] Vigabatrin, an irreversible inhibitor of GABA transaminase, is a particularly effective targeted therapy for infantile spasms in TSC patients.[146] Given the mechanism of disease, mTOR inhibitors (rapamycin, everolimus) have been implemented as precision therapy in the TSC population. Everolimus is selective for mTORC1, whereas rapamycin is less selective and has more side effects.[150] Long-term treatment with mTOR inhibitors has been found to reduce SEGA tumor volume and reduce severity of epilepsy in TSC patients.[146,151] Additionally, ketones generated subsequent to ketogenic diet are thought to downregulate mTOR activity[146] thus can be implemented as targeted therapy in patients with TSC syndrome. Cortical tubers can also be targeted for resection in patients with focal seizures and a "dominant" epileptiform tuber.[148,152] In a multicenter observational pediatric study, a larger area of resection was correlated with higher likelihood of seizure freedom, suggesting that the epileptogenic zone likely includes perituberal cortex.[153] Studies in the TSC population have shown that 57% of patients are able to achieve seizure freedom after resective surgery.[154]

2.3 New avenues of personalized treatment in genetic epilepsy

The ultimate goal for precision medicine in genetic epilepsy is either to correct the pathogenic gene mutation or to modulate its expression in order to compensate for the associated cascade impact. Several new avenues have been investigated recently as the state-of-the-art precision medicine approach, mainly in preclinical phases. Such interventions may be utilized in near future for epilepsy in clinical setting, and take the form of individualized rational drug design, including molecules such as anti-sense oligonucleotides that alter the transcriptional products,[155] or perhaps even editing the genome itself via technologies such as CRISPR.[156] A greater understanding of the nature of specific mutations, both on a clinical and a fundamental physiological level will be needed to herald such a future.

2.3.1 Human induced pluripotent stem cell (iPSC) based-disease model

iPSC derived neurons, created from patients with genetic epilepsies, are a novel tool to understand the molecular mechanisms underpinning epileptic phenotypes, thus enabling personalized drug-development.[157,158] It carries the advantage of allowing us to study the disease in the context of each patient's unique genetic constellation. iPSC models have been used to study numerous genetic epilepsy syndromes: Dravet syndrome (SCN1A),

Malignant Migrating Partial Seizures of Infancy (KCNT1), Female-Limited Epilepsy and Cognitive Impairment (PCDH19), Tuberous Sclerosis (TSC), Miller-Dieker Syndrome (17p13.3 deletion), Lissencephaly (TUBA1A, LIS1), Benign familial infantile epilepsy (PRRT2), Timothy Syndrome (CANCA1C).[158] Furthermore, there has been promising research regarding the transplantation of stem cells into the brain for personalized or targeted treatment of refractory epilepsy. A phase I study ($n=22$) has demonstrated safety and potential efficacy of employing autologous patient-derived mesenchymal stem cells as a personalized therapeutic in patients with refractory epilepsy.[159] Studies to investigate the use of GABA-producing inhibitory interneurons as a targeted therapeutic in drug-resistant unilateral mesial temporal lobe epilepsy are also on the horizon (ClinicalTrials.gov: NCT05135091).

2.3.2 Oligonucleotides (ASOs)

Antisense oligonucleotides (ASOs) are another future direction of precision therapy for genetic epilepsy syndromes. ASOs are short strands of DNA or RNA that bind to a complementary RNA sequence, thereby inhibiting its function. By inhibiting specific RNA sequences, ASOs can effectively downregulate or upregulate the production of certain downstream proteins and can theoretically be used with both select loss of function and gain of function mutations. ASOs are currently being studied in a few genetic epilepsy syndromes in preclinical animal models. Most notably, a study utilizing ASOs in a mouse model of Dravet syndrome demonstrated that ASOs that increased expression of productive SCN1A resulted in decreased incidence of seizures and sudden unexplained death in epilepsy (SUDEP).[99] There is a currently enrolling open-label early clinical trial evaluating the use of these ASOs in children with Dravet's syndrome (clinicaltrials.gov identifier: NCT04740476). Another preclinical study involving the use of ASOs in a SCN2A gain of function mutation mouse model also demonstrated reduced seizures and prolonged lifespan.[160] There is also some evidence of efficacy of ASOs in a preclinical model of KCNT1.[161] There are likely many additional genetic epilepsy syndromes that could be potentially treated by targeted ASOs.

2.3.3 Clustered regularly interspaced short palindromic repeats-associated protein 9 (CRISPR-Cas9)-derived model

CRISPR-Cas9 is a technology recently developed for the direct engineering of targeted sites within the genome. This was discovered as an endogenous defense mechanism in bacteria against viral attack – specifically, one which allows cleavage of specific sequences of genetic material. In brief, this, along with the utilization of a cell's own DNA repair system, has allowed the ability not only to cleave and therefore remove DNA at specific sites, but also to introduce new genetic material in this same region.[162] This has a range of potential applications, for example, removing deleterious mutations (and replacing with wild type sequences), or surveying the genome via deletion of key sequences to understand the importance of genomic regions in disease, or with regards to drug targeting or efficacy. Furthermore, a related reversible form of gene regulation known as CRISPR interference (CRISPRi) or CRISPR activation (CRISPRa), have been shown to both decrease and increase, respectively, gene transcription in selective regions of the genome.[162] Specific to epilepsy, an upregulation of KCNA1 via CRISPRa technology was shown in a mouse model of focal temporal lobe epilepsy to both significantly reduce seizure burden and to improve cognitive deficits related to spatial memory and novel object recognition.[156] Although CRISPR-Cas9 technology has been used on human stem cells to correct mutation in the SCN1A gene,[163] there are as of yet no clinical trials involving CRISPR-Cas9, CRISPRi, or CRISPRa in patients with epilepsy, and technical limitations with regards to delivery and safety remain to be overcome and demonstrated prior to human trials.[156,162]

3. Precision therapy with personalized pharmacogenomics

Pharmacogenomics refers to genetic factors that alter the pharmacokinetic, pharmacodynamic or other response of the body to a drug. Although the data remain somewhat limited in epilepsy, several well-known examples exist that involve metabolic enzymes, transporters, and drug target sites and will be discussed here.

The cytochrome P450 isozyme (CYP) family is responsible for the majority of anti-seizure medication metabolism. CYP2C9 and CYP2C19, in particular, exhibit polymorphism in which metabolism of antiseizure medications may differ significantly between individuals, resulting in varying drug serum levels.[164,165] For example, the alleles CYP2C9*2 (rs 1799853) and CYP2C9*3 (rs1057910(C)) result in slow metabolism of phenytoin and significantly higher serum levels in affected individuals.[166] CYP2C19 polymorphisms may also affect the conversion of clobazam to n-desmethylclobazam,[165] as well as possibly zonisamide metabolism.[164]

The best-known polymorphism related to adverse drug reaction is the HLA-B*15:02 genetic marker, showing a greater rate of Stevens-Johnson syndrome after carbamazepine use, particularly in the Han Chinese population. HLA-B*15:02 has also been found in other southern Asian populations, and has been linked to Stevens-Johnson syndrome with phenytoin, oxcarbazepine and lamotrigine. HLA-A*31:01 has also been associated with hypersensitivity reactions after carbamazepine.[164]

In addition, polymorphism in the genes which encode drug target molecules are shown to be associated with the efficacy of anti-seizure medications. For example, SCN1A, SCN2A and SCN3A encode for voltage-gated sodium channels, and their polymorphisms are associated with the efficacy of multiple anti-seizure medications, including valproic acid, carbamazepine and phenytoin.[167,168]

4. Precision therapy with neuromodulation

Patients with medically refractory epilepsy may consider surgical resection or neuromodulation to target the seizure onset zone or critical nodes in network propagation. Neuromodulation has emerged as a nondestructive and effective approach when compared to traditional resection. Since epilepsy patients have unique and yet usually stereotypical neurophysiological patterns at seizure onset, neuromodulation provides physicians a potential opportunity to deliver a tailored treatment, by utilizing neurostimulators to detect epileptiform activities as defined for each individual patient, which then deliver electrical stimulation to normalize the neuronal activities. In this section, we will discuss the responsive neurostimulation, and also deep brain stimulation as examples.

4.1 Responsive neurostimulation (RNS system)

Closed-loop responsive neurostimulation to the epileptogenic zone is deemed acceptably safe, and effectively decreases rate of disabling seizures, a category consisting of focal seizures with motor symptoms and/or impaired awareness as well as focal seizures that generalize to bilateral tonic-clonic.[169,170] The RNS system received FDA approval in 2013, to be used in addition to anti-epileptic medications in adult patients with refractory epilepsy characterized by focal-onset seizures generated by no more than two foci.[169] Its hardware consists of a programmable neurostimulator device, which is implanted in the skull underneath the scalp, and connected to one or two intracranial wires containing electrodes in the form of depth leads or cortical strip leads.[170] Lead targets are determined via extensive pre-surgical evaluation to localize seizure onset and propagation.

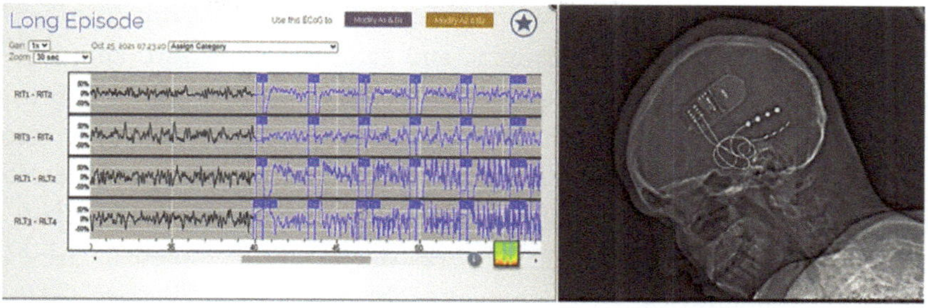

Fig. 1 RNS treatment for refractory epilepsy. (A) Example of RNS ECoG data depicting right temporal seizure pattern and subsequent electrical stimulation therapy administered in a patient with neurostimulator implanted onto skull connected to two intracranial RNS leads. (B) Right temporal cortical strips targeting seizure onset zones on scout CT head imaging.

As opposed to open-loop neuromodulation where electrical stimulation is constantly administered or delivered at fixed intervals, the RNS system in real-time collects data and detects seizure patterns triggering targeted therapeutic electrical stimulation.[169] Data can be used to inform personalized changes in the detection and stimulation parameters. Nair et al.[171] demonstrated a median of 75% reduction in seizure frequency at 9 years of neurostimulation. In addition to being a precision therapeutic, RNS is also a novel diagnostic tool, allowing for long-term recording of electrocorticography (Fig. 1). This unique dataset can be implemented to strategically guide further surgical options if epilepsy remains refractory.

4.2 Deep brain stimulation (DBS)

Deep brain stimulation (DBS) to select regions of the brain has been demonstrated to be another effective treatment for medically refractory epilepsy. DBS often is a form of open-loop neurostimulation, in which stimulation is applied continuously or intermittently, but irrespective of underlying epileptic activity. There are two modalities by which stimulation can be delivered—low frequency stimulation and high frequency stimulation. There is evidence that low frequency promotes normal neuronal electrical activity, while high frequency stimulation disrupts synchronized neuronal electrical activity[172]

When considering the DBS target structure for the treatment of epilepsy, there are a few different theories behind target selection. One approach is to directly stimulate the seizure onset zone, for example: the hypothalamus, amygdala, or hippocampus. Another approach is to stimulate the deep subcortical structures involved in epileptogenesis and propagation, for example: the thalamus, basal ganglia, and cerebellum. A third approach is to stimulate the deep connecting fibers such as the corpus callosum or fornix, to decrease the seizure threshold minimize seizure spread.[173] The most selected target for the treatment of medically refractory epilepsy is the anterior nucleus of the thalamus. The Stimulation of the Anterior Nucleus of the Thalamus for Epilepsy (SANTE) trial demonstrated a median decrease in seizure frequency of 56% in patients with medically refractory focal or secondarily generalized seizures who were implanted with bilateral DBS to the ATN.[174] The centromedian nucleus of the thalamus has shown to be a promising target for patients with generalized seizures, as well as in patients with Lennox-Gastaut Syndrome.[175] There are other less common neurostimulation targets for epilepsy, including the hippocampus, head of the caudate, subthalamic nucleus, cerebellar nuclei, and hypothalamus, and a recent review has highlighted some of these DBS targets for epilepsy[176] (Fig. 2).

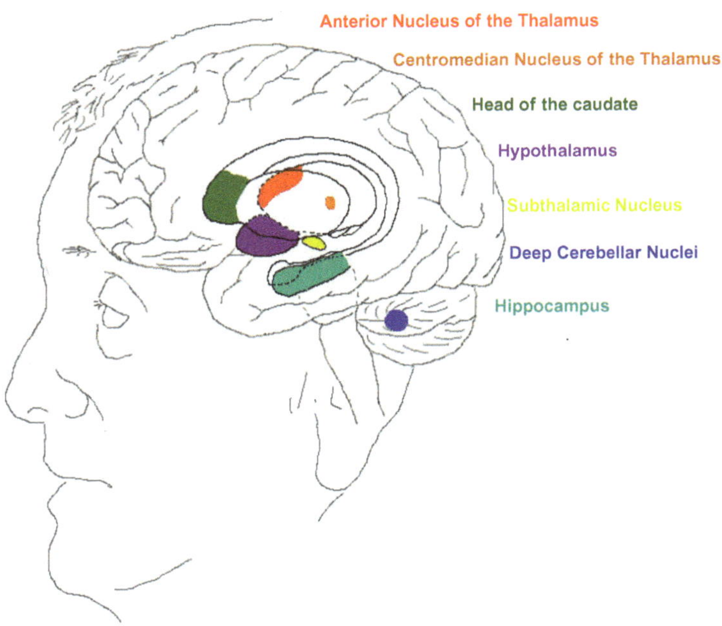

Fig. 2 Summary of deep brain stimulation targets within the brain with possible efficacy in epilepsy. Colored regions in the image match the written descriptions of the same color to the right. The only target for which there is evidence in the form of a randomized clinical trial and has approved by FDA is the anterior nucleus of the thalamus.

Closed-loop DBS presents another opportunity for precision treatment of medically refractory epilepsy. One of the newer DBS models, the Medtronic Percept™, has recently approved by FDA in 2020. It provides BrainSense™, a technique allows for recording of local field potential signals from deep brain structure.[177] This is the first closed-loop adaptive DBS device, which has the capacity to adjust stimulation based on recording of low frequency field potential oscillations. The implications of such oscillations within the context of epilepsy, as well as the manner in which stimulation should be adjusted based on these oscillations, are directions for future research.

5. Conclusion

Precision medicine in epilepsy is a growing field, now not only incorporating surgical decision-making on a patient-by-patient basis, but leveraging genetic information, developing precision genetic therapies, and adding neuromodulatory techniques to patient care. As in most fields of medicine, genetics has had a profound impact: it has shown that knowledge of patients' genomes may aid in disease diagnosis, drug choices (both for therapeutic advantage and avoidance of side effects and intolerance), rational drug design, and the development of personalized gene therapy, for which several trials are ongoing. In addition, neuromodulatory techniques include electrical stimulation of various brain regions to help improve seizure burden, and even to achieve seizure freedom in a portion of these refractory cases. No doubt these fields will continue to evolve. It may very well be that a cure for epilepsy will come from personalized treatment for the epilepsies, one-by-one, and shifting away from "one-size fits all."

Funding

None.

References

1. Ginsburg GS, Phillips KA. Precision medicine: from science to value. *Health Aff.* 2018;37(5):694–701. https://doi.org/10.1377/hlthaff.2017.1624.
2. Hansen TF, Møller RS. The first step towards personalized risk prediction for common epilepsies. *Brain*. 2019;142(11):3316–3318. https://doi.org/10.1093/brain/awz318.
3. Berkovic SF, Scheffer IE, Petrou S, et al. A roadmap for precision medicine in the epilepsies. *Lancet Neurol.* 2015;14(12):1219–1228. https://doi.org/10.1016/S1474-4422(15)00199-4.

4. Fan YS, Ouyang X, Peng J, et al. Frequent detection of parental consanguinity in children with developmental disorders by a combined CGH and SNP microarray. *Mol Cytogenet.* 2013;6(1):38. https://doi.org/10.1186/1755-8166-6-38.
5. Coppola A, Cellini E, Stamberger H, et al. Diagnostic implications of genetic copy number variation in epilepsy plus. *Epilepsia.* 2019;60(4):689–706. https://doi.org/10.1111/epi.14683.
6. Fernández IS, Loddenkemper T, Gaínza-Lein M, Sheidley BR, Poduri A. Diagnostic yield of genetic tests in epilepsy: a meta-analysis and cost-effectiveness study. *Neurology.* 2019;92(5):e418–e428. https://doi.org/10.1212/WNL.0000000000006850.
7. Poduri A, Sheidley BR, Shostak S, Ottman R. Genetic testing in the epilepsies-developments and dilemmas. *Nat Rev Neurol.* 2014;10(5):293–299. https://doi.org/10.1038/nrneurol.2014.60.
8. Dunn P, Albury CL, Maksemous N, et al. Next generation sequencing methods for diagnosis of epilepsy syndromes. *Front Genet.* 2018;295(3):751–763. https://doi.org/10.3389/fgene.2018.00020.
9. Johannesen KM, Nikanorova N, Marjanovic D, et al. Utility of genetic testing for therapeutic decision-making in adults with epilepsy. *Epilepsia.* 2020;61(6):1234–1239. https://doi.org/10.1111/epi.16533.
10. Choi M, Scholl UI, Ji W, et al. Genetic diagnosis by whole exome capture and massively parallel DNA sequencing. *Proc Natl Acad Sci U S A.* 2009;106(45):19096–19101. https://doi.org/10.1073/pnas.0910672106.
11. Yang Y, Muzny DM, Xia F, et al. Molecular findings among patients referred for clinical whole-exome sequencing. *JAMA.* 2014;312(18):1870–1879. https://doi.org/10.1001/jama.2014.14601.
12. Yang Y, Muzny DM, Reid JG, et al. Clinical Whole-Exome Sequencing for the Diagnosis of Mendelian Disorders. *N Engl J Med.* 2013;369(16):1502–1511. https://doi.org/10.1056/nejmoa1306555.
13. Valencia CA, Husami A, Holle J, et al. Clinical impact and cost-effectiveness of whole exome sequencing as a diagnostic tool: a pediatric center's experience. *Front Pediatr.* 2015;3:67. https://doi.org/10.3389/fped.2015.00067.
14. Belkadi A, Bolze A, Itan Y, et al. Whole-genome sequencing is more powerful than whole-exome sequencing for detecting exome variants. *Sources.* 2015;112(17):5473–5478. https://doi.org/10.2307/26462582.
15. Mattick JS, Dinger M, Schonrock N, Cowley M. Whole genome sequencing provides better diagnostic yield and future value than whole exome sequencing. *Med J Aust.* 2018;209(5):197. https://doi.org/10.5694/mja17.01176.
16. Palmer EE, Sachdev R, Macintosh R, et al. Diagnostic yield of whole genome sequencing after nondiagnostic exome sequencing or gene panel in developmental and epileptic encephalopathies. *Neurology.* 2021;96(13):e1770–e1782. https://doi.org/10.1212/WNL.0000000000011655.
17. Bruxel EM, Bruno DCF, do Canto AM, et al. Multi-omics in mesial temporal lobe epilepsy with hippocampal sclerosis: clues into the underlying mechanisms leading to disease. *Seizure.* 2021;90:34–50. https://doi.org/10.1016/j.seizure.2021.03.002.
18. Lovisari F, Simonato M. Gene networks and microRNAs: promises and challenges for treating epilepsies and their comorbidities. *Epilepsy Behav.* 2021;121(Pt B):106488. https://doi.org/10.1016/j.yebeh.2019.106488.
19. do Canto AM, Donatti A, Geraldis JC, Godoi AB, da Rosa DC, Lopes-Cendes I. Neuroproteomics in epilepsy: what do we know so far? *Front Mol Neurosci.* 2021;13:604158. https://doi.org/10.3389/fnmol.2020.604158.
20. Arnatkeviciute A, Fulcher BD, Bellgrove MA, Fornito A. Where the genome meets the connectome: understanding how genes shape human brain connectivity. *Neuroimage.* 2021;244:118570. https://doi.org/10.1016/j.neuroimage.2021.118570.

21. Woo YJ, Roussos P, Haroutunian V, et al. Comparison of brain connectomes by MRI and genomics and its implication in Alzheimer's disease. *BMC Med.* 2020;18(1):23–40. https://doi.org/10.1186/s12916-019-1488-1.
22. Mills PB, Footitt EJ, Mills KA, et al. Genotypic and phenotypic spectrum of pyridoxine-dependent epilepsy (ALDH7A1 deficiency). *Brain.* 2010;133(7):2148–2159. https://doi.org/10.1093/brain/awq143.
23. Mercimek-Mahmutoglu S, Cordeiro D, Cruz V, et al. Novel therapy for pyridoxine dependent epilepsy due to ALDH7A1 genetic defect: L-arginine supplementation alternative to lysine-restricted diet. *Eur J Paediatr Neurol.* 2014;18(6):741–746. https://doi.org/10.1016/j.ejpn.2014.07.001.
24. Coughlin CR, Swanson MA, Spector E, et al. The genotypic spectrum of ALDH7A1 mutations resulting in pyridoxine dependent epilepsy: a common epileptic encephalopathy. *J Inherit Metab Dis.* 2018;244:1–9. https://doi.org/10.1007/s10545-018-0219-7. Published online.
25. Basura GJ, Hagland SP, Wiltse AM, Gospe SM. Clinical features and the management of pyridoxine-dependent and pyridoxine-responsive seizures: Review of 63 North American cases submitted to a patient registry. *Eur J Pediatr.* 2009;168(6):697–704. https://doi.org/10.1007/s00431-008-0823-x.
26. Bok LA, Halbertsma FJ, Houterman S, et al. Long-term outcome in pyridoxine-dependent epilepsy. *Dev Med Child Neurol.* 2012;54(9):849–854. https://doi.org/10.1111/j.1469-8749.2012.04347.x.
27. Alghamdi M, Bashiri FA, Abdelhakim M, et al. Phenotypic and molecular spectrum of pyridoxamine-5′-phosphate oxidase deficiency: a scoping review of 87 cases of pyridoxamine-5′-phosphate oxidase deficiency. *Clin Genet.* 2021;99(1):99–110. https://doi.org/10.1111/cge.13843.
28. Mills PB, Surtees RAH, Champion MP, et al. Neonatal epileptic encephalopathy caused by mutations in the PNPO gene encoding pyridox(am)ine 5′-phosphate oxidase. *Hum Mol Genet.* 2005;14(8):1077–1086. https://doi.org/10.1093/hmg/ddi120.
29. Sudarsanam A, Singh H, Wilcken B, et al. Cirrhosis associated with pyridoxal 5′-phosphate treatment of pyridoxamine 5′-phosphate oxidase deficiency. *JIMD Rep.* 2014;17:67–70. https://doi.org/10.1007/8904_2014_338.
30. Darin N, Reid E, Prunetti L, et al. Mutations in PROSC disrupt cellular pyridoxal phosphate homeostasis and cause vitamin-B6-dependent epilepsy. *Am J Hum Genet.* 2016;99(6):1325–1337. https://doi.org/10.1016/j.ajhg.2016.10.011.
31. Kaler SG. ATP7A-related copper transport diseasesg-emerging concepts and future trends. *Nat Rev Neurol.* 2011;7(1):15–29. https://doi.org/10.1038/nrneurol.2010.180.
32. Verrotti A, Cusmai R, Darra F, et al. Epilepsy in Menkes disease: an electroclinical long-term study of 28 patients. *Epilepsy Res.* 2014;108(9):1597–1603. https://doi.org/10.1016/j.eplepsyres.2014.08.006.
33. Kaler SG. Neurodevelopment and brain growth in classic Menkes disease is influenced by age and symptomatology at initiation of copper treatment. *J Trace Elem Med Biol.* 2014;28(4):427–430. https://doi.org/10.1016/j.jtemb.2014.08.008.
34. Haddad MR, Choi EY, Zerfas PM, et al. Cerebrospinal fluid-directed rAAV9-rsATP7A plus subcutaneous copper histidinate advance survival and outcomes in a Menkes disease mouse model. *Mol Ther Methods Clin Dev.* 2018;10:165–178. https://doi.org/10.1016/j.omtm.2018.07.002.
35. Brunetti S, Malerba L, Giordano L, et al. Cerebral folate transporter deficiency syndrome in three siblings: why genetic testing for developmental and epileptic encephalopathies should be performed early and include the FOLR1 gene. *Am J Med Genet A.* 2021;185(8):2526–2531. https://doi.org/10.1002/ajmg.a.62345.
36. Delmelle F, Thöny B, Clapuyt P, Blau N, Nassogne MC. Neurological improvement following intravenous high-dose folinic acid for cerebral folate transporter deficiency

caused by FOLR-1 mutation. *Eur J Paediatr Neurol.* 2016;20(5):709–713. https://doi.org/10.1016/j.ejpn.2016.05.021.
37. Wyss M, Kaddurah-Daouk R. Creatine and creatinine metabolism. *Physiol Rev.* 2000;80(3):1107–1213. https://doi.org/10.1152/physrev.2000.80.3.1107.
38. Clark JF, Cecil KM. Diagnostic methods and recommendations for the cerebral creatine deficiency syndromes. *Pediatr Res.* 2015;77(3):398–405. https://doi.org/10.1038/pr.2014.203.
39. Stockler-Ipsiroglu S, Apatean D, Battini R, et al. Arginine:glycine amidinotransferase (AGAT) deficiency: clinical features and long term outcomes in 16 patients diagnosed worldwide. *Mol Genet Metab.* 2015;116(4):252–259. https://doi.org/10.1016/j.ymgme.2015.10.003.
40. Leuzzi V, Mastrangelo M, Battini R, Cioni G. Inborn errors of creatine metabolism and epilepsy. *Epilepsia.* 2013;54(2):217–227. https://doi.org/10.1111/epi.12020.
41. Battini R, Alessandrì MG, Leuzzi V, et al. Arginine:glycine amidinotransferase (AGAT) deficiency in a newborn: early treatment can prevent phenotypic expression of the disease. *J Pediatr.* 2006;148(6):828–830. https://doi.org/10.1016/j.jpeds.2006.01.043.
42. Stockler S, Schutz PW, Salomons GS. Cerebral creatine deficiency syndromes: clinical aspects, treatment and pathophysiology. *Subcell Biochem.* 2007;46:149–166. https://doi.org/10.1007/978-1-4020-6486-9_8.
43. Schulze A, Ebinger F, Rating D, Mayatepek E. Improving treatment of guanidinoacetate methyltransferase deficiency: reduction of guanidinoacetic acid in body fluids by arginine restriction and ornithine supplementation. *Mol Genet Metab.* 2001;74(4). https://doi.org/10.1006/mgme.2001.3257.
44. Schulze A, Hoffmann GF, Bachert P, et al. Presymptomatic treatment of neonatal guanidinoacetate methyltransferase deficiency. *Neurology.* 2006;67(4). https://doi.org/10.1212/01.wnl.0000230152.25203.01.
45. Stockler-Ipsiroglu S, van Karnebeek C, Longo N, et al. Guanidinoacetate methyltransferase (GAMT) deficiency: outcomes in 48 individuals and recommendations for diagnosis, treatment and monitoring. *Mol Genet Metab.* 2014;111(1):16–25. https://doi.org/10.1016/j.ymgme.2013.10.018.
46. Mercimek-Mahmutoglu S, Stoeckler-Ipsiroglu S, Adami A, et al. GAMT deficiency: features, treatment, and outcome in an inborn error of creatine synthesis. *Neurology.* 2006;67(3). https://doi.org/10.1212/01.wnl.0000234852.43688.bf.
47. Dhar SU, Scaglia F, Li FY, et al. Expanded clinical and molecular spectrum of guanidinoacetate methyltransferase (GAMT) deficiency. *Mol Genet Metab.* 2009;96(1). https://doi.org/10.1016/j.ymgme.2008.10.008.
48. van de Kamp JM, Betsalel OT, Mercimek-Mahmutoglu S, et al. Phenotype and genotype in 101 males with x-linked creatine transporter deficiency. *J Med Genet.* 2013;50(7). https://doi.org/10.1136/jmedgenet-2013-101658.
49. Kurosawa Y, DeGrauw TJ, Lindquist DM, et al. Cyclocreatine treatment improves cognition in mice with creatine transporter deficiency. *J Clin Investig.* 2012;122(8). https://doi.org/10.1172/JCI59373.
50. Trotier-Faurion A, Passirani C, Béjaud J, et al. Dodecyl creatine ester and lipid nanocapsule: a double strategy for the treatment of creatine transporter deficiency. *Nanomedicine (Lond).* 2015;10(2):185–191. https://doi.org/10.2217/nnm.13.205.
51. Li D, Yuan H, Ortiz-Gonzalez XR, et al. GRIN2D recurrent de novo dominant mutation causes a severe epileptic encephalopathy treatable with NMDA receptor channel blockers. *Am J Hum Genet.* 2016;99(4):802–816. https://doi.org/10.1016/j.ajhg.2016.07.013.

52. von Stülpnagel C, Ensslen M, Møller RS, et al. Epilepsy in patients with GRIN2A alterations: genetics, neurodevelopment, epileptic phenotype and response to anticonvulsive drugs. *Eur J Paediatr Neurol.* 2017;21(3):530–541. https://doi.org/10.1016/j.ejpn.2017.01.001.
53. XiangWei W, Kannan V, Xu Y, et al. Heterogeneous clinical and functional features of GRIN2D-related developmental and epileptic encephalopathy. *Brain.* 2019;142(10):3009–3027. https://doi.org/10.1093/brain/awz232.
54. Pierson TM, Yuan H, Marsh ED, et al. GRIN2A mutation and early-onset epileptic encephalopathy: personalized therapy with memantine. *Ann Clin Transl Neurol.* 2014;1(3):190–198. https://doi.org/10.1002/acn3.39.
55. Fitzgerald MP, Fiannacca M, Smith DM, et al. Treatment responsiveness in KCNT1-related epilepsy. *Neurotherapeutics.* 2019;16(3):848–857. https://doi.org/10.1007/s13311-019-00739-y.
56. Abdelnour E, Gallentine W, McDonald M, Sachdev M, Jiang YH, Mikati MA. Does age affect response to quinidine in patients with KCNT1 mutations? Report of three new cases and review of the literature. *Seizure.* 2018;55:1–3. https://doi.org/10.1016/j.seizure.2017.11.017.
57. Jia Y, Lin Y, Li J, et al. Quinidine therapy for Lennox-Gastaut syndrome with KCNT1 mutation. A case report and literature review. *Front Neurol.* 2019;10(FEB). https://doi.org/10.3389/fneur.2019.00064.
58. Corbett MA, Bellows ST, Melody L, Carroll R, et al. Dominant KCNA2 mutation causes episodic ataxia and pharmacoresponsive epilepsy. *Neurology.* 2016;87(19):1975–1984.
59. Canafoglia L, Castellotti B, Ragona F, et al. Progressive myoclonus epilepsy caused by a gain-of-function KCNA2 mutation. *Seizure.* 2019;65:106–108. https://doi.org/10.1016/j.seizure.2019.01.005.
60. Syrbe S, Hedrich UBS, Riesch E, et al. De novo loss-or gain-of-function mutations in KCNA2 cause epileptic encephalopathy. *Nat Genet.* 2015;47(4):393–399. https://doi.org/10.1038/ng.3239.
61. UB SH, Lauxmann S, Wolff M, et al. *E P I L E P S Y 4-Aminopyridine Is a Promising Treatment Option for Patients With Gain-of-Function KCNA2-Encephalopathy.* vol 13; 2021. https://www.science.org.
62. Kuersten M, Tacke M, Gerstl L, Hoelz H, Stülpnagel C, v., Borggraefe I. Antiepileptic therapy approaches in KCNQ2 related epilepsy: a systematic review. *Eur J Med Genet.* 2020;63(1). https://doi.org/10.1016/j.ejmg.2019.02.001.
63. Biervert C, Schroeder BC, Kubisch C, et al. A potassium channel mutation in neonatal human epilepsy. *Science.* 1998;279:403–406.
64. Barrese V, Miceli F, Virginia Soldovieri M, et al. Neuronal potassium channel openers in the management of epilepsy: role and potential of retigabine. *Clin Pharmacol Adv Appl.* 2010;2:225–236. https://doi.org/10.2147/CPAA.S15369.
65. Bock C, Surur AS, ristin Beirow K, et al. Sulfide analogues of flupirtine and retigabine with nanomolar K V 7.2/K V 7.3 channel opening activity. *ChemMedChem.* 2019;14:952–964. https://doi.org/10.1002/cmdc.201900112.
66. Zaki MS, Selim L, EL-Bassyouni HT, et al. Molybdenum cofactor and isolated sulphite oxidase deficiencies: clinical and molecular spectrum among Egyptian patients. *Eur J Paediatr Neurol.* 2016;20(5):714–722. https://doi.org/10.1016/j.ejpn.2016.05.011.
67. Hitzert MM, Bos AF, Bergman KA, et al. Favorable outcome in a newborn with molybdenum cofactor type A deficiency treated with CPMP CASE REPORT. *Pediatrics.* 2012;130(4):1005–1010. http://publications.aap.org/pediatrics/article-pdf/130/4/e1005/1088497/peds_2011-3330.pdf.

68. Schwahn BC, van Spronsen FJ, Belaidi AA, et al. Efficacy and safety of cyclic pyranopterin monophosphate substitution in severe molybdenum cofactor deficiency type A: a prospective cohort study. *Lancet.* 2015;386(10007):1955–1963. https://doi.org/10.1016/S0140-6736(15)00124-5.
69. Yakubovskaya E, Chen Z, Carrodeguas JA, Kisker C, Bogenhagen DF. Functional human mitochondrial DNA polymerase γ forms a heterotrimer. *J Biol Chem.* 2006;281(1):374–382. https://doi.org/10.1074/jbc.M509730200.
70. Ropp PA, Copeland WC. *Cloning and Characterization of the Human Mitochondrial DNA Polymerase, DNA Polymerase Gamma.* Vol 36.; 1996.
71. J. Hynynen, T. Pokka, J. Komulainen-Ebrahim, P. Myllynen, M. Kärppä, L. Pylvänen, R. Kälviäinen, A. Sokka, A. Jyrkilä, J. Lähdetie, L. Haataja, A. Mäkitalo, P. Ylikotila, K. Eriksson, P. Haapala, H. Ansakorpi, R. Hinttala, P. Vieira, K. Majamaa, H. Rantala, J. Uusimaa, 2018. Variants p.Q1236H and p.E1143G in mitochondrial DNA polymerase gamma POLG1 are not associated with increased risk for valproate-induced hepatotoxicity or pancreatic toxicity: a retrospective cohort study of patients with epilepsy. Epilepsia 59(11), 2125-2136, doi: 10.1111/epi.14568.
72. Horvath R, Hudson G, Ferrari G, et al. Phenotypic spectrum associated with mutations of the mitochondrial polymerase γ gene. *Brain.* 2006;129(7):1674–1684. https://doi.org/10.1093/brain/awl088.
73. Hakonen AH, Heiskanen S, Juvonen V, et al. Mitochondrial DNA polymerase W748S mutation: a common cause of autosomal recessive ataxia with ancient european origin. *Am J Hum Genet.* 2005;77:430–441.
74. van Goethem G, Luoma P, Rantamäki M, et al. POLG mutations in neurodegenerative disorders with ataxia but no muscle involvement. *Neurology.* 2004;63:1251–1257. www.neurology.org.
75. Davidzon G, Mancuso M, Ferraris S, et al. POLG mutations and Alpers syndrome. *Ann Neurol.* 2005;57(6):921–923. https://doi.org/10.1002/ana.20498.
76. Specchio N, Pietrafusa N, Calabrese C, et al. POLG1-related epilepsy: review of diagnostic and therapeutic findings. *Brain Sci.* 2020;10(11):1–13. https://doi.org/10.3390/brainsci10110768.
77. Stewart JD, Horvath R, Baruffini E, et al. Polymerase γGene POLG determines the risk of sodium valproate-induced liver toxicity. *Hepatology.* 2010;52(5):1791–1796. https://doi.org/10.1002/hep.23891.
78. Kahn-Kirby AH, Amagata A, Maeder CI, et al. Targeting ferroptosis: a novel therapeutic strategy for the treatment of mitochondrial disease-related epilepsy. *PLoS One.* 2019;14(3). https://doi.org/10.1371/journal.pone.0214250.
79. Chen WJ, Lin Y, Xiong ZQ, et al. Exome sequencing identifies truncating mutations in PRRT2 that cause paroxysmal kinesigenic dyskinesia. *Nat Genet.* 2011;43(12):1252–1255. https://doi.org/10.1038/ng.1008.
80. Heron SE, Grinton BE, Kivity S, et al. PRRT2 mutations cause benign familial infantile epilepsy and infantile convulsions with choreoathetosis syndrome. *Am J Hum Genet.* 2012;90(1):152–160. https://doi.org/10.1016/j.ajhg.2011.12.003.
81. Schwenk J, Harmel N, Brechet A, et al. High-resolution proteomics unravel architecture and molecular diversity of native AMPA receptor complexes. *Neuron.* 2012;74(4):621–633. https://doi.org/10.1016/j.neuron.2012.03.034.
82. Stelzl U, Worm U, Lalowski M, et al. A human protein-protein interaction network: A resource for annotating the proteome. *Cell.* 2005;122(6):957–968. https://doi.org/10.1016/j.cell.2005.08.029.
83. Ebrahimi-Fakhari D, Saffari A, Westenberger A, Klein C. The evolving spectrum of PRRT2-associated paroxysmal diseases. *Brain.* 2015;138(12):3476–3495. https://doi.org/10.1093/brain/awv317.

84. Bruno MK, Hallett M, Gwinn-Hardy K, et al. Clinical evaluation of idiopathic paroxysmal kinesigenic dyskinesia new diagnostic criteria. *Neurology*. 2004;63(12):2280–2287.
85. Li HF, Chen WJ, Ni W, et al. Clinical/scientific notes. *Neurology*. 2013;80(16): 1534–1535. https://doi.org/10.1212/WNL.0b013e31828cf7e1.
86. Lossin C. A catalog of SCN1A variants. *Brain Dev*. 2009;31(2):114–130. https://doi.org/10.1016/j.braindev.2008.07.011.
87. Catterall WA, Kalume F, Oakley JC. NaV1.1 channels and epilepsy. *J Physiol*. 2010;588 (11):1849–1859. https://doi.org/10.1113/JPHYSIOL.2010.187484.
88. Zuberi SM, Brunklaus A, Birch R, Reavey DE, Duncan J, Forbes GH. Genotype-phenotype associations in SCN1A-related epilepsies. *Neurology*. 2011;76(7):594–600. www.neurology.org.
89. Marini C, Scheffer IE, Nabbout R, et al. The genetics of Dravet syndrome. *Epilepsia*. 2011;52(suppl 2):24–29. https://doi.org/10.1111/j.1528-1167.2011.02997.x.
90. Escayg A, Goldin AL. Sodium channel SCN1A and epilepsy: mutations and mechanisms. *Epilepsia*. 2010;51(9):1650–1658. https://doi.org/10.1111/j.1528-1167.2010.02640.x.
91. Hirose S, Scheffer IE, Marini C, et al. SCN1A testing for epilepsy: application in clinical practice. *Epilepsia*. 2013;54(5):946–952. https://doi.org/10.1111/epi.12168.
92. Ishii A, Watkins JC, Chen D, Hirose S, Hammer MF. Clinical implications of SCN1A missense and truncation variants in a large Japanese cohort with Dravet syndrome. *Epilepsia*. 2017;58(2):282–290. https://doi.org/10.1111/epi.13639.
93. Guerrini R, Dravet C, Genton P, Belmonte A, Kaminska A, Dulac T. Lamotrigine and seizure aggravation in severe myoclonic epilepsy. *Epilepsia*. 1998;39(5):508–512.
94. Chiron C, Marchand MC, Tran A, et al. Stiripentol in severe myoclonic epilepsy in infancy: a randomised placebo-controlled syndrome-dedicated trial. *Lancet*. 2000;356 (9242):1638–1642. https://doi.org/10.1016/S0140-6736(00)03157-3.
95. Cho MJ, Kwon SS, Ko A, et al. Efficacy of stiripentol in dravet syndrome with or without SCN1A mutations. *J Clin Neurol*. 2018;14(1):22–28. https://doi.org/10.3988/jcn.2018.14.1.22.
96. Devinsky O, Cross JH, Laux L, et al. Trial of cannabidiol for drug-resistant seizures in the dravet syndrome. 2017;376(21):2011–2020. https://doi.org/10.1056/NEJMOA1611618.
97. Lagae L, Sullivan J, Knupp K, et al. Fenfluramine hydrochloride for the treatment of seizures in Dravet syndrome: a randomised, double-blind, placebo-controlled trial. *Lancet*. 2019;394(10216):2243–2254. https://doi.org/10.1016/S0140-6736(19)32500-0.
98. Wirrell EC, Laux L, Donner E, et al. Optimizing the diagnosis and management of dravet syndrome: recommendations from a North American consensus panel. *Pediatr Neurol*. 2017;68:18–34.e3. https://doi.org/10.1016/j.pediatrneurol.2017.01.025.
99. Han Z, Chen C, Christiansen A, et al. Antisense oligonucleotides increase Scn1a expression and reduce seizures and SUDEP incidence in a mouse model of Dravet syndrome. *Sci Transl Med*. 2020;12(558):6100. https://doi.org/10.1126/SCITRANSLMED.AAZ6100.
100. Sanders SJ, Campbell AJ, Cottrell JR, et al. Progress in understanding and treating SCN2A-mediated disorders. *Trends Neurosci*. 2018;41(7):442–456. https://doi.org/10.1016/j.tins.2018.03.011.
101. Carvill GL, Tambunan D, Mackay MT, et al. *SCN2A Encephalopathy A Major Cause of Epilepsy of Infancy with Migrating Focal Seizures*.; 2015.
102. Wolff M, Johannesen KM, Hedrich UBS, et al. Genetic and phenotypic heterogeneity suggest therapeutic implications in SCN2A-related disorders. *Brain*. 2017;140(5): 1316–1336. https://doi.org/10.1093/brain/awx054.

103. Foster LA, Johnson MR, MacDonald JT, et al. Infantile epileptic encephalopathy associated with SCN2A mutation responsive to oral mexiletine. *Pediatr Neurol.* 2017;66:108–111. https://doi.org/10.1016/j.pediatrneurol.2016.10.008.
104. Liu Y, Schubert J, Sonnenberg L, et al. Neuronal mechanisms of mutations in SCN8A causing epilepsy or intellectual disability. *Brain.* 2019;142(2):376–390. https://doi.org/10.1093/brain/awy326.
105. Larsen J, Carvill GL, Gardella E, et al. The phenotypic spectrum of SCN8A encephalopathy. *Neurology.* 2015;84(5):480–489. https://doi.org/10.1212/WNL.0000000000001211.
106. Johnson Jr J, Focken T, Khakh K, et al. NBI-921352, a first-in-class, NaV1.6 selective, sodium channel inhibitor that prevents seizures in Scn8a gain-of-function mice, and wild-type mice and rats. *Elife.* 2012;11:1–26, e724682. https://doi.org/10.1101/2021.08.29.458125.
107. Klepper J, Akman C, Armeno M, et al. Glut1 deficiency syndrome (Glut1DS): state of the art in 2020 and recommendations of the international Glut1DS study group. *Epilepsia Open.* 2020;5(3):354–365. https://doi.org/10.1002/epi4.12414.
108. Erro R, Stamelou M, Ganos C, et al. The clinical syndrome of paroxysmal exercise-induced dystonia: diagnostic outcomes and an algorithm. *Mov Disord Clin Pract.* 2014;1(1):57–61. https://doi.org/10.1002/mdc3.12007.
109. Klepper J. Glucose transporter deficiency syndrome (GLUT1DS) and the ketogenic diet. *Epilepsia.* 2008;49:46–49. https://doi.org/10.1111/j.1528-1167.2008.01833.x.
110. Kass HR, Winesett SP, Bessone SK, Turner Z, Kossoff EH. Use of dietary therapies amongst patients with GLUT1 deficiency syndrome. *Seizure.* 2016;35:83–87. https://doi.org/10.1016/j.seizure.2016.01.011.
111. de Giorgis V, Masnada S, Varesio C, et al. Overall cognitive profiles in patients with GLUT1 Deficiency Syndrome. *Brain Behav.* 2019;9(3). https://doi.org/10.1002/brb3.1224.
112. Klein P, Tyrlikova I, Mathews GC. Dietary treatment in adults with refractory epilepsy: a review. *Neurology.* 2014;83(21):1978–1985.
113. Johannesen KM, Gardella E, Linnankivi T, et al. Defining the phenotypic spectrum of SLC6A1 mutations. *Epilepsia.* 2018;59(2):389–402. https://doi.org/10.1111/epi.13986.
114. Goodspeed K, Pérez-Palma E, Iqbal S, et al. Current knowledge of SLC6A1-related neurodevelopmental disorders. *Brain. Communications.* 2020;2(2). https://doi.org/10.1093/braincomms/fcaa170.
115. Palmer S, Towne MC, Pearl PL, et al. SLC6A1 mutation and ketogenic diet in epilepsy with myoclonic-atonic seizures. *Pediatr Neurol.* 2016;64:77–79. https://doi.org/10.1016/j.pediatrneurol.2016.07.012.
116. Wang J, Wang J, Han X, et al. Report of the largest chinese cohort with SLC19A3 gene defect and literature review. *Front Genet.* 2021;12. https://doi.org/10.3389/fgene.2021.683255.
117. Tonduti D, Invernizzi F, Panteghini C, et al. SLC19A3 related disorder: treatment implication and clinical outcome of 2 new patients. *Eur J Paediatr Neurol.* 2018;22(2):332–335. https://doi.org/10.1016/j.ejpn.2017.11.012.
118. Haack TB, Klee D, Strom TM, et al. Infantile Leigh-like syndrome caused by SLC19A3 mutations is a treatable disease. *Brain.* 2014;137(9). https://doi.org/10.1093/brain/awu128.
119. Kimizu T, Takahashi Y, Oboshi T, et al. A case of early onset epileptic encephalopathy with de novo mutation in SLC35A2: clinical features and treatment for epilepsy. *Brain Dev.* 2017;39(3):256–260. https://doi.org/10.1016/J.BRAINDEV.2016.09.009.
120. Witters P, Tahata S, Barone R, et al. Clinical and biochemical improvement with galactose supplementation in SLC35A2-CDG. *Genet Med.* 2020;22(6):1102–1107. https://doi.org/10.1038/s41436-020-0767-8.

121. Kodera H, Nakamura K, Osaka H, et al. De novo mutations in SLC35A2 encoding a UDP-galactose transporter cause early-onset epileptic encephalopathy. *Hum Mutat.* 2013;34(12):1708–1714. https://doi.org/10.1002/HUMU.22446.
122. Dörre K, Olczak M, Wada Y, et al. A new case of UDP-galactose transporter deficiency (SLC35A2-CDG): molecular basis, clinical phenotype, and therapeutic approach. *J Inherit Metab Dis.* 2015;38:931–940. https://doi.org/10.1007/s10545-015-9828-6.
123. Abramov D, Guiberson NGL, Burré J. STXBP1 encephalopathies: clinical spectrum, disease mechanisms, and therapeutic strategies. *J Neurochem.* 2021;157(2):165–178. https://doi.org/10.1111/jnc.15120.
124. Stamberger H, Nikanorova M, Willemsen MH, et al. STXBP1 encephalopathy a neurodevelopmental disorder including epilepsy. *Neurology.* 2016;86:954–962. http://exac.broadinstitute.org/.
125. Lin L, Sohar I, Lackland H, Lobel P. The human CLN2 protein/tripeptidyl-peptidase I is a serine protease that autoactivates at acidic pH. *J Biol Chem.* 2001;276(3). https://doi.org/10.1074/jbc.M008562200.
126. Gardner E, Bailey M, Schulz A, Aristorena M, Miller N, Mole SE. Mutation update: review of TPP1 gene variants associated with neuronal ceroid lipofuscinosis CLN2 disease. *Hum Mutat.* 2019;40(11). https://doi.org/10.1002/humu.23860.
127. van Beersel G, Tihon E, Demine S, Hamer I, Jadot M, Arnould T. Different molecular mechanisms involved in spontaneous and oxidative stress-induced mitochondrial fragmentation in tripeptidyl peptidase-1 (TPP-1)-deficient fibroblasts. *Biosci Rep.* 2013;33 (2). https://doi.org/10.1042/BSR20120104.
128. Haltia M. The neuronal ceroid-lipofuscinoses. *J Neuropathol Exp Neurol.* 2003;62 (1):1–13. https://doi.org/10.1093/jnen/62.1.1.
129. Nickel M, Simonati A, Jacoby D, et al. Disease characteristics and progression in patients with late-infantile neuronal ceroid lipofuscinosis type 2 (CLN2) disease: an observational cohort study. *Lancet Child Adolesc Health.* 2018;2(8). https://doi.org/10.1016/S2352-4642(18)30179-2.
130. Fietz M, AlSayed M, Burke D, et al. Diagnosis of neuronal ceroid lipofuscinosis type 2 (CLN2 disease): expert recommendations for early detection and laboratory diagnosis. *Mol Genet Metab.* 2016;119(1–2). https://doi.org/10.1016/j.ymgme.2016.07.011.
131. Schulz A, Ajayi T, Specchio N, et al. Study of intraventricular cerliponase Alfa for CLN2 disease. *N Engl J Med.* 2018;378(20). https://doi.org/10.1056/nejmoa1712649.
132. Sondhi D, Hackett NR, Peterson DA, et al. Enhanced survival of the LINCL mouse following CLN2 gene transfer using the rh.10 rhesus Macaque-derived Adeno-associated virus vector. *Mol Ther.* 2007;15(3). https://doi.org/10.1038/sj.mt.6300049.
133. Sondhi D, Kaminsky SM, Hackett NR, et al. Slowing late infantile Batten disease by direct brain parenchymal administration of a rh.10 adeno-associated virus expressing CLN2. *Sci Transl Med.* 2020;12(572). https://doi.org/10.1126/SCITRANSLMED.ABB5413.
134. Walder RY, Shalev H, Brennan TMH, et al. Familial hypomagnesemia maps to chromosome 9q, not to the X chromosome: genetic linkage mapping and analysis of a balanced translocation breakpoint. *Hum Mol Genet.* 1997;6(9). https://doi.org/10.1093/hmg/6.9.1491.
135. Walder RY, Landau D, Meyer P, et al. Mutation of TPRM6 causes familial hypomagnesemia with secondary hypocalcemia. *Nat Genet.* 2002;31(2). https://doi.org/10.1038/ng901.
136. Schlingmann KP, Weber S, Peters M, et al. Hypomagnesemia with secondary hypocalcemia is caused by mutations in TPRM6, a new member of the TPRM gene family. *Nat Genet.* 2002;31(2). https://doi.org/10.1038/ng889.
137. Lomelino-Pinheiro S, Bastos M, de Lages AS. A novel TRPM6 variant (C.3179T > A) causing familial hypomagnesemia with secondary hypocalcemia. *Endocrinol Diabetes Metab Case Rep.* 2020;2020(1). https://doi.org/10.1530/EDM-20-0005.

138. Lainez S, Schlingmann KP, van der Wijst J, et al. New TRPM6 missense mutations linked to hypomagnesemia with secondary hypocalcemia. *Eur J Hum Genet.* 2014; 22(4). https://doi.org/10.1038/ejhg.2013.178.
139. Katayama K, Povalko N, Yatsuga S, et al. New TRPM6 mutation and management of hypomagnesaemia with secondary hypocalcaemia. *Brain Dev.* 2015;37(3). https://doi.org/10.1016/j.braindev.2014.06.006.
140. Anast CS, Mohs JM, Kaplan SL, Burns TW. Evidence for parathyroid failure in magnesium deficiency. *Science.* 1979;1972(177):4049. https://doi.org/10.1126/science.177.4049.606.
141. Hartnett KA, Stout AK, Rajdev S, Rosenberg PA, Reynolds IJ, Aizenman E. NMDA receptor-mediated neurotoxicity: a paradoxical requirement for extracellular Mg2+ in Na+/Ca2+-free solutions in rat cortical neurons in vitro. *J Neurochem.* 1997;68(5). https://doi.org/10.1046/j.1471-4159.1997.68051836.x.
142. Nardone R, Brigo F, Trinka E. Acute symptomatic seizures caused by electrolyte disturbances. *J Clin Neurol.* 2016;12(1). https://doi.org/10.3988/jcn.2016.12.1.21.
143. Schlingmann KP, Sassen MC, Weber S, et al. Novel TRPM6 mutations in 21 families with primary hypomagnesemia and secondary hypocalcemia. *J Am Soc Nephrol.* 2005; 16(10). https://doi.org/10.1681/ASN.2004110989.
144. Shalev H, Phillip M, Galil A, Carmi R, Landau D. Clinical presentation and outcome in primary familial hypomagnesaemia. *Arch Dis Child.* 1998;78(2). https://doi.org/10.1136/adc.78.2.127.
145. Astor MC, Løvås K, Wolff ASB, et al. Hypomagnesemia and functional hypoparathyroidism due to novel mutations in the Mg-channel TRPM6. *Endocr Connect.* 2015;4 (4). https://doi.org/10.1530/ec-15-0066.
146. Switon K, Kotulska K, Janusz-Kaminska A, Zmorzynska J, Jaworski J. Tuberous sclerosis complex: from molecular biology to novel therapeutic approaches. *IUBMB Life.* 2016;68(12):955–962. https://doi.org/10.1002/iub.1579.
147. Lam HC, Nijmeh J, Henske EP. New developments in the genetics and pathogenesis of tumours in tuberous sclerosis complex. *J Pathol.* 2017;241(2):219–225. https://doi.org/10.1002/path.4827.
148. Feliciano DM, Lin TV, Hartman NW, et al. A circuitry and biochemical basis for tuberous sclerosis symptoms: from epilepsy to neurocognitive deficits. *Int J Dev Neurosci.* 2013;31(7):667–678. https://doi.org/10.1016/j.ijdevneu.2013.02.008.
149. Jansen FE, van Huffelen AC, Algra A, van Nieuwenhuizen O. Epilepsy surgery in tuberous sclerosis: a systematic review. *Epilepsia.* 2007;48(8):1477–1484. https://doi.org/10.1111/j.1528-1167.2007.01117.x.
150. Franz DN, Belousova E, Sparagana S, et al. Long-term use of everolimus in patients with tuberous sclerosis complex: final results from the EXIST-1 study. *PLoS One.* 2016;11(6). https://doi.org/10.1371/journal.pone.0158476.
151. Ruffolo G, Iyer A, Cifelli P, et al. Functional aspects of early brain development are preserved in tuberous sclerosis complex (TSC) epileptogenic lesions. *Neurobiol Dis.* 2016;95:93–101. https://doi.org/10.1016/j.nbd.2016.07.014.
152. Fallah A, Weil AG, Wang S, Lewis E, Baca CB, Mathern GW. Cost-utility analysis of competing treatment strategies for drug-resistant epilepsy in children with Tuberous Sclerosis Complex. *Epilepsy Behav.* 2016;63:79–88. https://doi.org/10.1016/j.yebeh.2016.07.034.
153. Fallah A, Rodgers SD, Weil AG, et al. Resective epilepsy surgery for tuberous sclerosis in children: determining predictors of seizure outcomes in a multicenter retrospective cohort study. *Neurosurgery.* 2015;77(4):517–524. https://doi.org/10.1227/NEU.0000000000000875.
154. Jansen FE, Vincken KL, Algra A, et al. Cognitive impairment in tuberous sclerosis complex is a multifactorial condition. *Neurology.* 2007;70:916–923. www.neurology.org.

155. Morris G, O'Brien D, Henshall DC. Opportunities and challenges for microRNA-targeting therapeutics for epilepsy. *Trends Pharmacol Sci*. 2021;42(7):605–616. https://doi.org/10.1016/j.tips.2021.04.007.
156. Colasante G, Qiu Y, Massimino L, et al. In vivo CRISPRa decreases seizures and rescues cognitive deficits in a rodent model of epilepsy. *Brain*. 2020;143(3):891–905. https://doi.org/10.1093/brain/awaa045.
157. Hirose S, Tanaka Y, Shibata M, et al. Application of induced pluripotent stem cells in epilepsy. *Mol Cell Neurosci*. 2020;108. https://doi.org/10.1016/j.mcn.2020.103535.
158. Sterlini B, Fruscione F, Baldassari S, Benfenati F, Zara F, Corradi A. Progress of induced pluripotent stem cell technologies to understand genetic epilepsy. *Int J Mol Sci*. 2020;21(2). https://doi.org/10.3390/ijms21020482.
159. Hlebokazov F, Dakukina T, Ihnatsenko S, et al. Treatment of refractory epilepsy patients with autologous mesenchymal stem cells reduces seizure frequency: an open label study. *Adv Med Sci*. 2017;62(2):273–279. https://doi.org/10.1016/j.advms.2016.12.004.
160. Li M, Jancovski N, Jafar-Nejad P, et al. Antisense oligonucleotide therapy reduces seizures and extends life span in an SCN2A gain-of-function epilepsy model. *J Clin Investig*. 2021;131(23). https://doi.org/10.1172/JCI152079.
161. Burbano LE, Li M, Jancovski N, et al. Antisense oligonucleotide therapy for KCNT1 encephalopathy. *bioRxiv*. 2020;1–50. Published online.
162. Doudna JA, Charpentier E. The new frontier of genome engineering with. *CRISPR-Cas9*. 2002;346(6213):1–9. https://www.science.org.
163. Liu J, Gao C, Chen W, et al. CRISPR/Cas9 facilitates investigation of neural circuit disease using human iPSCs: mechanism of epilepsy caused by an SCN1A loss-of-function mutation. *Translational. Psychiatry*. 2016;6. https://doi.org/10.1038/tp.2015.203.
164. Balestrini S, Sisodiya SM. Pharmacogenomics in epilepsy. *Neurosci Lett*. 2018;667:27–39. https://doi.org/10.1016/j.neulet.2017.01.014.
165. Zhou SF, Liu JP, Chowbay B. Polymorphism of human cytochrome P450 enzymes and its clinical impact. *Drug Metab Rev*. 2009;41(2):89–295. https://doi.org/10.1080/03602530902843483.
166. Lee CR, Goldstein JA, Pieper JA. Cytochrome P450 2C9 polymorphisms: a comprehensive review of the in-vitro and human data. *Pharmacogenetics*. 2002;12(3):251–263. http://journals.lww.com/jpharmacogenetics.
167. Li M, Zhong R, Lu Y, Zhao Q, Li G, Lin W. Association between SCN1A rs2298771, SCN1A rs10188577, SCN2A rs17183814, and SCN2A rs2304016 polymorphisms and responsiveness to antiepileptic drugs: a meta-analysis. *Front Neurol*. 2021;11. https://doi.org/10.3389/fneur.2020.591828.
168. Bao Y, Liu X, Xiao Z. Association between two SCN1A polymorphisms and resistance to sodium channel blocking AEDs: a meta-analysis. *Neurol Sci*. 2018;39(6):1065–1072. https://doi.org/10.1007/s10072-018-3308-3.
169. Skarpaas TL, Jarosiewicz B, Morrell MJ. Brain-responsive neurostimulation for epilepsy (RNS® System). *Epilepsy Res*. 2019;153:68–70. https://doi.org/10.1016/J.EPLEPSYRES.2019.02.003.
170. Heck CN, King-Stephens D, Massey AD, et al. Two-year seizure reduction in adults with medically intractable partial onset epilepsy treated with responsive neurostimulation: final results of the RNS System Pivotal trial FULL-LENGTH ORIGINAL RESEARCH. *Epilepsia*. 2014;55(3):432–441. https://doi.org/10.1111/epi.12534.
171. Nair DR, Laxer KD, Weber PB, et al. Nine-year prospective efficacy and safety of brain-responsive neurostimulation for focal epilepsy. *Neurology*. 2020;95(9):e1244–e1256. https://doi.org/10.1212/WNL.0000000000010154.

172. Wyckhuys T, Raedt R, Vonck K, Wadman W, Boon P. Comparison of hippocampal Deep Brain Stimulation with high (130Hz) and low frequency (5Hz) on afterdischarges in kindled rats. *Epilepsy Res.* 2010;88(2–3):239–246. https://doi.org/10.1016/j.eplepsyres.2009.11.014.
173. Zangiabadi N, Ladino LDi, Sina F, Orozco-Hernández JP, Carter A, Téllez-Zenteno JF. Deep brain stimulation and drug-resistant epilepsy: a review of the literature. *Front Neurol.* 2019;10(JUN). https://doi.org/10.3389/fneur.2019.00601.
174. Fisher R, Salanova V, Witt T, et al. Electrical stimulation of the anterior nucleus of thalamus for treatment of refractory epilepsy. *Epilepsia.* 2010;51(5). https://doi.org/10.1111/j.1528-1167.2010.02536.x.
175. Velasco F, Velasco M, Jiménez F, et al. Predictors in the treatment of difficult-to-control seizures by electrical stimulation of the centromedian thalamic nucleus. *Neurosurgery.* 2000;47(2). https://doi.org/10.1097/00006123-200008000-00007.
176. Li MCH, Cook MJ. Deep brain stimulation for drug-resistant epilepsy. *Epilepsia.* 2018;59(2):273–290. https://doi.org/10.1111/epi.13964.
177. Goyal A, Goetz S, Stanslaski S, et al. The development of an implantable deep brain stimulation device with simultaneous chronic electrophysiological recording and stimulation in humans. *Biosens Bioelectron.* 2021;176. https://doi.org/10.1016/j.bios.2020.112888.

CHAPTER SEVEN

EGFR, NF-κB and noncoding RNAs in precision medicine

Yanli Li, Xiaomin Liu, and Zhongliang Ma*
Lab for Noncoding RNA & Cancer, School of Life Sciences, Shanghai University, Shanghai, China
*Corresponding author: e-mail address: zlma@shu.edu.cn

Contents

1. Part 1: Epidermal growth factor receptor (EGFR) 191
2. EGFR as a target in lung cancer treatment 192
3. Acquired resistance of EGFR TKIs 192
   3.1 First-generation EGFR TKIs 193
   3.2 Second-generation EGFR TKIs 193
   3.3 Third-generation EGFR TKIs 193
4. Precision medicine of EGFR in lung cancer 194
5. Part 2: Nuclear factor-κappa B signaling 195
6. NF-κB signaling pathway and its activation 195
   6.1 Canonical NF-κB pathway 196
   6.2 Non-canonical NF-κB pathway 196
   6.3 EGFR and NF-κB 196
   6.4 Noncoding RNAs 197
7. Part 3: Noncoding RNAs and EGFR in precision medicine 197
   7.1 miRNAs in EGFR-targeted therapy 197
   7.2 LncRNAs and EGFR-TKIs 199
   7.3 tRNA-derived RNA fragments and EGFR 200
   7.4 Noncoding RNAs and NF-κB signaling in cancer 200
   7.5 p50 and p65 200
   7.6 TNF-α 201
   7.7 IκB 203
   7.8 MyD88 203
   7.9 CBM complex 204
   7.10 TAK1 204
   7.11 Other ncRNAs and NF-κB signal 204
   7.12 NF-κB signaling in chronic inflammation to cancer 205
8. Conclusion and future perspectives 207
References 209

Progress in Molecular Biology and Translational Science, Volume 190
ISSN 1877-1173
https://doi.org/10.1016/bs.pmbts.2022.05.001

Abstract

Cancer has become a major disease that threatens human health because of its high morbidity and mortality. With the development of precision medicine, the efficiency of tumor treatment has been significantly improved. More attention has been paid to targeted therapy and immunotherapy as the key to precision treatment of cancer. Targeting epidermal growth factor receptor (EGFR) has become one of the most important targeted treatments for various cancers. NF-kb signal play an important role in carcinogenesis. Non-coding RNAs (ncRNAs) were found to a new player in carcinogenesis, treatment and regulation of the efficacy of anticancer drugs in the past few years. Therefore, in this chapter, we aim to summarize roles of EGFR, NF-κB and regulatory functions of non-coding RNAs in cancer treatment.

Abbreviations

EGFR	Epidermal growth factor receptor
AR	Amphiregulin
RTK	Receptor of tyrosine kinase
TKIs	Tyrosine kinases inhibitors
TK	Tyrosine kinases
TGF-α	Transforming growth factor-α
mPFS	median progression-free survival
EMT	Epithelial mesenchymal transition
NF-κB	Nuclear Factor-κappa B
RHD	Rel homology domain
IκB	inhibitor of κB
NLS	Nuclear localization signal
TADs	Transactivation domains
TNFR	Tumor necrosis factor receptor
TLR	Toll-like receptor
IL-1R	Interleukin-1 receptor
BAFR	B cell-activating factor receptor
LTβR	lymphotoxin β receptor
RANK	Receptor activator of NF-κB
NIK	NF-κB inducing kinase
STAT	Signal transducer and activator of transcription
miRNAs	MicroRNAs
lncRNAs	Long non-coding RNAs
tRFs	tRNA-derived RNA fragments
TFR	Transferrin receptors
TNF-α	Tumor necrosis factor-α
IKK	Inhibitors of NF-κB kinase
ESC	Endometrial stromal cells
IL-8	Interleukin-8
MyD88	myeloid differentiation factor 88
TAK1	TGFβ-activated kinase-1
HCC	Hepatocellular carcinoma
NKILA	NF-κB interacting lncRNA
ESCC	Esophageal squamous cell carcinoma

ciRS-7	miR-7 sponge circRNA
BBB	Blood-brain barrier
IL-1	Interleukin 1
MCP-1	Monocyte chemoattractant protein

1. Part 1: Epidermal growth factor receptor (EGFR)

Epidermal growth factor receptor (EGFR) is one of the four members of the HER family receptors, which comprise EGFR/HER1/erbB1, HER2/erbB2, HER3/erbB3, and HER4/erbB4. There are 11 species of growth factors in the HER family which can be broadly divided into those that specifically bind with EGFR (EGF, TGF-α, Amphiregulin (AR)), those that binding with EGFR and HER4 (BTC, HB-EGF, Epidermal regulators), and those that binding with HER3 and HER4 (Neuregulin). Although HER2 has no corresponding ligand, it usually binds to a ligand similar to the one that activates it, readily forming a dimer with other members of the HER family. Additionally, EGFR is a receptor of tyrosine kinase (RTK). It is consisted of a C-terminus intracellular region that possesses the kinase activity, and an N-terminus extracellular ligand-binding site, a hydrophobic transmembrane domain.[1] The EGFR signaling network plays a significant role in the epithelial tissues maintenance and growth, active EGFR signaling is frequently observed in lung cancer, and EGFR level is related with advanced stage of disease and bad prognosis.[2] HER family receptor related malformation from primary lung tumors and NSCLC brain metastases have some striking differences.[3] Therefore, in the development of new drugs for cancer treatment, EGFR and its signaling components can be used as targets, as with chimeric monoclonal antibodies (panitumumab and cetuximab)[4,5] and tyrosine kinases inhibitors (TKIs) (gefitinib, erlotinib, and afatinib).[6–8] However, to date, cancer heterogeneity and drug resistance greatly limit the usefulness of anti-EGFR agents.[9]

The EGFR gene is an oncogene-driven gene, with tyrosine kinases (TK) activity.[10] Meanwhile, discovered in 2004, EGFR mutation (L858R) was the first molecular alteration in lung cancer that is shown to confer sensitivity to specific targeted therapies, namely TKIs.[11–14] EGFR-TKIs can inhibit EGFR autophosphorylation activation and its downstream signaling pathways through competitive binding with the EGFR binding region, preventing the binding of ATP and EGFR receptors.

These molecular alterations could be influential in lung cancer and have important implications in lung cancer treatment.[15] Currently, none of the recurrent molecular alterations that are commonly altered in lung squamous cell carcinoma have proven to be as predictive for response to therapy as EGFR alterations in lung adenocarcinoma. Despite the establishment of tumor profiling for lung adenocarcinomas, its clinical benefits for other histologic subtypes of lung cancer, such as lung SCC and SCLC, remain unclear. Furthermore, a significant minority of patients with NSCLC have activating mutations in EGFR[16] and typically develop resistance within 9–12 months. However, anti-EGFR antibody therapy is used extensively in treatment cancer.[17]

2. EGFR as a target in lung cancer treatment

Many studies have highlighted the relationship between EGFR and cancers. High expression of EGFR correlates with poor survival in cancer of the head and neck, as well as in cervical and bladder carcinomas.[17–19] In NSCLC, EGFR as a prognostic factor remains disputable. Some studies have verified that EGFR over-expression is predictive of a poor result in NSCLC,[20–24] whereas others don't.[24–26]

In the progression of NSCLC, EGFR, regarded as a cancer driver gene,[27,28] shares numerous characteristics with other oncogenes, including stimulating DNA synthesis, cell cycle, cell proliferation, cell metastasis and invasion,[29,30] and has been proposed as an attractive and promising target for anti-cancer treatment.[31] The expression of high levels of EGFR can also be up-regulated by co-expression of receptor ligands (such as transforming growth factor-α (TGF-α) or EGF). The duration of EGFR signaling is stimulated by the positive feedback loop formed by EGFR, a ligand-releasing protease, and the RAS-MAPK signaling pathway. In spite of low levels expression of EGFR, the signaling response can be prolonged in a cell that is efficient in recapturing the endogenous ligand.[32] Research in breast cancer has shown that co-expression of TGF-α and EGFR had a substantial effect on survival compared to the co-expression of HER-2 and EGFR.[33] In lung cancer, one study has reported that synchronous expression of EGFR and HER-2 is predictive of increased recurrence risk. Moreover, success with PD-L1 blockers suggests that activation of the immune response may be strategy for treatment of NSCLC.[34]

3. Acquired resistance of EGFR TKIs

In lung cancer, especially in NSCLC, a variety of EGFR mutations exist that are closely linked to tumor development. The discovery of activating mutations in EGFR and the eventual approval of TKIs are milestones in the history of NSCLC treatment.

3.1 First-generation EGFR TKIs

Gefitinib and Erlotinib have greatly improved progression-free survival versus standard chemotherapy for EGFR-positive NSCLC. Yet, due to the emergence of resistance, disease progression eventually occurs in almost all patients. Quinazoline has the core skeleton of gefitinib and erlotinib combined with EGFR reversibility. However, almost all patients who received gefitinib/erlotinib treatment eventually and inevitably acquired resistance, with a median progression-free survival (mPFS) of approximately 9–11 months.[35] The most common mechanism of acquired resistance is point mutation in exon 20 of T790M, with an incidence of approximately 50–60%.[36] The threonine (T) at site 790 of EGFR is located at the ATP binding site of the C-terminus. The T790M results in the substitution of the relatively small, polar Thr residue with the much larger, nonpolar Met residue. This results in Coulombic changes and steric hindrance, which interferes with binding of EGFR TKIs to the receptor. Other mechanisms of resistance include MET gene amplification,[37] small cell transformation,[38] and epithelial mesenchymal transition (EMT).[39]

3.2 Second-generation EGFR TKIs

The second-generation EGFR TKIs were designed to overcome gefitinib and erlotinib resistance, and mainly included BIBW 2992 (afatinib), PF00299804 (dacomitinib), HKI-272 (neratinib), and CI-1033 (canertinib). These have greater potential to overcome or delay resistance[40] because they have better binding constants and inhibitor geometry, so they fit the receptor better. Currently it is unclear whether or not they can overcome resistance to first-generation EGFR TKIs entirely. Recent discoveries based on these principles continue to inspire the next generation of innovative clinical trials in these diseases.

3.3 Third-generation EGFR TKIs

To date, acquired resistance to EGFR-TKI is an unavoidable process and usually appears after 10–12 months of therapy. New EGFR-TKIs with specific capability bind to T790M mutated receptor have been developed and successfully tested in patients with acquired resistance.[41,42] Moreover, the emerging third-generation EGFR TKIs have demonstrated high tolerability.. The most famous of the third generation TKIs is AZD9291 (osimertinib). AZD9291 overcomes T790M drug resistance. However, some patients showed resistance to this drug because of a C797S mutation. Reasons for drug resistance also include C-MET amplification, small cell lung cancer transformation, and activation of downstream genes (KRAS or BRAF). It is possible that efficacious treatment may be attainable by using a combination of the first and third generation EGFR-TKIs. In fact, Osimertinib, HM61713 (olmutinib), CO-1686 (rociletinib) and others (ASP8273, EGF816) are the subjects of several clinical trials and AZD9291 has already obtained FDA and EMA approval for the therapy of EGFR mutant T790M-positive NSCLC.[43]

In the future, when EGFR-TKI drug resistance occurs, genetic testing could be used to select the treatment method corresponding to the resistance mechanism.

4. Precision medicine of EGFR in lung cancer

Precision medicine in the treatment of lung cancer has dramatically impacted diagnostic pathology, and precision medicine provides a better understanding of the mechanisms of disease at the molecular level.

In 2004 and 2007, EGFR mutations and ALK-rearrangements that confer sensitivity to TKIs were the first molecular alterations in lung adenocarcinoma found and heralded the initiation of the era of precision medicine for lung cancer.[12,44] Among EGFR mutations, the $EGFR^{T790M}$ mutation was a clear target for drug development. The remarkable responses to TKIs observed in patients, as well as the discoveries made studying these molecular subsets of lung cancer, served as catalysts for further exploration of the lung cancer genome, leading to the incorporation of molecular testing in routine clinical practice. Clinical trials have revealed that treatment of advanced EGFR lung cancers is superior to chemotherapy, using the appropriate TKIs.[16,45] Conversely, it has also been shown that patients with mutant EGFR lung cancers rarely respond to EGFR TKIs and are more likely to benefit from chemotherapy, underlining the importance of matching tumor genotype to therapy.[45] Here, we show the timeline of the history of EGFR development. Continuing efforts to find new types of precision medicine for lung cancers is necessary to improve outcomes for patients with lung cancer, as well as biomarker-driven clinical trials. In lung cancer, the impact of precision medicine has resulted in considerable changes, as well as challenges, in diagnostic pathology.[46]

5. Part 2: Nuclear factor-κappa B signaling

Cancer is a malignancy that involves aberrant oncogenes or activation of pre-existing oncogenes.[47,48] Many studies have shown that the occurrence and development of tumors are related to inflammation, and the activation of inflammation-related signaling pathways plays vital roles in tumor microenvironment.[49] NF-κB signaling pathway activation promotes the transcription of inflammatory factors, chemokines, adhesion molecules, growth factors and other related genes such as EGFR, which in turn leads to tumor development.[50–52]

The transcription factor NF-κB, a nuclear factor that binds to the enhancer of the immunoglobulin kappa light-chain of activated B cell tumors, was discovered in 1986.[53,54] NF-κB also has been found to be closely linked with cancer development and progression. It induces expression of many genes involved in developmental processes, the inflammatory response, immunoreaction, as well as cellular proliferation, apoptosis, and angiogenesis.[55–57]

Non-coding RNA and NF-κB act as crucial factors in the development of cancer, both of which can be applied as tumor markers and drug targets in clinical therapy.[58]

6. NF-κB signaling pathway and its activation

The NF-κB transcription factor family comprises five main elements, viz. Rel (c-Rel), RelB, p65 (RelA, NF-κB3), p50 (NF-κB1) and p52 (NF-κ B2), all of which have an N-terminal region of ~300 amino acids termed the Rel homology domain (RHD). RHD contains the binding domain ofDNA, dimerization domain and the inhibitor of κB (IκB) binding domain besides the nuclear localization signal (NLS). p65, c-Rel and RelB possess C-terminal transactivation domains (TADs) that endow the proteins with the ability to activate NF-κB, thereby positively modulating its expression. While p50 and p52 lack the transcription activation domain, their homodimers can inhibit transcription.[59] These family members can form many homodimers or heterodimers, among which p50/p65 heterodimers, that play a major role in physical function, are the earliest found.[60,61] Different NF-κB/Rel dimers have different binding sites, giving them diverse properties. As a consequence, different target genes are distinctively induced by different NF-κB dimers and play different functions.

Activation of NF-κB has three forms: release from IκB molecules, cleavage of the inhibitory ankyrin repeat domains of p100 and p105, activation of IKK through the CBM (CARMA1, BCL10 and MALT1) complex. In general, NF-κB signaling is split into canonical and the non-canonical pathways.

6.1 Canonical NF-κB pathway

(1) In the cytoplasm, NF-κB will bind to IκB as trimers when inert. (2) Upon recognition of external stimulation such as bacteria, viruses, parasites, inflammatory cytokines and fibroblast growth factors, the tumor necrosis factor receptor (TNFR), Toll-like receptor (TLR) and interleukin-1 receptor (IL-1R) will be activated, to which MyD88 and IRAK will bind and pass the activation signal to downstream proteins as important adaptors. Consequently, the downstream proteins TRAF2, NIK and RIP activate IKK, leading to ubiquitination, IκB phosphorylation, and further release of NF-κB dimers with the dissociated IκB molecules being degraded by proteasomes. (3) The NF-κB dimers enter the nucleus and bind to the κB domains of target genes, altering gene expression.

6.2 Non-canonical NF-κB pathway

The non-canonical NF-κB pathway is composed of IκB family members, especially p100 and p105. Upon stimulation, the B cell-activating factor receptor (BAFR), lymphotoxin β receptor (LTβR), receptor activator of NF-κB (RANK) and CD40 will be activated, as a consequence of which p100 will be phosphorylated by the downstream NF-κB inducing kinase (NIK), eventually leading to the ubiquitination of p100 and its degradation into p52. The free p52-RelB dimers will enter the nucleus and bind to the target genes of the κB domain.

6.3 EGFR and NF-κB

As a vital regulator, EGFR has a pivotal effect on the NF-κB signaling pathway and cancer. In some ways, EGFR and NF-kB are partners in cancer.[62,63] For example, EGFR can indirectly or directly activate signal transducer and activator of transcription (STAT) signal.[64] NF-κB-mediated ERK reactivation is observed in EGFR-mutant NSCLC patients who gained gefitinib resistance.[65] Studies have elucidated that NF-κB is activated by EGFR activation, as well as dependent on NF-κB pathways underlying resistance to EGFR inhibitors.[66–68] In addition, environmental factors also can affect EGFR expression through regulating NF-κB.[69] We and others have proved that noncoding RNAs play important roles in carcinogenesis and are involved in regulating EGFR and NF-κB.

6.4 Noncoding RNAs

MicroRNAs (miRNAs) are a class of endogenous regulatory non-coding RNAs involved in cell carcinogenesis that are about 20–25 nucleotide (nt) in length.[70] They can bind to the 3′-UTR of protein coding mRNAs, typically leading to mRNA degradation. Alternatively, they can also cause repression of target mRNA translation.[71] miRNA cleavage is related to cancer progression and clinical features, so as an important regulator of life processes, miRNAs have an intimate relationship with tumors.[72]

LncRNA is a type of non-coding RNAs with a length >200 nt. It has been shown that lncRNA participate in a variety of biological processes, such as gene regulation, epigenetic regulation, cell cycle regulation and cell differentiation regulation.[73,74] Therefore, a more in-depth study of the functions and mechanisms involving lncRNA will help reveal their role in tumorigenesis and their importance as a primary tool for diagnosis and cancer treatment.

7. Part 3: Noncoding RNAs and EGFR in precision medicine

7.1 miRNAs in EGFR-targeted therapy

miRNAs can functionally impact cell fate determination through the regulation of critical protein expression, thus playing a pivotal role in the diverse processes of human cancer, acting as either tumor oncogenes or suppressors.[75–78] Recently, increasing numbers of miRNAs have been correlated with the drug resistance oflung cancer cells to anti-EGFR agents, indicating that miRNAs may serve as novel targets or promising predictive biomarkers for anti-EGFR therapy. Moreover, miRNA-based therapy has been suggested to be a rational and potentially effective approach for the therapeutic targeting of EGFR.[79] Many miRNAs, such as miR-200a, miR-27a/27b, miR-133a, and miR-134 have been verified to directly target EGFR[79–83] (Table 1). These EGFR-miRNAs regulation network studies demonstrated that miRNA-based therapy could possibly be utilized to target EGFR, except for TKIs and classical mAbs for EGFR-targeted therapies.[79]

Our lab has confirmed that miR-34a, miR-181a-5p, miR-32, and miR-486-5p play vital roles in the progression of NSCLC.[102–104] Our

Table 1 miRNAs that target EGFR signaling pathway involved in cancer.

miRNAs	Carcinomas	Biological effect	Regulation	Citations
miR-200	Anaplastic thyroid cancer/bladder cancer	Regulate EMT and reverse resistance of EGFR therapy	Down	84,85
miR-23b/27b	Bladder cancer	Regulate EGFR and suppress cancer	Down	80
miR-27a	renal cell carcinoma	Suppress human RCC cell proliferation and induce cell apoptosis	Down	81
miR-133a	NSCLC	Suppresses multiple oncogenic membrane receptors and cell invasion	Down	82
miR-134	NSCLC	Inhibit proliferation	Down	83
miR-7	Various cancer cells	Inhibit EGFR-PI3K-AKT signaling and reverse radio resistance	Down	86
miR-34a	Solid cancer	Regulate Axl receptor tyrosine kinase by targeting SIRT1 and MEK1	Down	87
miR-145	Lung cancer	Negatively regulate EGFR expression	Down	88
miR-146a	NSCLC	Inhibit EGFR in NSCLC cancer cells	Down	89,90
miR-146b-5p	Glioblastoma	Suppress EGFR expression	Down	91
miR-206	Squamous lung cancer	Suppress EGFR signaling	Down	92
miR-135a-1	Prostate cancer	Inhibit cell growth and migration	Down	93
miR-133a	NSCLC	Suppress EGFR signaling	Down	82
miR-133b	NSCLC	Suppress EGFR pathway signaling and enhance susceptibility to EGFR-TKI	Down	94
miR-1203,1237,541,542-5p	Human lung cancer	Downregulate EGFR	Down	86,95
miR-199a-3p	Prostate cancer	Suppress the expansion and tumor	Down	96
miR-2861	Cervical cancer	Inhibit tumor growth	Down	97
miR-25	Lung cancer	Upregulate EGFR	Up	98
miR-24		Activates EGFR signaling	Up	99
miR-21	Glioblastoma	Regulate the EGFR/AKT pathway in a PTEN independent manner	Up	100
miR-34a	NSCLC	EGFR	Down	101

recent studies demonstrated that miR-34a and miR-146-5p can suppress NSCLC by directly targeting EGFR *in vitro* and *in vivo*. These findings demonstrate that altered miRNA expression may be related to the oncogenesis of lung cancer. Some studies have verified that EGFR mutations can be regulated by miRNA in cancer therapies. At present, small RNA possesses the best potential as a diagnostic biomarkers and therapeutic drug for cancer, which is the most mature miRNA.[105] For instance, because of its pivotal role in lung cancer, liver cancer, and breast cancer among others, miR-34a is considered to be the most likely of the miRNAs to become a diagnostic marker and target of drugs.[106–108]

In conclusion, the emerging role of miRNAs as regulators could not only regulate EGFR signaling, but also the lung cancer cells resistance to anti-EGFR therapy. Additionally, miRNAs could also be employed as novel therapeutic targets to circumvent the resistance of lung cancer cells to EGFR inhibitors and as biomarkers for response to anti-EGFR agents.

7.2 LncRNAs and EGFR-TKIs

Long non-coding RNAs (lncRNAs) are involved in a number of biological processes, such as chromatin modification, gene regulation, transcription activation and interference, and cellular processes, including cell apoptosis, migration, tumor invasion, metastasis, and drug resistance.[109–111] They play vital roles in tumor occurrence and development.

lncRNAs can bind to transcription sites and regulate the expression of genes, whereas coding genes and miRNAs have no such functions.[112] This suggests that lncRNAs may be better epigenetic regulators of gene expression regulation. Research has shown that some lncRNAs, including lncRNA UCA1, H19, BC200, and BC087858, were increased in gefitinib-resistant human lung cancer cells, as determined by lncRNA microarray analysis.[113] Studies have also demonstrated that lncRNA UCA1 may stimulate non-T790M acquired resistance for EGFR-TKIs by activating the AKT/mTOR pathway and EMT.[114] Moreover, over-expression of lncRNA BC087858 could act via a novel mechanism by which acquired resistance for EGFR-TKIs can develop in EGFR-mutant NSCLC patients without T790M mutation.[115] Another study showed that stimulation of the PI3K/AKT and MEK/ERK pathways, as well as EMT, could be involved in the resistance to EGFR-TKIs.[116,117]

In conclusion, further exploration of the function and mechanism of lncRNAs will reveal their critical role in the process of lung cancer generation, and their significance as a diagnostic tool as well as in the treatment of cancer.

7.3 tRNA-derived RNA fragments and EGFR

A close connection has been established between cancers and a variety of small non-coding RNAs, such as miRNAs, piRNAs, and circRNAs, but not including tRNA-derived RNA fragments (tRFs). tRFs, a class of small RNAs, are non-coding single-stranded RNAs 14–35 nt in length that always derive from the 5′ end or 3′ end of tRNA in a particular cleavage site.[118,119] The length and the generation of tRFs are very similar to those of miRNAs. Through the action of anticodon-cleaving enzymes, some tRNA cleavages can be generated from mature tRNAs. The biological function of tRFs is still unclear. Some studies have shown that, in biological processes, tRFs have the capacity to regulate some cellular processes, including translational efficiency under stress conditions,[120] oncogenic transformation,[121] and mitochondrial-mediated apoptosis.[122] EGFR and transferrin receptors (TFR) are known to be involved in cell growth and to be expressed in normal human epidermis. Currently, our understanding of EGFR and TRF expression in human cancer is incomplete.

However, the biological functions of tRFs have been the focus of many research efforts. Some studies have shown that, in biological processes, the function of 5′ tRFs might extend much farther than that of 3′ tRFs.[123,124] In addition, it has been found that tRFs suppress breast cancer progression via YBX1 displacement.[125] Furthermore, some tRFs can active cellular functions through argonaute engagement, such as cell proliferation and RNA silencing.[126–128] TRF expression is not associated with cellular proliferation in embryonic and fetal epidermis, whereas EGFR appears to be correlated with proliferation and undifferentiated cells.[129]

7.4 Noncoding RNAs and NF-κB signaling in cancer

In tumorigenesis, abnormal expression of noncoding RNA is usually accompanied by the activation of NF-κB signaling pathway. Noncoding RNAs can regulate the NF-κB pathway through translocating themselves to some upstream signal molecules or NF-κB members. Meanwhile, NF-κB pathway can also regulate the expression level of noncoding RNAs by synthesis of proteins.[130] Below, we briefly review key aspects of miRNAs and NF-κB.

7.5 p50 and p65

The precursor of p50, p105, was a target protein of miR-9. Expression of miR-9 is controlled by TNF-a and IL-1B. It also suggests that miR-9 will function as a trimming agent in the anti-inflammatory responses to hinder the negative control by p50 homodimers in some monocytes. miR-9 can inhibit the development of ovarian cancer and gastric cancer through regulation of the NF-κB pathway.[131–133] Moreover, researchers have discovered that miR-143 can function as the transcription target of NF-κB and promote metastasis in lung cancer.[74,134] It also is know that over-expression of miR-143 will decrease the viability of the cells and have an effect on the expression of p65 in the NF-κB pathway.[135] Many miRNAs have a direct or indirect regulatory function on p50/p65. This can explain why NF-κB doesn't work alone, but instead forms part of a network with a variety of miRNAs.

Studies have shown that miR-145-5p was a tumor inhibitor,[136] by virtue of regulation of the NF-κB/YY1 pathway, whose aberrant expression can lead to the development of rhabdomyosarcoma.[137] In addition, Mott *et al.* also verified that the inhibitory effect NF-κB has on miR-29b exists in biliary duct and cholangiocarcinoma.[138] Certainly, a series of miRNAs such as miR-125b-1, miR-125a-5p, miR-494, miR-130b are under the regulation of NF-κB[139–143] (Table 2), which determines the mode of its effects on the expression of several other genes, and thus its function.

7.6 TNF-α

Tumor necrosis factor-α (TNF-α) is a secreted proinflammatory cytokine. Chen *et al.* found that miR-27a could protect human mitral valve interstitial cells via TNF-α.[139–143] Over-expression of miR-27a protected hMVICs from TNF-α-induced cellular injury, indicating that miR-27a plays as a vital role in regulation of TNF-α release. To date, miR-27a is considered as an important negative regulatory factor in human inflammation, which has a close connection with TNF-α.

Additionally, miR-146a has been shown to have a connection with TNF-α in gastric diseases and psoriasis patients.[164,165] MiR-146a can increase the secretion of TNF-α, which is confirmed by experiments. Compared with the control group, the overexpression of miR-146a significantly was up-regulated the expression level of TNF-α in a gastritis group, while that of the gastric cancer group was significantly reduced.

Thus, TNF-α and NF-κB have an intimate interaction and miRNAs have an essential role in regulation. Some miRNAs interact with TNF-α in inflammation, while TNF-α is a vital component in NF-κB pathway regulation.[166,167] Some miRNAs regulate the expression and secretion of TNF-α indirectly through interacting with important members of NF-κB pathway, like p50 and p65.[168]

Table 2 miRNAs and NF-κB signaling pathway.

miRNA	Target gene	Disease/cancer	Function	Reference
miR-302e	RelA	Allergic inflammation	Anti-inflammatory	144
miR-194 miR-195	TRAF6 IKKα	Intervertebral disc degeneration Wilms tumor	Anti-inflammatory Proliferation, apoptosis	145 146
miR-199a-3p	IKKβ	Cystic fibrosis	Chronic pulmonary inflammation	147
miR-23a	IKKβ	Articular chondrocytes	Anti-inflammatory	148
miR-429	IKKβ	Cervical cancer	Proliferation, apoptosis	149
miR-7	Myd88	Eriocheir sinensis	Enhance white spot syndrome	150
let-7b	IKKε	Human glioma	Invasion, migration	151
miR-342-3p	IKK-γ	Hepatocellular carcinoma	Proliferation	152
miR-503	IKKβ	NSCLC	Metastasis	108
miR-451	IKKβ	Hepatocellular carcinoma	Proliferation	153
miR-199a	IKKβ	Endometrial stromal cell (ESC)	Cell adhesion, migration, invasion	154
miR-194 miR-502-5p	TRAF6 TRAF2	Nucleus pulposus cells of the intervertebral disc Chondrocyte injury	Anti-inflammatory Proliferation, apoptosis	145 155
miR-18a-5p	IRF2	NSCLC	Proliferation, metastasis, apoptosis	156
miR-708-3p	STAT3	Idiopathic pulmonary fibrosis	Inhibit fibrosis	57
miR-411	STAT3	Cervical cancer	Cell proliferation and invasion	157
miR-29a	STAT3	Human retinoblastoma	Cell proliferation, migration, invasion, apoptosis	158
miR-124	STAT3	NSCLC	Proliferation, apoptosis	159
miR-181b	STAT3	Atherosclerosis	Proliferation, apoptosis, cell cycle	160
miR-26b	TAK1	Hepatocellular carcinoma	Apoptosis	161
miR-20a	TAK1	Osteosarcoma	Proliferation	162
miR-23b	TRAF1	Sepsis	Proliferation Immunosuppression	163

7.7 IκB

The regulation of IκB by miRNAs is not direct, but through the activation of inhibitors of NF-κB kinase (IKK). The IKK family members mainly include IKKα, IKKβ, IKKγ and IKKε. Firstly, miRNAs bind with the IKK complex to activate IKK, causing phosphorylation of IκB, which then dissociates from the p50/p65/IκB trimers and is ubiquitinated. Then, NF-κB switches from an inert to an activated state, activating a series of subsequent reactions.[169]

Studies have shown that during the differentiation of human macrophages, the expression levels of miR-223, miR-23a, miR-15a and miR-16 decreased rapidly, which directly lead to the increase of IKKα expression and cause elevated expression level of p52.

Over-expression of p52 will strongly inhibits the level of expression of target genes in the NF-κB pathway. When miRNAs expression decreased, the activation of macrophages was apparently inhibited.[35,148] In tumors with NF-κB activity, miR-199a targets IKKβ and inhibits its expression in the endometrial stromal cells (ESC).[154] As the IKKβ expression decreases, miR-199a suppresses the activation of NF-κB pathway and the synthesis of interleukin-8 (IL-8), which means miR-199a can reduce the invasiveness of ESCs through inhibiting the NF-κB pathway.[154,170] IKKε also played a contrasting role in the activation of NF-κB pathway. A series of experiments have verified that miR-155 is involved in the negative-feedback loop through down-regulation of IKKε, IKKβ and other target genes that produce transcription factors for NF-κB.[171] Additionally, through building the regulatory feedback loop of IKBKε/YAP1/miR-Let-7b/I, IKKε promotes glioblastoma progression.[172]

7.8 MyD88

In ovarian cancer, d'Adhemar *et al.* found that miR-146a and miR-21 bind to the mRNA of myeloid differentiation factor 88 (MyD88), influencing the regulatory function of MyD88 in TLR4 pathway, whose activation does not directly promote the activation of downstream NF-κB. Meanwhile, they found that the regulatory function of miR-146a and miR-21 on MyD88 is not limited only to the activation of TLR4 and NF-κB pathway, but also in regulating the expression level of MyD88 to directly influence the sensitivity of ovarian cancer cells to pharmacotherapy.[173] miR-200b and miR-200c are important in the activation of the TLR4 and NF-κB pathways via binding with MyD88.[174] They observed that after transfecting miR-200b and miR-200c into HEK293-TLR4 cells, the expression of MyD88 decreases, while other target genes like IRAK-1 and TRAF-6 have no effect. This showed that miR-200b and miR-200c can influence the TLR4 pathway through regulation of MyD88, which has a vital role in defenses against microbial pathogens.[174,175]

In short, miRNAs can affect NF-κB efficiently through MyD88, which has a crucial role in the biogenesis and development of inflammation and tumors.

7.9 CBM complex

The CBM complex modulates nuclear transcription factor activation such as NF-κB, which is initiated by antigen receptor activation. CARMA1 is a cytoskeletal protein that changes its conformation after antigen receptor activation and recruits downstream signaling pathway components BCL10 and MALT1. The mutual recognition of CARMA1 and the linker protein BCL10 is mediated by the CARD domain. CARD, BCL10, and MALT1 form a stable complex, termed CBM complex, which also has a regulatory effect on miRNAs. The CBM complex also has a regulatory effect on miRNA. Recent studies have shown that miR-181d acts as a tumor suppressor through targeting MALT1 to suppress glioblastoma growth through NF-κB pathways.[176] Additionally, miR-26 restrains TNF-α/NF-κB signaling and IL-6 expression by silencing MALT1.[177] Gu and Sun found that miR-539 can regulate migration and invasion through targeting CARMA1 in human thyroid cancer.[178]

7.10 TAK1

Zhao *et al.* showed that miR-26b can inhibit the activation of the NF-κB pathway by targeting TGFβ-activated kinase-1 (TAK1) and TAB3. They observed that in the hepatocellular carcinoma (HCC) cell line QGY-7703 and MHCC-97H, miR-26b can bind directly to the 3′-UTR of TAK1 and TAB3 to suppress their expression, inhibiting the TNF-α induced NF-κB pathway, eventually enhancing the sensitivity of lung cancer cells to pharmacotherapy.[161] TAK1 has a special and important position in the NF-κB pathway because it can bind to TAB1 to activate NF-κB-induced kinase and IκB kinase.[179]

7.11 Other ncRNAs and NF-κB signal

LncRNA has been found to play a vital part in pathological and physiological processes, containing tumor formation and metastasis. Different lncRNAs have different molecular mechanisms that play different biological functions.[180]

Tumor-associated NF-κB activation and lncRNA over-expression are most directly related to the inhibition of IκB, which acts as a negative regulator of NF-κB signaling. Liu *et al.* found that NKILA (NF-κB interacting lncRNA) binds to NF-κB, which is up-regulated by inflammatory factors in breast cancer. NKILA inhibits the activation of NF-κB signaling pathway by masking the location phosphorylation of IκB for IKK phosphorylation suppression. Further studies revealed that there are two hairpin structures A and B at 300-500 nt of NKILA. Hairpin A binds to the DNA-binding region of NF-κB, and hairpin B binds to S51-R73 of p65, preventing IκBα detachment that forming a stable NKILA/NF-κB/IκBα complex.[181] Yang *et al.* study found that FTH1P3 (long non-protein coding RNA ferritin heavy chain 1 pseudogene 3) regulated metastasis and invasion through SP1 (specificity protein 1)/NF-κB (p65) in esophageal squamous cell carcinoma (ESCC).[182] The mechanism of lncRNAs has been studied. A deeper understanding of the molecular mechanisms and biological functions of lncRNA will help to find new effective anticancer strategies in tumorigenesis.

MiRNAs and lncRNAs have been demonstrated to play an irreplaceable role in NF-κB signaling pathway. In addition, other type ncRNAs, such as circRNA and piRNA, are also involved in the regulation of tumor-associated signaling pathways. Studies have shown that miR-7 sponge circRNA (ciRS-7) has more than 70 conserved miR-7 binding sites, which can effectively sponge miR-7 and suppress the inhibition of miR-7 target genes, promoting correlation gene expression and activating NF-κB signaling pathways.[183,184] There are few reports about piRNA in cancer. *Leng et al.* found that piR-DQ590027, miR-377and miR-153 in GECs had lower expression. PiR-DQ590027 bound to MIR17HG. FOXR2, was a downstream target of miR-377 and miR-153. This study proved that piR-DQ590027/MIR17HG/miR-153 (miR-377)/FOXR2 pathway plays a crucial role in regulating the permeability of glioma-conditioned normal blood-brain barrier (BBB).[185] Another study shown that piR-021285 can induce genes methylation in breast cancer.[186] The role of these ncRNAs in tumors remains to be further explored.

7.12 NF-κB signaling in chronic inflammation to cancer

Since the discovery of NF-κB 50 years ago, researchers have demonstrated that NF-κB is the "master switch" of gene expression of various pro-inflammatory mediators.[187] It is expressed in many cancers, including hepatocellular carcinoma (HCC), colon cancer, breast cancer, prostate cancer, cervical cancer, and chronic lymphocytic leukemia.[188–191] NF-κB has an important function in inflammation by regulating the expression of pro-inflammatory cytokines such as TNF, interleukin 1 (IL-1) and monocyte chemoattractant protein (MCP-1).[192] Drugs such as aminosalicylate and lipopolysaccharide that are used clinically to combat inflammation work by inhibiting NF-κB. At the same time, a large number of studies have verified that these drugs can inhibit the growth of some cancers.[193,194] Although NF-κB activation can prevent self-activated apoptosis, it acts as a transcription factor that regulates the production of cytotoxic chemicals such as nitric oxide, which in turn may indirectly lead to apoptosis in other cells.[195] Environmental factors affect NF-κB activation and inflammation reaction that induce to promote tumor growth in lung cancer.[196] Particularly, in glioblastoma, it produces neurotoxic reactive oxygen species, which are induces by cytokines through NF-κB activation. Moreover, NF-κB and STAT3 synergistically regulate a number of target genes affecting growth, angiogenesis and immune-suppression.[197] In addition, they also collaboratively control a common set of genes encoding cytokines.[198] This means that NF-κB activity is regulated by STAT3 and plays a significant role in cancer.

It has been shown that the occurrence and development of cancers are closely related to inflammation, and the activation of inflammation-related signaling pathways is important in the formation of the tumor microenvironment. It is worth noting that cancers also have abnormal activation of the inflammatory signaling pathway independent of external stimuli. Activation of the NF-κB signaling pathway promotes the transcription of inflammatory factors, chemokines, adhesion molecules, growth factors and other related genes, which in turn leads to tumor development.[50]

Abnormal activation of inflammatory signaling pathways in tumors can promote the expression of cellular inflammatory factors or activate genes associated with tumor progression. At the same time, inflammatory factors can accelerate the development of malignant cells through paracrine effects. For instance, in breast cancer, low expression of miR-892b leads to aberrant activation of the NF-κB signaling pathway, increasing tumor cell proliferation, metastasis and angiogenesis.[199] In addition, lncRNA-SRLR binds to the IL-6 promoter region, activating the STAT3 signaling pathway in tumor cells by autocrine or paracrine secretion.[200] Tumor cell inflammation is similar to a series of phenotypes in traditional inflammatory cells that participate in immune responses. It has the characteristics of rapid proliferation, anti-apoptosis, invasion and migration and secretion of inflammatory factors. At present, most of the evidence indicates that inflammation in tumors is one of the causes of tumor deterioration.

8. Conclusion and future perspectives

The evolution of modern medical technology has enabled advancements in research to reveal the mechanism of the development of a variety of incurable disease. In the case of cancer, the emerging view is that cancer is a "genetic disease." With next-generation sequencing technology and promotion of the human genome project, the treatment of cancer has been gradually moving towards the era of precision therapy. Under this condition, new therapies take advantage of RNA and other powerful features targeting genes that play a critical role in tumorigenesis.

The importance of miRNA regulatory networks in tumors is increasingly recognized. Meanwhile, we should combine molecular mechanisms to overcome the emergence of resistance. As described in this review, several molecular pathways underlying the mechanisms of this disease have been elaborated in part, among them the EGFR pathway is one of the well-known signal cascades that play a pivotal role in oncogenesis. In lung cancer, dysregulation of EGFR signaling is frequently found. Strategies to effectively inhibit the EGFR signaling pathway have been mounted for developing anti-cancer therapeutic agents. However, because of the development of drug-resistance, most anti-EGFR-targeted agents are unable to prevent cancer progression. Therefore, studies of the mechanisms underpinning the resistance toward anti-EGFR agents may enable the development of efficacious anti-EGFR therapies in lung cancer treatment.

Finally, when determining the applicability of EGFR-TKI in practice, clinical benefit should be carefully analyzed based on clinical background and the prediction of the presence or absence of EGFR mutations. Furthermore, immune-based cancer prevention can also influence pre-malignant biology. It has been shown that cancer vaccines reprogram the immune response to prevent, reject, and detect pre-malignant cells, which might be applicable in EGFR therapy.[201] Going forward, in lung cancer, EGFR may be the rising star in the era of precision medicine.

In cancer, activation of NF-κB signaling pathway is ubiquitous. Its activation is not caused by stimulation of external pathogens, but rather by the abnormal regulation of the cells. Previous studies have found that ncRNA may be involved in this regulation process. Compared with protein synthesis, the production of RNA products in cells is faster. When cells are stimulated by some factors, DNA rapidly transcribes RNA and undergoes splice processing, without the need for protein modification to function. Different RNA products appear in different spliced states and perform different functions. Meanwhile, RNA has more modifications than proteins, which have an effect on the subcellular localization, activity and structure of RNA that makes RNA more diverse in functioning. These properties of RNA determine their status in regulating signaling pathways. The regulation of NF-κB signaling by RNA requires more in-depth research and is of great significance for finding new cancer therapies.

Activation of NF-κB signaling can cause cells to be resistant to chemotherapeutic drugs and inhibit cell apoptosis.[202] Combined with the effects of NF-κB signaling on tumor formation and drug resistance, it can be seen that targeting NF-κB may effectively inhibit the development of tumors and alleviate the tolerance to chemotherapeutic drugs caused by the activation of NF-κB signaling.

As vital transcriptional regulators, miRNAs, non-coding single-chain small RNA molecules, are involved in the development of various tumors and thus may have potential therapeutic value.[203] miRNA has great clinical significance for the research in tumors. Among those pathways in which miRNAs are involved, the NF-κB pathway is one of the most important. It performs an important regulatory function in immune reactions, inflammations and carcinogenesis.[130,204] The main connection of miRNAs and NF-κB lies in miRNAs directly or indirectly regulating the expression level of the target genes in the NF-κB pathway. Currently, the most commonly used method for inhibiting the function of miRNA is using anti-miR oligonucleotides. After the endogenous mature miRNA is inhibited by the anti-miR oligonucleotide, the downstream gene that is inhibited by the miRNA is released. Therefore, anti-miR oligonucleotides are effective in inhibiting miRNAs that promote the NF-κB signaling pathways in cancer.[205] In general, the clinical application of miRNA-related products requires more in-depth experiments.[206] In addition, lncRNA is a potential therapeutic target, but it is essential to find an effective way to inhibit it. With the study of lncRNA in the NF-κB signaling pathway, it is believed that there will be drugs or treatments designed for lncRNA.

Finally, research on the relationship between non-coding RNA and NF-κB signaling pathways is increasingly deepening in tumors. Undoubtedly, the study of the relationship between these has extremely important clinical significance for the mechanism of tumor development and the provision of new drug target sites for tumor treatment. Further efforts also are required to determine if members of the pathway are clinically significant as biomarkers or therapeutic targets.

References

1. Yoshida T, Zhang G, Haura EB. Targeting epidermal growth factor receptor: central signaling kinase in lung cancer. *Biochem Pharmacol.* 2010;80(5):613–623.
2. Janne PA, Engelman JA, Johnson BE. Epidermal growth factor receptor mutations in non-small-cell lung cancer: implications for treatment and tumor biology. *J Clin Oncol.* 2005;23(14):3227–3234.
3. Sun M, Behrens C, Feng L, et al. HER family receptor abnormalities in lung cancer brain metastases and corresponding primary tumors. *Clin Cancer Res.* 2009; 15(15):4829–4837.
4. Wang Y, Deng G, Liu X, et al. Monoclonal antibodies in lung cancer. *Expert Opin Biol Ther.* 2013;13(2):209–226.
5. Chai B, Ma Z, Wang X, et al. Functions of non-coding RNAs in regulating cancer drug targets. *Acta Biochim Biophys Sin.* 2022;54(3):279–291.
6. Perez R, Crombet T, de Leon J, et al. A view on EGFR-targeted therapies from the oncogene-addiction perspective. *Front Pharmacol.* 2013;4:53.
7. Remon J, Moran T, Majem M, et al. Acquired resistance to epidermal growth factor receptor tyrosine kinase inhibitors in EGFR-mutant non-small cell lung cancer: a new era begins. *Cancer Treat Rev.* 2014;40(1):93–101.
8. Yu HA, Riely GJ, Lovly CM. Therapeutic strategies utilized in the setting of acquired resistance to EGFR tyrosine kinase inhibitors. *Clin Cancer Res.* 2014;20(23):5898–5907.
9. Lin Y, Wang X, Jin H. EGFR-TKI resistance in NSCLC patients: mechanisms and strategies. *Am J Cancer Res.* 2014;4(5):411–435.
10. Yun CH, Boggon TJ, Li Y, et al. Structures of lung cancer-derived EGFR mutants and inhibitor complexes: mechanism of activation and insights into differential inhibitor sensitivity. *Cancer Cell.* 2007;11(3):217–227.
11. Rikova K, Guo A, Zeng Q, et al. Global survey of phosphotyrosine signaling identifies oncogenic kinases in lung cancer. *Cell.* 2007;131(6):1190–1203.
12. Lynch TJ, Bell DW, Sordella R, et al. Activating mutations in the epidermal growth factor receptor underlying responsiveness of non-small-cell lung cancer to gefitinib. *N Engl J Med.* 2004;350(21):2129–2139.
13. Paez JG, Janne PA, Lee JC, et al. EGFR mutations in lung cancer: correlation with clinical response to gefitinib therapy. *Science.* 2004;304(5676):1497–1500.
14. Pao W, Miller V, Zakowski M, et al. EGF receptor gene mutations are common in lung cancers from "never smokers" and are associated with sensitivity of tumors to gefitinib and erlotinib. *Proc Natl Acad Sci U S A.* 2004;101(36):13306–13311.
15. Gandara DR, Hammerman PS, Sos ML, et al. Squamous cell lung cancer: from tumor genomics to cancer therapeutics. *Clin Cancer Res.* 2015;21(10):2236–2243.
16. Riely GJ, Yu HA. EGFR: the paradigm of an oncogene-driven lung cancer. *Clin Cancer Res.* 2015;21(10):2221–2226.
17. Herbst RS, Kim ES, Harari PM. IMC-C225, an anti-epidermal growth factor receptor monoclonal antibody, for treatment of head and neck cancer. *Expert Opin Biol Ther.* 2001;1(4):719–732.
18. Colquhoun AJ, Mellon JK. Epidermal growth factor receptor and bladder cancer. *Postgrad Med J.* 2002;78(924):584–589.
19. Yong BK, Gwi EK, Nam HC, et al. Overexpression of cyclooxygenase-2 is associated with a poor prognosis in patients with squamous cell carcinoma of the uterine cervix treated with radiation and concurrent chemotherapy. *Cancer.* 2002;95(3): 531–539.
20. Ohsaki Y, Tanno S, Fujita Y, et al. Epidermal growth factor receptor expression correlates with poor prognosis in non-small cell lung cancer patients with p53 over-expression. *Oncol Rep.* 2000;7(3):603–607.
21. Pastorino U, Andreola S, Tagliabue E, et al. Immunocytochemical markers in stage I lung cancer: relevance to prognosis. *J Clin Oncol.* 1997;15(8):2858–2865.

22. Selvaggi G, Novello S, Torri V, et al. Epidermal growth factor receptor overexpression correlates with a poor prognosis in completely resected non-small-cell lung cancer. *Ann Oncol.* 2004;15(1):28–32.
23. Harpole Jr DH, Richards WG, Herndon Ii JE, et al. Angiogenesis and molecular biologic substaging in patients with stage I non-small cell lung cancer. *Ann Thorac Surg.* 1996;61(5):1470–1476.
24. Pfeiffer P, Clausen PP, Andersen K, et al. Lack of prognostic significance of epidermal growth factor receptor and the oncoprotein p185(HER-2) in patients with systemically untreated non-small-cell lung cancer: an immunohistochemical study on cryosections. *Br J Cancer.* 1996;74(1):86–91.
25. Rusch V, Baselga J, Cordon-Cardo C, et al. Differential expression of the epidermal growth factor receptor and its ligands in primary non-small cell lung cancers and adjacent benign lung. *Cancer Res.* 1993;53(10):2379–2385.
26. Cox G, Jones JL, O'Byrne KJ. Matrix metalloproteinase 9 and the epidermal growth factor signal pathway in operable non-small cell lung cancer. *Clin Cancer Res.* 2000; 6(6):2349–2355.
27. Dias F, Teixeira AL, Santos JI, et al. Renal cell carcinoma development and miRNAs: a possible link to the EGFR pathway. *Pharmacogenomics.* 2013;14(14):1793–1803.
28. Stumm G, Eberwein S, Rostock-Wolf S, et al. Concomitant overexpression of the EGFR and erbB-2 genes in renal cell carcinoma (RCC) is correlated with dedifferentiation and metastasis. *Int J Cancer.* 1996;69(1):17–22.
29. Lui VWY, Grandis JR. EGFR-mediated cell cycle regulation. *Anticancer Res.* 2002;22 (1 A):1–11.
30. Shelton JG, Steelman LS, Abrams SL, et al. Conditional EGFR promotes cell cycle progression and prevention of apoptosis in the absence of autocrine cytokines. *Cell Cycle.* 2005;4(6):822–830.
31. Kari C, Chan TO, Rocha de Quadros M, et al. Targeting the epidermal growth factor receptor in cancer: apoptosis takes center stage. *Cancer Res.* 2003;63(1):1–5.
32. Shvartsman SY, Hagan MP, Yacoub A, et al. Autocrine loops with positive feedback enable context-dependent cell signaling. *Am J Phys Cell Phys.* 2002;282(3 51-3): C545–C559.
33. Umekita Y, Ohi Y, Sagara Y, et al. Co-expression of epidermal growth factor receptor and transforming growth factor-α predicts worse prognosis in breast-cancer patients. *Int J Cancer.* 2000;89(6):484–487.
34. Akbay EA, Koyama S, Carretero J, et al. Activation of the PD-1 pathway contributes to immune escape in EGFR-driven lung tumors. *Cancer Discovery.* 2013;3(12):1355–1363.
35. Paz-Ares L, Soulières D, Melezínek I, et al. Clinical outcomes in non-small-cell lung cancer patients with EGFR mutations: pooled analysis. *J Cell Mol Med.* 2010;14 (1-2):51–69.
36. Gazdar AF. Activating and resistance mutations of EGFR in non-small-cell lung cancer: role in clinical response to EGFR tyrosine kinase inhibitors. *Oncogene.* 2009;28: S24–S31.
37. Engelman JA, Zejnullahu K, Mitsudomi T, et al. MET amplification leads to gefitinib resistance in lung cancer by activating ERBB3 signaling. *Science.* 2007;316 (5827):1039–1043.
38. Popat S, Wotherspoon A, Nutting CM, et al. Transformation to "high grade" neuroendocrine carcinoma as an acquired drug resistance mechanism in EGFR-mutant lung adenocarcinoma. *Lung Cancer.* 2013;80(1):1–4.
39. Suda K, Tomizawa K, Fujii M, et al. Epithelial to mesenchymal transition in an epidermal growth factor receptor-mutant lung cancer cell line with acquired resistance to erlotinib. *J Thorac Oncol.* 2011;6(7):1152–1161.
40. Metro G, Cappuzzo F. Emerging drugs for small-cell lung cancer. *Expert Opin Emerg Drugs.* 2009;14(4):591–606.

41. Jänne PA, Chih-Hsin Yang J, Kim DW, et al. AZD9291 in EGFR inhibitor-resistant non-small-cell lung cancer. *N Engl J Med.* 2015;372(18):1689–1699.
42. Kim ES. Olmutinib: first global approval. *Drugs.* 2016;76(11):1153–1157.
43. Minari R, Bordi P, Tiseo M. Third-generation epidermal growth factor receptor-tyrosine kinase inhibitors in T790M-positive non-small cell lung cancer: review on emerged mechanisms of resistance. *Transl Lung Cancer Res.* 2016;5(6):695–708.
44. Soda M, Choi YL, Enomoto M, et al. Identification of the transforming EML4-ALK fusion gene in non-small-cell lung cancer. *Nature.* 2007;448(7153):561–566.
45. Mok TS, Wu YL, Thongprasert S, et al. Gefitinib or carboplatin-paclitaxel in pulmonary adenocarcinoma. *N Engl J Med.* 2009;361(10):947–957.
46. Politi K, Herbst RS. Lung cancer in the era of precision medicine. *Clin Cancer Res.* 2015;21(10):2213–2220.
47. Antoni S, Ferlay J, Soerjomataram I, et al. Bladder cancer incidence and mortality: a global overview and recent trends. *Eur Urol.* 2017;71(1):96–108.
48. Siegel RL, Miller KD, Fuchs HE, et al. Cancer statistics, 2021. *CA Cancer J Clin.* 2021;71(1):7–33.
49. Zhang X, Wang X, Chai B, et al. Downregulated miR-18a and miR-92a synergistically suppress non-small cell lung cancer via targeting Sprouty 4. *Bioengineered.* 2022;13(4):11281–11295.
50. Pikarsky E, Porat RM, Stein I, et al. NF-κB functions as a tumour promoter in inflammation-associated cancer. *Nature.* 2004;431(7007):461–466.
51. Liu X, Shao Y, Zhou J, et al. Nuclear factor kappaB signaling and its related non-coding RNAs in cancer therapy. *Mol Ther Nucleic Acids.* 2020;19:208–217.
52. Wang X, Li W, Lu S, et al. Modulation of the Wound Healing through Noncoding RNA Interplay and GSK-3beta/NF-kappaB Signaling Interaction. *Int J Genomics.* 2021;2021:9709290.
53. Sen R, Baltimore D. Multiple nuclear factors interact with the immunoglobulin enhancer sequences. *Cell.* 1986;46(5):705–716.
54. Sen R, Baltimore D. Multiple nuclear factors interact with the immunoglobulin enhancer sequences. *Cell.* 1986;46:705–716. *J Immunol.* 2006;177(11):7485–7496.
55. Tian B, Liu Z, Yang J, et al. Selective antagonists of the bronchiolar epithelial NF-κB-bromodomain-containing protein 4 pathway in viral-induced airway inflammation. *Cell Rep.* 2018;23(4):1138–1151.
56. Wu MY, Li CJ, Yiang GT, et al. Molecular regulation of bone metastasis pathogenesis. *Cell Physiol Biochem.* 2018;46(4):1423–1438.
57. Liu B, Li R, Zhang J, et al. MicroRNA-708-3p as a potential therapeutic target via the ADAM17-GATA/STAT3 axis in idiopathic pulmonary fibrosis. *Exp Mol Med.* 2018;50(3):e465.
58. Fang J, Zhang H, Jin S. Epigenetics and cervical cancer: from pathogenesis to therapy. *Tumor Biol.* 2014;35(6):5083–5093.
59. Hayden MS, Ghosh S. NF-κB, the first quarter-century: remarkable progress and outstanding questions. *Genes Dev.* 2012;26(3):203–234.
60. Kim JW, Jang SM, Kim CH, et al. New molecular bridge between RelA/p65 and NF-κB target genes via histone acetyltransferase TIP60 cofactor. *J Biol Chem.* 2012;287(10):7780–7791.
61. Caamaño J, Hunter CA. NF-κB family of transcription factors: central regulators of innate and adaptive immune functions. *Clin Microbiol Rev.* 2002;15(3):414–429.
62. Soldevilla MM, de Caso DMC, Menon AP, et al. Aptamer-iRNAs as therapeutics for cancer treatment. *Pharmaceuticals.* 2018;11(4).
63. Shostak K, Chariot A. EGFR and NF-κB: partners in cancer. *Trends Mol Med.* 2015;21(6):385–393.
64. Clay TD, Russell PA, Do H, et al. EGFR and KRAS mutations do not enrich for the activation of IL-6, JAK1 or phosphorylated STAT3 in resected lung adenocarcinoma. *Med Oncol.* 2017;34(10).

65. Cho JH, You YM, Il Yeom Y, et al. RNF25 promotes gefitinib resistance in EGFR-mutant NSCLC cells by inducing NF-κB-mediated ERK reactivation article. *Cell Death Dis.* 2018;9(6).
66. Ma J, Lyu H, Huang J, et al. Targeting of erbB3 receptor to overcome resistance in cancer treatment. *Mol Cancer.* 2014;13(1).
67. Jiang C, Zhu Y, Zhou Z, et al. TMEM43/LUMA is a key signaling component mediating EGFR-induced NF-κ B activation and tumor progression. *Oncogene.* 2017;36(20):2813–2823.
68. Chen CYA, Chang JT, Ho YF, et al. MiR-26 down-regulates TNF-α/NF-κB signalling and IL-6 expression by silencing HMGA1 and MALT1. *Nucleic Acids Res.* 2016;44(8):3772–3787.
69. Wang J, Zhu M, Wang L, et al. Amphiregulin potentiates airway inflammation and mucus hypersecretion induced by urban particulate matter via the EGFR-PI3Kα-AKT/ERK pathway. *Cell Signal.* 2019;53:122–131.
70. Croce CM. Causes and consequences of microRNA dysregulation in cancer. *Nat Rev Genet.* 2009;10(10):704–714.
71. Bartel DP. MicroRNAs: genomics, biogenesis, mechanism, and function. *Cell.* 2004;116(2):281–297.
72. Hayes J, Peruzzi PP, Lawler S. MicroRNAs in cancer: biomarkers, functions and therapy. *Trends Mol Med.* 2014;20(8):460–469.
73. Bhan A, Soleimani M, Mandal SS. Long noncoding RNA and cancer: a new paradigm. *Cancer Res.* 2017;77(15):3965–3981.
74. Li Y, Zhang H, Guo J, et al. Downregulation of LINC01296 suppresses non-small-cell lung cancer via targeting miR-143-3p/ATG2B. *Acta Biochim Biophys Sin Shanghai.* 2021;53(12):1681–1690.
75. Inui M, Martello G, Piccolo S. MicroRNA control of signal transduction. *Nat Rev Mol Cell Biol.* 2010;11(4):252–263.
76. Hoshino I, Matsubara H. MicroRNAs in cancer diagnosis and therapy: from bench to bedside. *Surg Today.* 2013;43(5):467–478.
77. Wang F, Chan LWC, Law HKW, et al. Exploring microRNA-mediated alteration of EGFR signaling pathway in non-small cell lung cancer using an mRNA: MiRNA regression model supported by target prediction databases. *Genomics.* 2014;104(6): 504–511.
78. Gomez GG, Wykosky J, Zanca C, et al. Therapeutic resistance in cancer: microRNA regulation of EGFR signaling networks. *Cancer Biol Med.* 2013;10(4): 192–205.
79. Zhen Q, Liu J, Gao L, et al. MicroRNA-200a targets EGFR and c-Met to inhibit migration, invasion, and gefitinib resistance in non-small cell lung cancer. *Cytogenetic Genome Res.* 2015;146(1):1–8.
80. Chiyomaru T, Seki N, Inoguchi S, et al. Dual regulation of receptor tyrosine kinase genes EGFR and c-Met by the tumor-suppressive microRNA-23b/27b cluster in bladder cancer. *Int J Oncol.* 2015;46(2):487–496.
81. Li Y, Li J, Sun X, et al. MicroRNA-27a functions as a tumor suppressor in renal cell carcinoma by targeting epidermal growth factor receptor. *Oncol Lett.* 2016;11 (6):4217–4223.
82. Wang LK, Hsiao TH, Hong TM, et al. MicroRNA-133a suppresses multiple oncogenic membrane receptors and cell invasion in non-small cell lung carcinoma. *PLoS One.* 2014;9(5).
83. Qin Q, Wei F, Zhang J, et al. miR-134 inhibits non-small cell lung cancer growth by targeting the epidermal growth factor receptor. *J Cell Mol Med.* 2016;20(10): 1974–1983.
84. Adam L, Zhong M, Choi W, et al. miR-200 expression regulates epithelial-to-mesenchymal transition in bladder cancer cells and reverses resistance to epidermal growth factor receptor therapy. *Clin Cancer Res.* 2009;15(16):5060–5072.

85. Zhang Z, Liu ZB, Ren WM, et al. The miR-200 family regulates the epithelial-mesenchymal transition induced by EGF/EGFR in anaplastic thyroid cancer cells. *Int J Mol Med*. 2012;30(4):856–862.
86. Webster RJ, Giles KM, Price KJ, et al. Regulation of epidermal growth factor receptor signaling in human cancer cells by microRNA-7. *J Biol Chem*. 2009;284(9):5731–5741.
87. Mudduluru G, Ceppi P, Kumarswamy R, et al. Regulation of Axl receptor tyrosine kinase expression by miR-34a and miR-199a/b in solid cancer. *Oncogene*. 2011;30(25):2888–2899.
88. Cho WC, Chow AS, Au JS. MiR-145 inhibits cell proliferation of human lung adenocarcinoma by targeting EGFR and NUDT1. *RNA Biol*. 2011;8(1):125–131.
89. Chen G, Umelo IA, Lv S, et al. miR-146a inhibits cell growth, cell migration and induces apoptosis in non-small cell lung cancer cells. *PLoS One*. 2013;8(3):e60317.
90. Xu B, Wang N, Wang X, et al. MiR-146a suppresses tumor growth and progression by targeting EGFR pathway and in a p-ERK-dependent manner in castration-resistant prostate cancer. *Prostate*. 2012;72(11):1171–1178.
91. Katakowski M, Zheng X, Jiang F, et al. MiR-146b-5p suppresses EGFR expression and reduces in vitro migration and invasion of glioma. *Cancer Investig*. 2010;28 (10):1024–1030.
92. Mataki H, Seki N, Chiyomaru T, et al. Tumor-suppressive microRNA-206 as a dual inhibitor of MET and EGFR oncogenic signaling in lung squamous cell carcinoma. *Int J Oncol*. 2015;46(3):1039–1050.
93. Xu B, Tao T, Wang Y, et al. hsa-miR-135a-1 inhibits prostate cancer cell growth and migration by targeting EGFR. *Tumour Biol*. 2016.
94. Liu L, Shao X, Gao W, et al. MicroRNA-133b inhibits the growth of non-small-cell lung cancer by targeting the epidermal growth factor receptor. *FEBS J*. 2012;279 (20):3800–3812.
95. Yamaguchi G, Takanashi M, Tanaka M, et al. Isolation of miRNAs that target EGFR mRNA in human lung cancer. *Biochem Biophys Res Commun*. 2012;420(2): 411–416.
96. Liu R, Liu C, Zhang D, et al. miR-199a-3p targets stemness-related and mitogenic signaling pathways to suppress the expansion and tumorigenic capabilities of prostate cancer stem cells. *Oncotarget*. 2016.
97. Xu J, Wan X, Chen X, et al. miR-2861 acts as a tumor suppressor via targeting EGFR/ AKT2/CCND1 pathway in cervical cancer induced by human papillomavirus virus 16 E6. *Sci Rep*. 2016;6:28968.
98. Dacic S, Kelly L, Shuai Y, et al. miRNA expression profiling of lung adenocarcinomas: correlation with mutational status. *Mod Pathol*. 2010;23(12):1577–1582.
99. Du WW, Fang L, Li M, et al. MicroRNA miR-24 enhances tumor invasion and metastasis by targeting PTPN9 and PTPRF to promote EGF signaling. *J Cell Sci*. 2013;126(Pt 6):1440–1453.
100. Zhou X, Ren Y, Moore L, et al. Downregulation of miR-21 inhibits EGFR pathway and suppresses the growth of human glioblastoma cells independent of PTEN status. *Lab Investig*. 2010;90(2):144–155.
101. Li YL, Liu XM, Zhang CY, et al. MicroRNA-34a/EGFR axis plays pivotal roles in lung tumorigenesis. *Oncogenesis*. 2017;6(8):e372.
102. Ma ZL, Hou PP, Li YL, et al. MicroRNA-34a inhibits the proliferation and promotes the apoptosis of non-small cell lung cancer H1299 cell line by targeting TGFβR2. *Tumor Biol*. 2015;36(4):2481–2490.
103. Ma Z, Qiu X, Wang D, et al. MiR-181a-5p inhibits cell proliferation and migration by targeting Kras in non-small cell lung cancer A549 cells. *Acta Biochim Biophys Sin*. 2015;47(8):630–638.

104. Shao Y, Shen YQ, Li YL, et al. Direct repression of the oncogene CDK4 by the tumor suppressor miR-486-5p in non-small cell lung cancer. *Oncotarget.* 2016;7(23):34011–34021.
105. Yang C, Jia X, Zhou J, et al. The MiR-17-92 gene cluster is a blood-based marker for cancer detection in non-small-cell lung cancer. *Am J Med Sci.* 2020;360(3):248–260.
106. Del Vescovo V, Denti MA. MicroRNA and lung cancer. *Adv Exp Med Biol.* 2015;889:153–177.
107. Yu T, Liu L, Li J, et al. MiRNA-10a is upregulated in NSCLC and may promote cancer by targeting PTEN. *Oncotarget.* 2015;6(30):30239–30250.
108. Yang Y, Liu L, Zhang Y, et al. MiR-503 targets PI3K p85 and IKK-β and suppresses progression of non-small cell lung cancer. *Int J Cancer.* 2014;135(7):1531–1542.
109. Chen J, Wang R, Zhang K, et al. Long non-coding RNAs in non-small cell lung cancer as biomarkers and therapeutic targets. *J Cell Mol Med.* 2014;18(12):2425–2436.
110. Esteller M. Non-coding RNAs in human disease. *Nat Rev Genet.* 2011;12(12): 861–874.
111. Cheetham SW, Gruhl F, Mattick JS, et al. Long noncoding RNAs and the genetics of cancer. *Br J Cancer.* 2013;108(12):2419–2425.
112. Li T, Mo X, Fu L, et al. Molecular mechanisms of long noncoding RNAs on gastric cancer. *Oncotarget.* 2016;7(8):8601–8612.
113. Cheng N, Li X, Zhao C, et al. Microarray expression profile of long non-coding RNAs in EGFR-TKIs resistance of human non-small cell lung cancer. *Oncol Rep.* 2015;33(2): 833–839.
114. Cheng N, Cai W, Ren S, et al. Long non-coding RNA UCA1 induces non-T790M acquired resistance to EGFR-TKIs by activating the AKT/mTOR pathway in EGFR-mutant non-small cell lung cancer. *Oncotarget.* 2015;6(27):23582–23593.
115. Alves ND, Correia JS, Patricio P, et al. Adult hippocampal neuroplasticity triggers susceptibility to recurrent depression. *Transl Psychiatry.* 2017;7.
116. Hampton KK, Craven RJ. Pathways driving the endocytosis of mutant and wild-type EGFR in cancer. *Oncoscience.* 2014;1(8):504–512.
117. Rolfo C, Giovannetti E, Hong DS, et al. Novel therapeutic strategies for patients with NSCLC that do not respond to treatment with EGFR inhibitors. *Cancer Treat Rev.* 2014;40(8):990–1004.
118. Ma Z, Zhou J, Shao Y, et al. Biochemical properties and progress in cancers of tRNA-derived fragments. *J Cell Biochem.* 2020;121(3):2058–2063.
119. Lee YS, Shibata Y, Malhotra A, et al. A novel class of small RNAs: tRNA-derived RNA fragments (tRFs). *Genes Dev.* 2009;23(22):2639–2649.
120. Ivanov P, Emara MM, Villen J, et al. Angiogenin-induced tRNA fragments inhibit translation initiation. *Mol Cell.* 2011;43(4):613–623.
121. Marshall L, Kenneth NS, White RJ. Elevated tRNA(iMet) synthesis can drive cell proliferation and oncogenic transformation. *Cell.* 2008;133(1):78–89.
122. Mei Y, Yong J, Liu H, et al. tRNA binds to cytochrome c and inhibits caspase activation. *Mol Cell.* 2010;37(5):668–678.
123. Dhahbi JM, Spindler SR, Atamna H, et al. 5' tRNA halves are present as abundant complexes in serum, concentrated in blood cells, and modulated by aging and calorie restriction. *BMC Genomics.* 2013;14:298.
124. Andersen KL, Collins K. Several RNase T2 enzymes function in induced tRNA and rRNA turnover in the ciliate Tetrahymena. *Mol Biol Cell.* 2012;23(1):36–44.
125. Goodarzi H, Liu X, Nguyen HC, et al. Endogenous tRNA-derived fragments suppress breast cancer progression via YBX1 displacement. *Cell.* 2015;161(4):790–802.
126. Gebetsberger J, Polacek N. Slicing tRNAs to boost functional ncRNA diversity. *RNA Biol.* 2013;10(12):1798–1806.
127. Haussecker D, Huang Y, Lau A, et al. Human tRNA-derived small RNAs in the global regulation of RNA silencing. *RNA.* 2010;16(4):673–695.

128. Garcia-Silva MR, Cabrera-Cabrera F, Guida MC, et al. Hints of tRNA-derived small RNAs role in RNA silencing mechanisms. *Genes (Basel).* 2012;3(4):603–614.
129. Zambruno G, Girolomoni G, Manca V, et al. Epidermal growth factor and transferrin receptor expression in human embryonic and fetal epidermal cells. *Arch Dermatol Res.* 1990;282(8):544–548.
130. Olarerin-George AO, Anton L, Hwang YC, et al. A functional genomics screen for microRNA regulators of NF-kappaB signaling. *BMC Biol.* 2013;11.
131. Guo LM, Pu Y, Han Z, et al. MicroRNA-9 inhibits ovarian cancer cell growth through regulation of NF-kappaB1. *FEBS J.* 2009;276(19):5537–5546.
132. Wan HY, Guo LM, Liu T, et al. Regulation of the transcription factor NF-kappaB1 by microRNA-9 in human gastric adenocarcinoma. *Mol Cancer.* 2010;9:16.
133. Wang Q, Sun ZX, Allgayer H, et al. Downregulation of E-cadherin is an essential event in activating beta-catenin/Tcf-dependent transcription and expression of its target genes in Pdcd4 knockdown cells. *Oncogene.* 2010;29(1):128–138.
134. Zhang X, Liu S, Hu T, et al. Up-regulated microRNA-143 transcribed by nuclear factor kappa B enhances hepatocarcinoma metastasis by repressing fibronectin expression. *Hepatology.* 2009;50(2):490–499.
135. Borralho PM, Kren BT, Castro RE, et al. MicroRNA-143 reduces viability and increases sensitivity to 5-fluorouracil in HCT116 human colorectal cancer cells. *FEBS J.* 2009;276(22):6689–6700.
136. Mei LL, Wang WJ, Qiu YT, et al. miR-145-5p suppresses tumor cell migration, invasion and epithelial to mesenchymal transition by regulating the Sp1/NF-kappaB signaling pathway in esophageal squamous cell carcinoma. *Int J Mol Sci.* 2017;18(9).
137. Wang H, Garzon R, Sun H, et al. NF-kappaB-YY1-miR-29 regulatory circuitry in skeletal myogenesis and rhabdomyosarcoma. *Cancer Cell.* 2008;14(5):369–381.
138. Mott JL, Kurita S, Cazanave SC, et al. Transcriptional suppression of mir-29b-1/mir-29a promoter by c-Myc, hedgehog, and NF-kappaB. *J Cell Biochem.* 2010;110(5):1155–1164.
139. Zhou R, Hu G, Liu J, et al. NF-kappaB p65-dependent transactivation of miRNA genes following Cryptosporidium parvum infection stimulates epithelial cell immune responses. *PLoS Pathog.* 2009;5(12):e1000681.
140. Zhou R, Hu G, Gong AY, et al. Binding of NF-kappaB p65 subunit to the promoter elements is involved in LPS-induced transactivation of miRNA genes in human biliary epithelial cells. *Nucleic Acids Res.* 2010;38(10):3222–3232.
141. Tang ZP, Zhao W, Du JK, et al. miR-494 Contributes to estrogen protection of cardiomyocytes against oxidative stress via targeting (NF-kappaB) repressing factor. *Front Endocrinol (Lausanne).* 2018;9:215.
142. Fan X, Liang Y, Liu Y, et al. The upregulation of TMPRSS4, partly ascribed to the downregulation of miR125a5p, promotes the growth of human lung adenocarcinoma via the NFkappaB signaling pathway. *Int J Oncol.* 2018;53(1):148–158.
143. Datta C, Subuddhi A, Kumar M, et al. Genome-wide mRNA-miRNA profiling uncovers a role of the microRNA miR-29b-1-5p/PHLPP1 signalling pathway in Helicobacter pylori-driven matrix metalloproteinase production in gastric epithelial cells. *Cell Microbiol.* 2018; e12859.
144. Xiao L, Jiang L, Hu Q, et al. MiR-302e attenuates allergic inflammation in vitro model by targeting RelA. *Biosci Rep.* 2018;38(3).
145. Kong L, Sun M, Jiang Z, et al. MicroRNA-194 inhibits lipopolysaccharide-induced inflammatory response in nucleus pulposus cells of the intervertebral disc by targeting TNF receptor-associated factor 6 (TRAF6). *Med Sci Monit.* 2018;24:3056–3067.
146. Zhu S, Fu W, Zhang L, et al. LINC00473 antagonizes the tumour suppressor miR-195 to mediate the pathogenesis of Wilms tumour via IKKalpha. *Cell Prolif.* 2018;51(1).

147. Bardin P, Marchal-Duval E, Sonneville F, et al. Small RNA and transcriptome sequencing reveal the role of miR-199a-3p in inflammatory processes in cystic fibrosis airways. *J Pathol.* 2018.
148. Hu J, Zhai C, Hu J, et al. MiR-23a inhibited IL-17-mediated proinflammatory mediators expression via targeting IKKalpha in articular chondrocytes. *Int Immunopharmacol.* 2017;43:1–6.
149. Fan JY, Fan YJ, Wang XL, et al. miR-429 is involved in regulation of NF-kappaB activity by targeting IKKbeta and suppresses oncogenic activity in cervical cancer cells. *FEBS Lett.* 2017;591(1):118–128.
150. Huang Y, Wang W, Xu Z, et al. Eriocheir sinensis microRNA-7 targets crab Myd88 to enhance white spot syndrome virus replication. *Fish Shellfish Immunol.* 2018;79:274–283.
151. Tian Y, Hao S, Ye M, et al. MicroRNAs let-7b/i suppress human glioma cell invasion and migration by targeting IKBKE directly. *Biochem Biophys Res Commun.* 2015;458 (2):307–312.
152. Zhao L, Zhang Y. miR-342-3p affects hepatocellular carcinoma cell proliferation via regulating NF-kappaB pathway. *Biochem Biophys Res Commun.* 2015;457(3):370–377.
153. Li HP, Zeng XC, Zhang B, et al. miR-451 inhibits cell proliferation in human hepatocellular carcinoma through direct suppression of IKK-beta. *Carcinogenesis.* 2013;34 (11):2443–2451.
154. Dai L, Gu L, Di W. MiR-199a attenuates endometrial stromal cell invasiveness through suppression of the IKKbeta/NF-kappaB pathway and reduced interleukin-8 expression. *Mol Hum Reprod.* 2012;18(3):136–145.
155. Zhang G, Sun Y, Wang Y, et al. MiR-502-5p inhibits IL-1beta-induced chondrocyte injury by targeting TRAF2. *Cell Immunol.* 2016;302:50–57.
156. Liang C, Zhang X, Wang HM, et al. MicroRNA-18a-5p functions as an oncogene by directly targeting IRF2 in lung cancer. *Cell Death Dis.* 2017;8(5):e2764.
157. Shan D, Shang Y, Hu T. MicroRNA-411 inhibits cervical cancer progression by directly targeting STAT3. *Oncol Res.* 2018.
158. Liu S, Zhang X, Hu C, et al. miR-29a inhibits human retinoblastoma progression by targeting STAT3. *Oncol Rep.* 2018;39(2):739–746.
159. Wang M, Meng B, Liu Y, et al. MiR-124 inhibits growth and enhances radiation-induced apoptosis in non-small cell lung cancer by inhibiting STAT3. *Cell Physiol Biochem.* 2017;44(5):2017–2028.
160. Zhong X, Ma X, Zhang L, et al. MIAT promotes proliferation and hinders apoptosis by modulating miR-181b/STAT3 axis in ox-LDL-induced atherosclerosis cell models. *Biomed Pharmacother.* 2018;97:1078–1085.
161. Zhao N, Wang R, Zhou L, et al. MicroRNA-26b suppresses the NF-kappaB signaling and enhances the chemosensitivity of hepatocellular carcinoma cells by targeting TAK1 and TAB3. *Mol Cancer.* 2014;13:35.
162. Yuan G, Zhao Y, Wu D, et al. miRNA-20a upregulates TAK1 and increases proliferation in osteosarcoma cells. *Future Oncol.* 2018;14(5):461–469.
163. Zhang H, Li H, Shaikh A, et al. Inhibition of MicroRNA-23b attenuates immunosuppression in late sepsis through NIK, TRAF1 and XIAP. *J Infect Dis.* 2018.
164. Zabaglia LM, Sallas ML, Santos MPD, et al. Expression of miRNA-146a, miRNA-155, IL-2, and TNF-alpha in inflammatory response to Helicobacter pylori infection associated with cancer progression. *Ann Hum Genet.* 2018;82(3):135–142.
165. Mensa E, Recchioni R, Marcheselli F, et al. MiR-146a-5p correlates with clinical efficacy in psoriasis patients treated with the TNF alpha inhibitor Adalimumab. *Br J Dermatol.* 2018.
166. Zhang C, Qian D, Zhao H, et al. MiR17 improves insulin sensitivity through inhibiting expression of ASK1 and anti-inflammation of macrophages. *Biomed Pharmacother.* 2018;100:448–454.

167. Feng Y, Chen L, Luo Q, et al. Involvement of microRNA-146a in diabetic peripheral neuropathy through the regulation of inflammation. *Drug Des Devel Ther.* 2018;12:171–177.
168. Lorente-Cebrian S, Mejhert N, Kulyte A, et al. MicroRNAs regulate human adipocyte lipolysis: effects of miR-145 are linked to TNF-alpha. *PLoS One.* 2014;9(1):e86800.
169. Neumann M, Naumann M. Beyond IkappaBs: alternative regulation of NF-kappaB activity. *FASEB J.* 2007;21(11):2642–2654.
170. Zhou HJ, Wang LQ, Xu QS, et al. Downregulation of miR-199b promotes the acute spinal cord injury through IKKbeta-NF-kappaB signaling pathway activating microglial cells. *Exp Cell Res.* 2016;349(1):60–67.
171. Tili E, Michaille JJ, Cimino A, et al. Modulation of miR-155 and miR-125b levels following lipopolysaccharide/TNF-alpha stimulation and their possible roles in regulating the response to endotoxin shock. *J Immunol.* 2007;179(8):5082–5089.
172. Zhang Z, Lu J, Guo G, et al. IKBKE promotes glioblastoma progression by establishing the regulatory feedback loop of IKBKE/YAP1/miR-Let-7b/i. *Tumour Biol.* 2017;39 (7):1010428317705575.
173. d'Adhemar CJ, Spillane CD, Gallagher MF, et al. The MyD88+ phenotype is an adverse prognostic factor in epithelial ovarian cancer. *PLoS One.* 2014;9(6): e100816.
174. Wendlandt EB, Graff JW, Gioannini TL, et al. The role of microRNAs miR-200b and miR-200c in TLR4 signaling and NF-kappaB activation. *Innate Immun.* 2012;18 (6):846–855.
175. Feng Y, Zou L, Yan D, et al. Extracellular microRNAs induce potent innate immune responses via TLR7/MyD88-dependent mechanisms. *J Immunol.* 2017;199(6): 2106–2117.
176. Yang F, Liu X, Liu Y, et al. miR-181d/MALT1 regulatory axis attenuates mesenchymal phenotype through NF-kappaB pathways in glioblastoma. *Cancer Lett.* 2017; 396:1–9.
177. Chen CY, Chang JT, Ho YF, et al. MiR-26 down-regulates TNF-alpha/NF-kappaB signalling and IL-6 expression by silencing HMGA1 and MALT1. *Nucleic Acids Res.* 2016;44(8):3772–3787.
178. Gu L, Sun W. MiR-539 inhibits thyroid cancer cell migration and invasion by directly targeting CARMA1. *Biochem Biophys Res Commun.* 2015;464(4):1128–1133.
179. Morioka S, Broglie P, Omori E, et al. TAK1 kinase switches cell fate from apoptosis to necrosis following TNF stimulation. *J Cell Biol.* 2014;204(4):607–623.
180. Quinodoz S, Guttman M. Long noncoding RNAs: an emerging link between gene regulation and nuclear organization. *Trends Cell Biol.* 2014;24(11):651–663.
181. Liu B, Sun L, Liu Q, et al. A cytoplasmic NF-kappaB interacting long noncoding RNA blocks IkappaB phosphorylation and suppresses breast cancer metastasis. *Cancer Cell.* 2015;27(3):370–381.
182. Yang L, Sun K, Chu J, et al. Long non-coding RNA FTH1P3 regulated metastasis and invasion of esophageal squamous cell carcinoma through SP1/NF-kB pathway. *Biomed Pharmacother.* 2018;106:1570–1577.
183. Hansen TB, Jensen TI, Clausen BH, et al. Natural RNA circles function as efficient microRNA sponges. *Nature.* 2013;495(7441):384–388.
184. Hansen TB, Kjems J, Damgaard CK. Circular RNA and miR-7 in cancer. *Cancer Res.* 2013;73(18):5609–5612.
185. Leng X, Ma J, Liu Y, et al. Mechanism of piR-DQ590027/MIR17HG regulating the permeability of glioma conditioned normal BBB. *J Exp Clin Cancer Res.* 2018;37(1):246.
186. Fu A, Jacobs DI, Hoffman AE, et al. PIWI-interacting RNA 021285 is involved in breast tumorigenesis possibly by remodeling the cancer epigenome. *Carcinogenesis.* 2015;36(10):1094–1102.

187. Tsoulfas G, Geller DA. NF-kappaB in transplantation: friend or foe? *Transpl Infect Dis.* 2001;3(4):212–219.
188. Ehedego H, Mohs A, Jansen B, et al. Loss of Cyclin E1 attenuates hepatitis and hepatocarcinogenesis in a mouse model of chronic liver injury. *Oncogene.* 2018.
189. Liu H, Li G, Zhang B, et al. Suppression of the NFkappaB signaling pathway in colon cancer cells by the natural compound Riccardin D from Dumortierahirsute. *Mol Med Rep.* 2018;17(4):5837–5843.
190. Romagny S, Bouaouiche S, Lucchi G, et al. S-Nitrosylation of cIAP1 switches cancer cell fate from TNFalpha/TNFR1-mediated cell survival to cell death. *Cancer Res.* 2018;78(8):1948–1957.
191. Deswaerte V, Nguyen P, West A, et al. Inflammasome adaptor ASC suppresses apoptosis of gastric cancer cells by an IL18-mediated inflammation-independent mechanism. *Cancer Res.* 2018;78(5):1293–1307.
192. Mattson MP, Camandola S. NF-kappaB in neuronal plasticity and neurodegenerative disorders. *J Clin Invest.* 2001;107(3):247–254.
193. Roebuck KA, Carpenter LR, Lakshminarayanan V, et al. Stimulus-specific regulation of chemokine expression involves differential activation of the redox-responsive transcription factors AP-1 and NF-kappaB. *J Leukoc Biol.* 1999;65(3):291–298.
194. Qureshi AA, Zuvanich EG, Khan DA, et al. Proteasome inhibitors modulate anticancer and anti-proliferative properties via NF-kB signaling, and ubiquitin-proteasome pathways in cancer cell lines of different organs. *Lipids Health Dis.* 2018;17(1):62.
195. Ahmadian E, Khosroushahi AY, Eftekhari A, et al. Novel angiotensin receptor blocker, azilsartan induces oxidative stress and NFkB-mediated apoptosis in hepatocellular carcinoma cell line HepG2. *Biomed Pharmacother.* 2018;99:939–946.
196. Pozniak J, Nsengimana J, Laye JP, et al. Genetic and environmental determinants of immune response to cutaneous melanoma. *Cancer Res.* 2019.
197. Silva TCC, de Jesus M-FN, de Faria Lopes GP, et al. Specific cytostatic and cytotoxic effect of dihydrochelerythrine in glioblastoma cells: role of NF-kB/b-catenin and STAT3/IL-6 pathways. *Anti Cancer Agents Med Chem.* 2018.
198. Yang J, Liao X, Agarwal MK, et al. Unphosphorylated STAT3 accumulates in response to IL-6 and activates transcription by binding to NFkappaB. *Genes Dev.* 2007;21 (11):1396–1408.
199. Zhu H, Zhao Y, Wu H, et al. Remarkable alterations of Nav1.6 in reactive astrogliosis during epileptogenesis. *Sci Rep.* 2016;6:38108.
200. Xu Z, Yang F, Wei D, et al. Long noncoding RNA-SRLR elicits intrinsic sorafenib resistance via evoking IL-6/STAT3 axis in renal cell carcinoma. *Oncogene.* 2017;36(14):1965–1977.
201. Spira A, Disis ML, Schiller JT, et al. Leveraging premalignant biology for immune-based cancer prevention. *Proc Natl Acad Sci U S A.* 2016.
202. Yu L, Li L, Medeiros LJ, et al. NF-kappaB signaling pathway and its potential as a target for therapy in lymphoid neoplasms. *Blood Rev.* 2017;31(2):77–92.
203. Rupaimoole R, Slack FJ. MicroRNA therapeutics: towards a new era for the management of cancer and other diseases. *Nat Rev Drug Discov.* 2017;16(3):203–222.
204. Zhao N, Wang R, Zhou L, et al. MicroRNA-26b suppresses the NF-κB signaling and enhances the chemosensitivity of hepatocellular carcinoma cells by targeting TAK1 and TAB3. *Mol Cancer.* 2014;13(1).
205. Soldevilla MM, Meraviglia-Crivelli de Caso D, Menon AP, et al. Aptamer-iRNAs as therapeutics for cancer treatment. *Pharmaceuticals (Basel).* 2018;11(4).
206. Catela Ivkovic T, Voss G, Cornella H, et al. microRNAs as cancer therapeutics: a step closer to clinical application. *Cancer Lett.* 2017;407:113–122.

CHAPTER EIGHT

Evolution of biomarker research in autoimmunity conditions for health professionals and clinical practice

Arthur Silverstein[a], Anton Dudaev[b], Maria Studneva[b,c], John Aitken[d], Sofya Blokh[e], Andrew David Miller[f], Sofia Tanasova[g], Noel Rose[am,†], John Ryals[h], Christoph Borchers[i], Anders Nordstrom[j], Marina Moiseyakh[b], Arturo Solís Herrera[k], Nikita Skomorohov[b], Trevor Marshall[l,m,n], Alan Wu[o,p], R. Holland Cheng[q], Ksenia Syzko[b], Philip D. Cotter[p,r], Marianna Podzyuban[b], William Thilly[s], Paul David Smith[t], Paul Barach[u], Khaled Bouri[v], Yehuda Schoenfeld[w], Eiji Matsuura[x], Veronika Medvedeva[y], Ilya Shmulevich[z], Liang Cheng[aa], Paul Seegers[ab], Yekaterina Khotskaya[ac], Keith Flaherty[ad,ae], Steven Dooley[af], Eric J. Sorenson[ag], Michael Ross[ah], and Sergey Suchkov[b,c,g,ai,aj,ak,al,*]

[a]Department of the History of Medicine, Johns Hopkins University, Baltimore, MD, United States
[b]MGUPP, Moscow, Russia
[c]EPMA (European Association for Predictive, Preventive and Personalized Medicine), Brussels, Belgium
[d]Otakaro Pathways, Christchurch, New Zealand
[e]Filatov's Moscow Clinical Pediatric Hospital, Moscow, Russia
[f]Mendel University and Veterinary Research Institute, Brno, Czech Republic
[g]A.I. Evdokimov Moscow State University of Medicine & Dentistry, Moscow, Russia
[h]Mathematics Department at Central Georgia Technical College, Warner Robins, GA, United States
[i]Lady Davis Institute, Segal Cancer Proteomics Centre, McGill University, Montreal, QC, Canada
[j]Center for Proteomics, Faculty of Medicine, Department of Molecular Biology, Umea University, Umea, Sweden
[k]University of Florida, Gainesville, FL, United States
[l]University of Adelaide, Adelaide, SA, Australia
[m]Murdoch University, Murdoch, WA, Australia
[n]The Autoimmunity Research Foundation, Thousand Oaks, CA, United States
[o]San Francisco General Hospital, San Francisco, CA, United States
[p]UCSF, San Francisco, CA, United States
[q]College of Biological Sciences, UC Davis, Davis, CA, United States
[r]ResearchDx and PacificDx, Irvine, CA, United States
[s]Massachusetts Institute of Technology, MIT, Cambridge, MA, United States
[t]Newberry College, Newberry, SC, United States
[u]Wayne State University, Detroit, IL, United States
[v]Georgetown University, Washington, DC, United States
[w]Sheba Medical Center in Tel Hashomer, Sackler Faculty of Medicine, Tel-Aviv University, Tel-Aviv, Israel

† Deceased

Progress in Molecular Biology and Translational Science, Volume 190
ISSN 1877-1173
https://doi.org/10.1016/bs.pmbts.2022.02.004

[x]Neutron Therapy Research, Center Collaborative Research Center for Okayama Medical Innovation Center & Department of Cell Chemistry, Okayama University, Okayama, Japan
[y]I.M. Sechenov First Moscow State Medical University (Sechenov University), Moscow, Russia
[z]Institute of Systems Biology (ISB), Seattle, WA, United States
[aa]Genitourinary Pathology Service, The Urologic Pathology Fellowship, and Molecular Diagnostics and Molecular Pathology Laboratories, Indiana University School of Medicine, Indianapolis, IN, United States
[ab]Stichting PALGA (Pathologisch Anatomisch Landelijk Geautomatiseerd Archief), Houten, The Netherlands
[ac]Clinical Decision Support Management, The Curematch, Inc, San Diego, CA, United States
[ad]Harvard Medical School, Boston, MA, United States
[ae]Mass General Cancer Center, Massachusetts General Hospital, Boston, MA, United States
[af]Medical Faculty Mannheim, Heidelberg University, Heidelberg, Germany
[ag]Mayo Clinic, Rochester, MN, United States
[ah]UCLA Medical Center Obstetrics & Gynecology, Los Angeles, CA, United States
[ai]ISPM (International Society for Personalized Medicine), Tokyo, Japan
[aj]PMC (Personalized Medicine Coalition), Washington, DC, United States
[ak]New York Academy of Sciences, New York, NY, United States
[al]Association for Medical Education in Europe (AMEE), Dundee, United Kingdom
[am]Johns Hopkins Bloomberg School of Public Health, Baltimore, MD, United States
*Corresponding author: e-mail address: ssuchkov57@gmail.com

Contents

1. Introductory 221
2. Abs with catalytic activity (catAbs or abzymes) 226
3. Progressing toward personalized and precision healthcare services and challenges in applying biomarkers for clinical autoimmunity conditions 231
   3.1 Classical technological approaches to autoAb-based biomarker discovery and validation 236
4. Specific autoimmunity conditions subareas 243
   4.1 Biomarkers in clinical practice: Biomarkers as applicable to autoimmune myocarditis 243
   4.2 Biomarkers in clinical practice: Biomarkers as applicable to multiple sclerosis 245
   4.3 Biomarkers in clinical practice: Biomarkers as applicable to type 1 diabetes 246
   4.4 Biomarkers in clinical practice: Biomarkers as applicable to SLE 258
   4.5 Biomarkers in clinical practice: Biomarkers as applicable to RA 264
   4.6 Biomarkers in clinical practice: Biomarkers as applicable to systemic sclerosis 267
5. Biomarkers in clinical practice: Biomarkers as applicable to personalized dietary protocols in autoimmune conditions 270
6. Conclusion 273
References 274

Abstract

Medical abzymology has made a great contribution to the development of general autoimmunity theory: it has put the autoantibodies (Ab) as the key brick of the theory to the level of physiological functionality by providing such Ab with the ability to catalyze and mediate direct and independent cytotoxic effect on cellular and molecular targets. Natural catalytic autoantibodies (abzymes) while being a pool of canonical Abs and possessing catalytic activity belong to the new group of physiologically active substances whose features and properties are evolutionary consolidated in one functionally active biomolecule. Therefore, further studies on Ab-mediated autoAg degradation and other targeted Ab-mediated proteolysis may provide biomarkers of newer generations and thus a supplementary tool for assessing the disease progression and predicting disability of the patients and persons at risks.

This chapter is a summary of current knowledge and prognostic perspectives toward catalytic Abs in autoimmunity and thus some autoimmune clinical cases, their role in pathogenesis, and the exploitation of both whole molecules and their constituent parts in developing highly effective targeted drugs of the future to come, and thus the therapeutic protocols being individualized.

1. Introductory

Personalized and precision medicine (PPM) is a particularly novel and exciting topic in medicine and healthcare services and bioindustries illustrating a concept that has the potential to transform medical interventions by providing effective, tailored therapeutic strategies based on the genome- and phenome-related profiles of an individual, while remaining mindful of the individualized situation, as well as on exposome-related profile (Fig. 1) illustrating the microenvironment.

PPM is also an emergent and rapidly developing method of daily clinical practice that uses upgraded clinical philosophy and innovative technologies to provide evidence-based and clinically valuable decisions in regard to the diagnosis and treatment, prediction and prognostication, prevention, and prophylaxis of disease. Increased utilization of molecular stratification of patients and persons at risks will provide medical professionals with evidence upon which to base canonical therapeutic strategies for individual patients or preventive and prophylactic manipulations for individual persons at risk. PPM thus has the potential to offer improved medication selection and targeted therapy being biomarker-based, reduce adverse effects, increase patient compliance, shift the goal of medicine from reaction to prevention, improve cost effectiveness, and increase patient confidence postmarketing by approving novel therapeutic strategies and altering the perception of medicine in the healthcare system.

The concept and armamentarium of PPM emphasize innovations that enable the physician and other practitioners to tailor disease prediction, subclinical and clinical diagnosis, prevention, and treatment to the individual patient or a person at risk, based on a personalized data-driven and biomarker-based precision approach.

Among the main challenges to implementation of PPM in medical practice is knowledge gap of professionals about biomarkers (Fig. 2).

f (genome, exposome) → phenome

Fig. 1 Multibiomarker-based profiles.

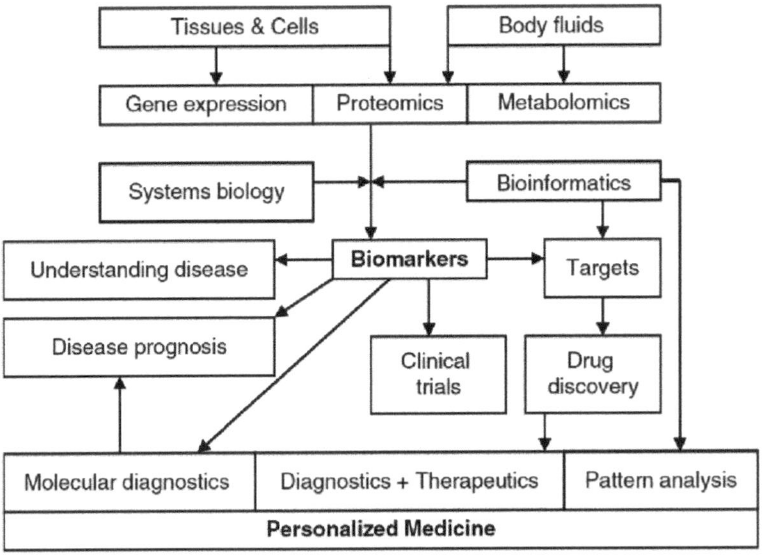

Fig. 2 Relation of biomarkers to PPM-related technology platforms and healthcare services.

Biomarkers can be measured and evaluated as indicators of normal or pathological processes or of pharmacological response to a therapeutic intervention. Classical biomarkers are measurable alterations in blood pressure, blood lactate levels following exercise, and blood glucose in diabetes mellitus. Any specific molecular alteration of DNA, RNA, metabolite, or protein level may be a molecular biomarker. Biomarkers usually mean molecular biomarkers and can be divided into three broad categories:

(i) those that track disease progression over time and correlate with known clinical measures;
(ii) those that detect the effect of a drug;
(iii) those that serve as surrogate end points in clinical trials.

PPM is highly dependent upon biomarkers, molecular health indicators that can be used to improve diagnoses and thus help to determine appropriate treatments. PPM would tailor biomarker-driven multimodal treatments to individual or subgroups of patients based on the use of disease-specific biomarkers.

Biomarkers applicate to improve patient outcomes in clinical practice and clinical trials, which is rapidly being adopted into daily medical care. Biomarkers can be anything that can be measured in a person (a patient or a person at risk) and may, in particular, include OMICS-related components measured from blood and other tissues as well as images from digital pathology and radiology.

A unique and multistep journey from the discovery of a biomarker to its implementation in clinical practice is long and complex. It involves surmounting many obstacles, including inadequate validation of the biomarker and evidence of its clinical utility, operational barriers to clinical implementation, and inadequate evidence of operational effectiveness or impact in daily clinical care.

Biomarkers can be anything that can be measured in a person and can range from genomic, phenomic, and exposomic sources measured from samples to imaging such as digital pathology, radiology, and molecular imaging with special contrast agents.

Quantitative biomarkers measure something using a continuous scale and will usually have associated values that represent "cut-offs" above (or sometime below) which signifies abnormality (biomarker positive). Quantitative, molecular phenotypes are becoming the mainstay of PPM. *Qualitative* biomarkers are either present or not (Fig. 3).

As an example, for patients with specific autoimmune conditions, a predictive biomarker that measures mechanisms of sensitivity or resistance would be used, in combination with a prognostic biomarker, to select the

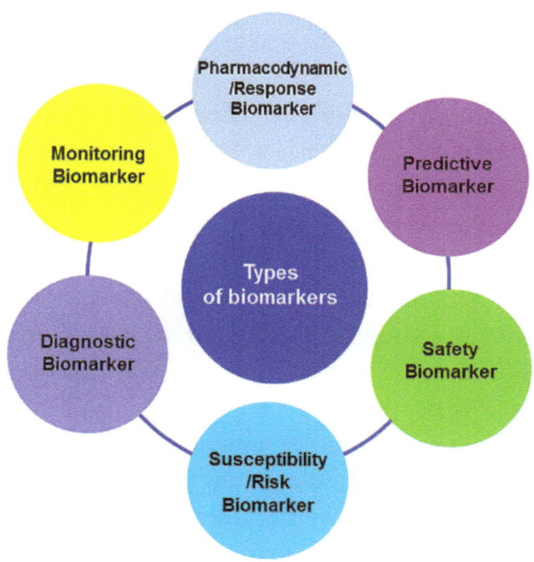

Fig. 3 Classification of biomarkers based on its main clinical applications. A canonical *diagnostic* biomarker is applied daily to identify individuals with a disease or condition of interest or to define a subset of the disease. A *prognostic* biomarker is used to estimate the outcome for a patient in the absence of a treatment. A *predictive* biomarker is used to estimate the benefit for a specific treatment, while being used to monitor the effectiveness of a prescribed treatment.

correct chemo- or immunotherapy. Application of this principle has helped increase the impact of treatments for autoimmune disorders and advanced PPM-based health care.

When a predictive biomarker is registered with a regulatory body (such as the FDA) along with an associated drug to select appropriate patients, it is referred to as a *companion* diagnostic and is on-label for the drug. Creation of panels of predictive biomarkers is needed for subclinical and early diagnostic and prognostic purposes, along with upgraded guidelines for biomarker-based predictive diagnostics and for preventive therapy.

Biomarkers can be used in clinical settings to facilitate drug *repurposing* and inform patient care decisions, and can be incorporated into drug development through the drug approval process, scientific community consensus followed by regulatory acceptance, and biomarker qualification. A new generation of biomarkers is required that define all aspects of the variability of unified system indicators. This would offer a new way to optimize treatment, decrease rehabilitation costs, and facilitate building new products and services in this area, viz. *multimarker-based companion diagnostics*.

These diagnostics are expected to identify individuals with a higher chance to benefit from a particular treatment, and in this way biomarkers can increase the success rate of drug development programs and accelerate the availability of new therapeutics.

For instance, in inflammatory bowel diseases (IBDs), C-reactive protein (CRP), fecal calprotectin, and fecal lactoferrin are evolving as a multimarker diagnostic panel for differential diagnosis, to monitor IBD progression, and to predict response to therapy.

In addition, having a complete understanding of a drug's pathway, interactome, and network interactions could expedite the identification of sensitizing mutations, drug interactions, or the risks of drug combinations to guide biomarker discovery, including simple, combinatorial, and network-based biomarkers (NBBs) (Fig. 4).

Clinically, those features can be prognostic, diagnostic, predictive, or response markers and clinicians could use any combination of those categorical biomarkers to describe those aspects of disease. In fact, clinicians are already using biomarkers in clinical decision making, e.g., integrating *biomarker as indicators of responses to therapeutic intervention* into pharmacogenetic tests.

The field of pharmacogenomics and nutrigenomics overlaps with the application areas, but expands the notion of biomarker-guided decisions to all drugs and uniquely emphasizes the role of genetic features.

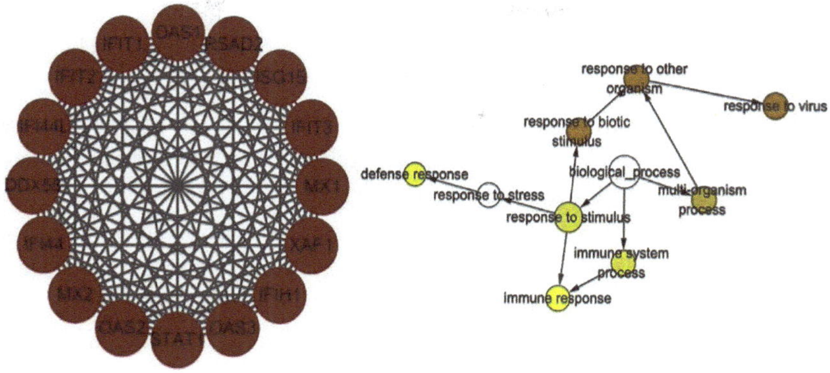

Fig. 4 Potential NBB-related protein biomarkers for systemic lupus erythematosus (SLE) determined by bioinformatics analysis. SLE is a heterogeneous autoimmune disorder, featuring with 90 (82 up- and 8 downregulated) differentially expressed genes (DEGs) common to female LN−, female LN+, and male LN+ using the GSE65391 and GSE49454 gene expression datasets from Gene Expression Omnibus database. The protein–protein interaction (PPI) network of 70 DEGs was constructed using STRING and cytoscape, and the Gene ontology and Kyoto Encyclopedia of Genes and Genomes pathway enrichment analysis showed that the PPI network was significantly enriched in defense response to virus, cytosol, protein binding, and measles. Sixteen hubgenes were identified from this PPI network, and Literature Mining Gene Networks molecular of GenCLiP 2.0 showed strong interaction between STAT1, DDX58, and IFIT1. Enrichment analysis of hubgenes in published literature showed the involvement of immune response and interferon-related genes in the pathogenesis of SLE. In addition, the transcription factors STAT1 and 2 and IRF6 and 9 had high Normalized Enrichment Score. The 70 DEGs with PPI network and 16 hubgenes are potential biomarkers of SLE, and can help improve diagnosis and develop individualized therapies. *NBBs*, network-based biomarkers.

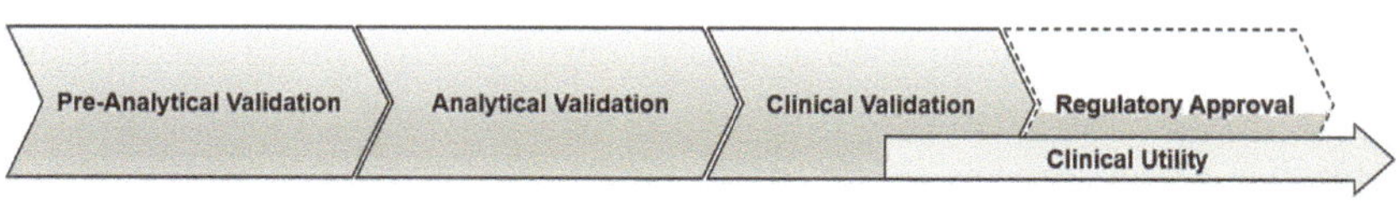

Fig. 5 The biomarker development process.

The FDA and NIH provided specific definitions for diagnostic, monitoring, pharmacodynamics/response, predictive, prognostic, safety, and susceptibility/risk biomarkers, and reasonably likely surrogate and validated surrogate endpoints.

The biomarker development process can be schematically divided into sequential phases, including laboratory and clinical validation, regulatory approval, and demonstration of clinical utility (Fig. 5).

Biomarkers as a target: The global promise of PPM is underpinned by large-scale biologic databases, OMICS-portfolios and diverse assays, and digital health technology. On this background circulating biomarkers, some of them already available as being preapproved by FDA/CE, may be at the forefront of PPM being used as a biotarget while representing a mediator of disease, or be tightly linked to a mediator.

Using clinical trans-omics, networks-related pathways, and IT algorithms to integrate clinical phenomes with molecular multiomics, autoimmune disease-specific biomarkers, and therapeutic targets can be identified and validated to find the causes of diseases and improve precise diagnosis, treatment, and prevention for specific patients or persons at risk.

Most current biomarkers do not satisfy the required characteristics for use among the spectrum of diseases. Better knowledge of the molecular history of diseases is necessary if biomarker development is to proceed.

2. Abs with catalytic activity (catAbs or abzymes)

Along with canonical Abs, some of the families proven to occur are Abs possessing with catalytic activity (catAbs or abzymes) and thus to belong to Abs with a feature of functionality!

CatAbs (or abzymes) are multivalent Igs, presumably of IgG isotype, endowed with a capacity to hydrolyze Ags. The enzymatic activity is located in the Fab fragment of the Ig molecule, which endows such antibodies with the ability to bind to specific antigens and hydrolyze them. Proteolytic Abs (or Ab-proteases) represent a significant portion of the family of abzymes that PPM uses to target specific Ags. Because of their Ag specificity, Ab-proteases also may be used as biomarkers able to control autoimmune disease progression to transform from subclinical into clinical stages, and to predict complications. Moreover, sequence-specific Ab-proteases have proved to be greatly informative and thus valuable as biomarkers to monitor autoimmune diseases at both subclinical and clinical stages while demonstrating their predictive value for the development of the disorder.

We may also identify specific pathways and interactome-based networks involved in diseases for which drugs have not yet been explored in appropriately designed trials.

NGBs help determine the probability of developing chronic pathologies or autoimmunity-predisposed condition. Key factors contributing to the growth of the global NGBs market include high prevalence of chronic autoimmune diseases; rising adoption of biomarkers for diagnostic, predictive, and prognostic applications; and increasing application in drug discovery and development.

For instance, the biomarkers of the future would be used for: (i) screening the general population or individuals at risk (panels of screening and predisposition biomarkers); (ii) the detection of the presence of a particular type of cancer (panels of diagnostic and prognostic biomarkers); (iii) monitoring the progression of autoimmune inflammation, and predicting the complications and outcome (panels of prognostic biomarkers); (iv) understanding whether a patient will benefit from a specific drug treatment (panels of predictive biomarkers); and (v) evaluating the drug's efficacy and optimizing the treatment, providing the tool to tailor treatment for individual autoimmunity-related patients or persons at risk (panels of pharmacodynamics biomarkers).

According to the FDA, an ideal biomarker (Fig. 6) must be specifically associated with a particular disease or disease state and be able to differentiate between similar physiological conditions. In this sense, characteristics of an ideal biomarker and basic statistical methods for evaluation of the latter mentioned would have to include:

- the biomarker should be easy to measure;
- the cost of follow-up tests should be relatively low and there should be proven treatments resulting in changes in biomarker levels;

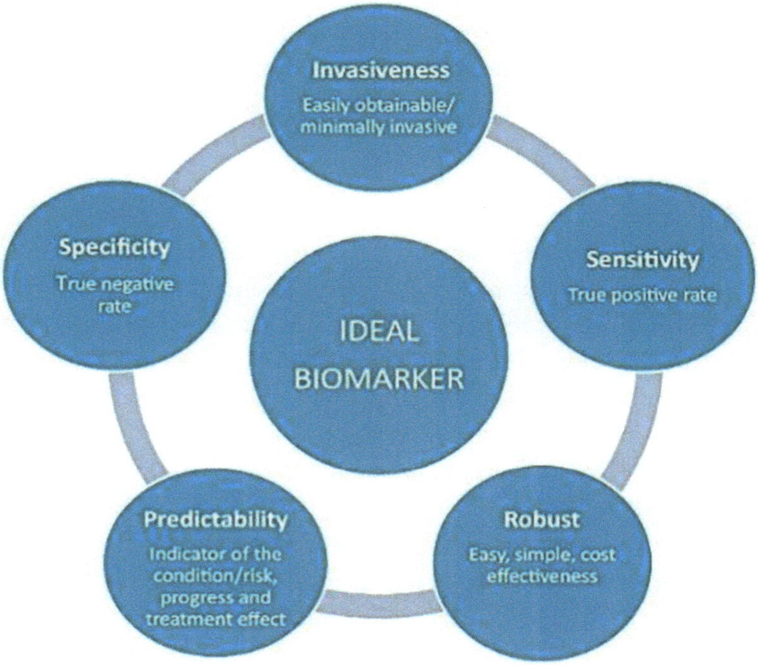

Fig. 6 Characteristics and features of an ideal biomarker.

- a rapid, simple, accurate, and inexpensive detection of the relevant marker should be available, together with a measurable and standard baseline as a reference point;
- standard biological sources, such as serum and urine, would be used for identifying biomarkers;
- it should be consistent across genders and ethnic groups.

The aspect of PPM that involves the measurement of biomarkers in clinical samples would involve understanding how biomarkers are defined, how they are used in clinical trials, and most importantly, how they are used in conjunction with drug treatment.

Those features have made clinical decisions more precise and personalized while improving health outcomes, but have a high bar for validation. Tremendous efforts have been made to date to discover novel biomarkers for use in clinical practice including:

(i) *biomarkers to facilitate drug development*

Increased focus on biomarker-based PPM is causing a paradigm shift in drug development.

(ii) *biomarkers as a guide for targeted therapy optimization* (Fig. 7)

Trials have looked at biomarkers as a guide for targeted therapy optimization and a generic trigger for having the targeted therapies titrated, rather than a one-marker/one-drug approach. Designing trials of biomarker-guided therapy has many challenges and most trials have important design issues. In addition to revisiting the strategy in the event of new effective drugs, optimal biomarker targets remain to be defined and would need to be studied in depth.

(iii) *Clinical validation of biomarker sets* (Fig. 8)

The process usually involves the application of the biomarker to a patient population that is entirely different (the validation dataset) to that used for the purpose of discovery and development (the training dataset). The true ability of the biomarker to guide PPM-related autoimmune disease regulations can be estimated in terms of sensitivity and specificity. This approach demonstrates the biomarker working very well in this defined group of patients and is referred to as "overfitting."

A *valid biomarker* is defined as a biomarker that is measured in an analytical test system with well-established performance characteristics and for which there is an established scientific framework or body of evidence that elucidates the physiologic, toxicologic, pharmacologic, or clinical significance of test results. The clinical utility (e.g., predict toxicity, effectiveness, or dosing) and use of epidemiology/population data (e.g., strength of

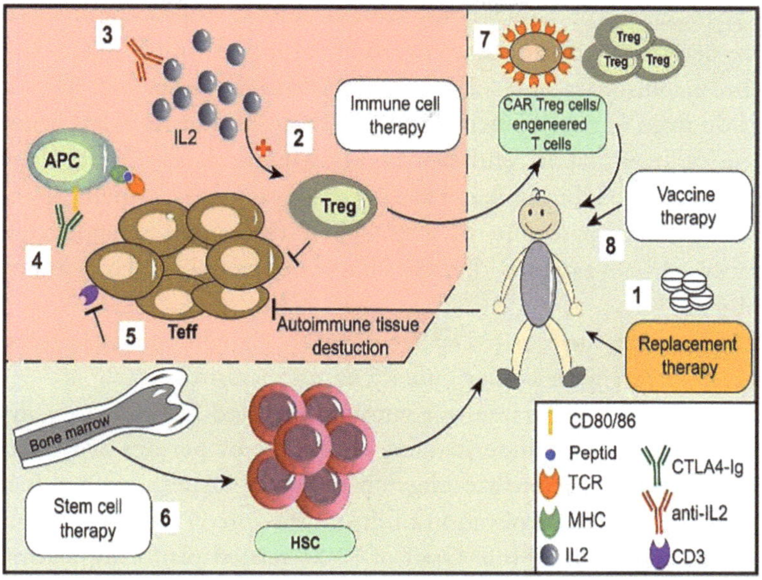

Fig. 7 Optimization of current and future biomarker-driven targeted therapy for autoimmune diseases. There are multiple immune-based therapeutics available for some of the most prevalent autoimmune diseases, but for others, there are few or no approved immune therapies. This dichotomy poses discrete challenges. First, for diseases in which multiple therapy choices exist, a rational decision tree is required to select the best therapy. Second, we must devise new strategies for the autoimmune diseases that have the highest unmet clinical need. This commentary outlines new strategies for designing more efficient and selective approaches for immune therapy of autoimmune diseases.

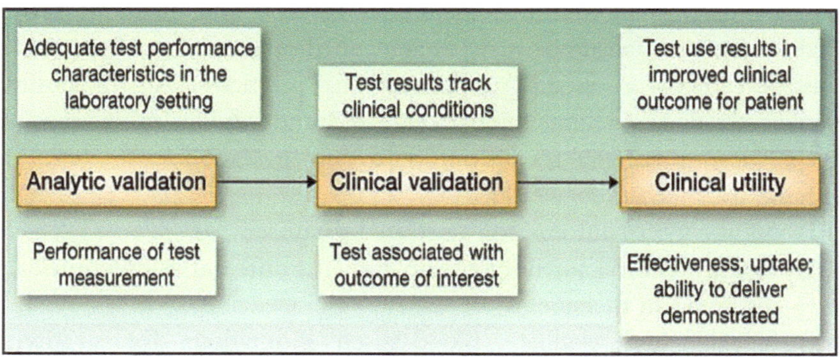

Fig. 8 Process for obtaining clinical utility with predictive/prognostic molecular biomarkers and other diagnostic tests. Analytical validation consists of demonstration of assay performance (accuracy, precision, specificity, etc.); clinical validation is the association between the test result and a clinical outcome, including the variability of a marker in subjects associated with a disease condition. Both analytical and clinical validations precede demonstration of clinical utility. Clinical utility is the evidence of improved, measurable clinical outcomes directly related to the use of the test (see text for further discussion).

genotype–phenotype associations) are examples of approaches that can be used to determine the necessary criteria for validation.

Chronic disorder and autoimmunity-related specialists today are in transition from standard practice to PPM, but this transition has been slow because of insufficient fundamental knowledge, lack of confidence in modern advanced biomarker-based testing, as well as medical inertia. The latter is in part due to entrenched resistance to application of biomarker-driven targeted therapies deeply engrained in clinical practice guidelines.

Autoimmune diseases (ADs) are a family of complex heterogeneous chronic disorders with similar underlying mechanisms.

New preventive therapies for autoimmune and inflammatory diseases would require greater understanding of patient or persons-at-risk subsets and the ability to personalize targeted therapies for either. In this sense, the identification and application of diagnostic, predictive, and prognostic biomarkers remain the Holy Grail of PPM-related platforms, algorithms, and protocols. We must identify new biomarkers and create simultaneously a fundamentally new strategy based upon subclinical recognition of the bio-indicators long before the disease manifests clinically.

Among the best validated autoimmunity-related predictive biomarkers are those that predict risks of chronic disabling disease. This is because chronic autoimmune diseases are preceded by a long subclinical (symptom-free) phase (Fig. 9).

Meanwhile, the utility of predictive biomarkers (including autoAbs) to monitor chronic autoimmune conditions (usually inflammation) and to predict stepwise transitions is dependent on three key parameters, which must be carefully assessed: (i) sensitivity, (ii) specificity, and (iii) positive predictive value. Accurate biomarker-based and -driven prediction and prognostication are vital for prevention of autoaggression.

Specifically, biomarkers would:

(i) build a global autoimmunity-related database;

(ii) identify clinical phenotypic subsets with differential outcomes and responses to therapies;

(iii) identify and evaluate blood-based biomarkers for stratifying autoimmunity-related individuals into clinically, biologically, and therapeutically meaningful clusters;

(iv) evaluate novel biomarkers and stratification strategies identified within the principal database and affiliated targeted datasets;

(v) define the most effective and practical stratification strategy for use in clinical trials;

Disease	Autoantibodies
Type 1 diabetes mellitus	Anti-GAD
	IA-2
	IAA
RA	RF (anti-IgG)
	Anti-CCP
SLE	Anti-PL
	Anti-Ro
	Anti-La
	Anti-dsDNA
	Anti-Sm
	Antinuclear ribonucleoprotein
	Anti-HS
	Antinucleosome
	Antihistone
	Antiribosomal P protein
Sjögren's disease	Anti-Ro
	Anti-La
APS	Anticardiolipin
	Lupus anticoagulant
PBC	Anti-gp120
	Anti-PDC
AIH	ANA
Crohn's disease	ASCA
UC	pANCA
Autoimmune Addison's disease	ACA (anti-21-hydroxylase)
AITD	Anti-TG
	Anti-TPO
Pemphigus	Antidesmoglein 1

GAD = glutamic acid decarboxylase; IA-2 = protein tyrosine phosphatase-like molecule; IAA = insulin autoantibodies; RF = rheumatoid factor; CCP = cyclic citrullinated peptide; PL = phospholipids; HS = heparan sulphate; PDC = pyruvate dehydrogenase complex; ANA = antinuclear antibodies; ASCA = anti-*Saccharomyces cerevisiae* antibodies; pANCA = perinuclear antineutrophil cytoplasmic antibodies; ACA = adrenal cortex cells antibodies; TG = thyroglobulin; TPO = thyroid peroxidase.

Fig. 9 Predictive biomarkers in the practice of autoimmunity-related conditions.

(vi) use a combination of systems biology, data-driven, and mechanistic modeling to identify biomarkers that correlate with major organ manifestations and long-term outcomes;

(vii) evaluate the performance of the identified biomarkers using the samples generated by the NECESSITY clinical trial.

3. Progressing toward personalized and precision healthcare services and challenges in applying biomarkers for clinical autoimmunity conditions

Recent advances in multiple "omics" (multiomics) approaches in combination with bioinformatics and biostatistics have made it possible to accelerate the discovery and development of specific biomarkers for complex chronic diseases.

Furthermore, effective translation of the preclinical biomarkers into the clinic will pave the way toward effective execution of personalized therapies across complex disease areas for the benefit of patients, healthcare providers, and the biopharmaceutical industry.

(i) *Genomics-related biomarkers*

Differential gene expression results in the carefully controlled (or regulated) expression of functional proteins or small RNA molecules which might serve as predictive and prognostic biomarkers. In this sense, a substantial need does exist for developing and validating a range of genomic biomarkers that would address a number of important unmet clinical needs in the autoimmunity field.

The journey mentioned above begins based on preexisting genetic equipment followed by the onset of a self-directed autoimmune response, which has a genetic predisposition to provoke and dictate the abnormal immune response depending upon interactions among susceptibility genes (Fig. 10) to surveil immune control, and various environmental (exposomal) factors.

Autoimmune disease	*MHC antigen*	*Relative risk*
Ankylosing spondylitis	B27	69.1
Reiter's syndrome	B27	37.0
Rheumatoid arthritis	Dw4/DR4	3.8
Sjögren's syndrome	Dw3	5.7
Systemic lupus	DR3 (Caucasian)	2.6
erythematosus	DR2 (Japanese)	5.3
	DQ3 (Chinese)	11.5
Celiac disease	DR3	11.6
Pernicious anemia	DR5	5.4
Pemphigus vulgaris	DR4 (Jews)	14.6
Type 1 diabetes mellitus	DR4	3.6
	DR3	4.8
	DR2	0.2
	BfF1	15.0
Graves disease	DR3	3.7
Addison's disease	Dw3	10.5
Hashimoto's thyroiditis	DR5	3.2
Multiple sclerosis	DR2	6.0
Goodpasture's syndrome	DR2	15.9

Fig. 10 The MHC/HLA biomarkers are the most valuable ones for assessing the susceptibility to autoimmunity and the autoimmunity risks. Different HLA alleles are linked to different autoimmune diseases. Among those alleles are: (i) A1 and DQW1 associated with SLE; (ii) DQ3-DR4, DQ3-DR9, DQ5-DR1, DQ5-DR10 associated with RA; and (iii) DR3 and B8 associated with RA, SLE, autoimmune thyroiditis, celiac, multiple sclerosis, myasthenia gravis, etc. Other genetic risk factors confirmed are polymorphisms in specific genes encoding molecules involved in antigen presentation, such as Tap-1 and proteosomes, as well as many other candidate genes.

(ii) *SNP-related (single nucleotide polymorphisms)* and *haplotype biomarkers* are important as they help in understanding the genetic basis of common autoimmune diseases. Until WGS testing (Whole Genome Sequencing test) of individual patients becomes feasible clinically, the identification of SNPs and haplotypes will prove instrumental in efforts to individualize health care in clinical autoimmunity practice. When an extensive inventory of genome-wide SNP scans has been assembled across diverse population samples, LD maps (linkage disequilibrium maps) using SNP or haplotypes will make it unnecessary to identify the precise genes involved in determining therapeutic efficacy or an adverse reaction.

Mitochondria have their own genome and there is growing evidence that defects of mitochondrial DNA (mtDNA) cause multisystem or tissue-specific autoimmune diseases. Thus, there is a need to assess DNA damage, modifications, or postreshuffled shifts because of the impact that different insults on genetic material may have on the autoimmunity predisposition and/or human health as a whole.

Profiling of mRNA in serum has been found to be useful for the development of predictive biomarkers for autoimmune disease and treatment outcome. With advances in biotechnology, more biological functions of circRNAs (circular RNA) have been found in several signaling pathways that are related to autoimmunity being expressed abnormally and play important roles in the progression of autoimmune diseases. From a clinical perspective, further investigations of circRNAs/sncRNAs/miRNAs have the potential to aid our understanding of the phenotypic diversity related to the autoimmune disease pathogenesis, subtype, or clinical parameters. It also will allow to elucidate mechanisms underpinning autoimmune disease and to improve diagnosis, prognosis, and prediction of response, while serving as a promising tool for PPM.

(iii) *Metabolomic biomarkers*, like the proverbial canaries in coal mines, can act as indicators of autoimmunity conditions and human health as a whole because the metabolome is highly sensitive to food, lifestyle, environment, seasons, and even mood. A single base change in DNA can lead to a 100,000-fold change in metabolite levels. Unfortunately, less than 1% of known metabolites are being used in autoimmunity-related practice as specific metabolic biomarkers.

(iv) *Cellular biomarkers* can be used for assessing autoimmune disease progression and therapeutic intervention and for the association of disease-related biomarkers on the protein level. Moreover, identification of biomarkers associated with particular cell types can in some cases be exploited and developed into indicators of certain types of autoimmune disease.

(v) *Antibody (Ab) biomarkers*

Notwithstanding the underlying mechanism of disease, virtually all autoimmune diseases are associated with circulating autoAbs, which bind self-protein. Furthermore, for many diseases, these autoAbs are found in serum samples many years before disease onset. AutoAbs, in addition to their contributions to disease pathogenesis, can be markers of future disease in presently healthy individuals. Tests that detect autoAbs could warn of the need to take preventive action, as for example, do antinuclear antibodies in the detection lupus erythematosus.

Abs against the family of Argonaute proteins (mainly AGO1 and AGO2), being reported in systemic autoimmunity, are detected in patients with a high suspicion of autoimmune disorders.

For T1D model, about half of the total risk is genetically predisposed, and about half of the risk is in the HLA and other regions to be useful for gene-based predictive testing (Fig. 11).

More precisely, autoimmunity results from disruption of different combinations of genes participating in the same immunologic function designated as tolerance. Meanwhile, some susceptibility loci are shared by several autoimmune diseases, whereas others are more disease specific. Nevertheless, due to the highly multifactorial nature of most autoimmune diseases, individual loci do not seem significant enough to be used as biomarkers or multimarker-based panels for susceptibility or as diagnostic tools.

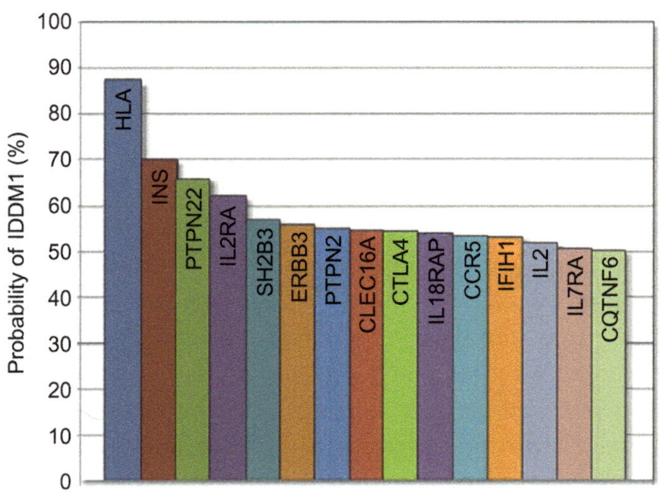

Fig. 11 Genomic biomarkers as applicable to T1D model.

(vi) *Proteomics-related biomarkers*

The autoAgs/Ab are not only hallmarks but also involved in pathogenesis of autoimmune diseases. The development of proteomic technologies provides an unprecedented ability to identify novel biosignatures to diagnose, classify, and guide therapeutic decision making in patients with autoimmune disease. The necessity of adopting existing technologies of protein chemistry, predisposition testing, targeted monitoring, and prevention of diseases through pharmacotherapy and nutrition will be focused as modern diagnostic tools in realizing the changeover from canonical medicine to PPM. Biological fluids can be used for serum/plasma proteomics to identify low-abundance proteins as biomarkers of autoimmune diseases.

Current proteomic research follows at least two pathways including: (i) the identification of proteins in patients' samples to be used as diagnostic or prognostic disease markers, and (ii) the discovery of cellular proteins related to the response to various therapies. In the future, the ability to routinely identify thousands of proteins in the body fluids will be available.

Proteomic techniques will pave the way for: (a) new biomarker discovery as diagnostic or prognostic markers, and (b) development of new ways to manage the disease, i.e., targeted intervention to modulate metabolic machinery in the cell or tissue early in the disease. The range of autoAbs and spectrum of their action in different autoimmune diseases at different stages of disease will guide a new understanding and managing of these diseases, which so far have been proved to be unsatisfactory and in certain cases elusive.

The developing proteomics technologies that search for disease biomarkers will eventually solve problems in pathogenesis and treatment of autoimmune diseases. Multiple proteomics technologies provide the potential to identify biomarkers with utility for PPM and targeted therapy. For effective Ab-based analysis, simultaneous determination of several Abs with multiparametric confirmatory assays will accelerate serological diagnosis compared with conventional monoparametric testing. This novel concept, the CytoBead assay, is applicable for the determination of various multiparametric antiAbs (Fig. 12), e.g., the determination of antinuclear Abs (ANA) and the corresponding autoAbs in molecular cytoplasmic and nuclear autoAg structures, revealing determinants of celiac disease or ANCA-associated vascular pathologies.

Ultimately, the development and use of such proteomic biomarkers for diagnosis, assessing prognosis, and guiding therapy will revolutionize the care for autoimmune disease patients.

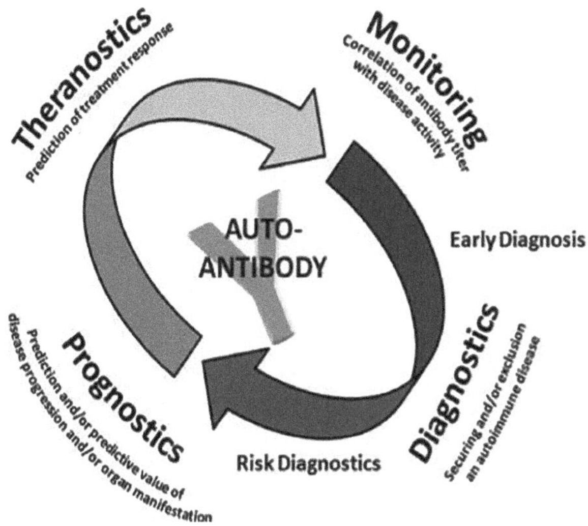

Fig. 12 The CytoBead assay as a novel approach of multiparametric autoantibody analysis in the diagnostics of systemic autoimmune diseases.

PPM applied to autoimmune disease will need paradigm shifts in the use and application of autoantibodies (*autoAbs*) and other biomarkers. For example, autoAbs combined with other multianalyte "OMICS" profiles will form the basis of disease prediction allowing for earlier intervention linked to disease prevention strategies, as well as earlier, effective, and personalized interventions for established disease. In this sense, newer diagnostic and predictive (biomarker-based) platforms that utilize emerging deep learning (DL) and artificial intelligence (AI) algorithms and close the gaps in autoAb-related diagnostics will benefit from paradigm shifts, thereby facilitating the personalized and precision health agenda.

3.1 Classical technological approaches to autoAb-based biomarker discovery and validation

(i) *Western blotting* allows detection of specific proteins to the level of single isotypes. However, it is associated with poor reproducibility, limited mass resolution, lack of accurate quantitation, low throughput, and lengthy time to result, while nonspecific cross-reactivity of mono- and polyclonal primary and secondary antibodies is common.

(ii) *Enzyme-linked immunosorbent assays (ELISAs)*, unlike western blotting, are adaptable to higher throughput of samples as are typically performed in microtiter plates whereby plate handling and detection systems can be automated. ELISA can be used to determine the exact amount of a specific protein in a sample, making it more readily quantitative as a technique compared to western blotting. The signals are usually produced by chromogenic reaction that generate a colored product, which is quantified by spectrophotometry. There are four types of ELISAs—sandwich, direct, indirect, and competitive—which differ in how the antigen and capture antibody are immobilized on a surface or in what order they are used.

In the context of autoAb detection, the direct and indirect ELISA formats are most commonly used, but are better suited to the analysis of a larger number of samples against a small number of antigens in screening, verification, and validation applications rather than as a primary discovery platform.

(iii) *Mass spectrometry* for serum biomarker discovery is in theory straightforward since the results are obtained in the form of identified and quantified proteins that are then compared between pathological and control groups. Recent advances in mass spectrometry instrumentation have significantly improved the depth, breadth, and reproducibility of protein identifications in many biological samples, which in turn has aided the identification of meaningful signatures that have diagnostic potential. Furthermore, mass spectrometry-based proteomics remains heavily constrained today in its ability to differentiate and assign function to individual Ab sequences within a large collection of immunoglobulins. This is partly because the affinity-matured antibody sequences are not germline encoded (and therefore do not appear in the proteome databases that underpin tandem mass spectrometry-based protein identifications) and is partly because both light and heavy chains are required for Ag specificity in an immunoglobulin, yet that pairing between light and heavy chain sequences (as well as the connectivity between the complementary determining regions with each light and heavy chain) is lost during proteolytic digest before mass spectrometry analysis. Moreover, antigen specificity cannot yet be predicted de novo from the primary immunoglobulin sequence. As a result, mass spectrometry is currently not well suited to the challenge of autoAb biomarker discovery; alternative technological platforms are therefore required to unravel the complexity of the human immunoglobulin repertoire and to detect and quantify novel autoantibody/autoantigen pairs in biological samples.

(iv) *Serologic Proteome Analysis (SERPA)* is a classical immunoproteomics approach to autoAg discovery that provides a robust way of screening Ab reactivity profiles in sera from patients with various diseases. The method, which is essentially an adaption of western blotting, involves separating proteins from a biological sample using two-dimensional electrophoresis on large format gels and then immunoblotting with patient or control sera. Unique protein spots identified by following blotting with patient but not control sera are excised from the gels and identified by mass spectrometry. However, the inherently high gel-to-gel variability and relatively low resolving power of individual gels impact on the accuracy of spot picking and impose a limitation due to comigrating proteins, which are especially problematic for low-abundance protein targets. This technology is thus less widely used for autoAg discovery now and has been largely supplanted by newer technologies that are better able to overcome these limitations.

(v) *Serological analysis of recombinant cDNA expression libraries (SEREX)* initially involves generation of a cDNA library from a cancer tissue or tumor cell line, followed by cloning of that library into a suitable expression vector, clonal separation of the library, and expression of the encoded proteins in *Escherichia coli* cells grown on solid media. Colonies are then transferred to a nitrocellulose or PVDF membrane, lysed, and the expressed recombinant proteins blotted with sera from patients and healthy controls. Seroreactive proteins are then identified by sequencing the cDNA from positive colonies [1], which makes it more sensitive than the SERPA method since the latter relies on direct protein identification and is therefore limited by absolute protein abundance. Furthermore, the clonal separation of the members of the cDNA expression library provides greater resolving power than the gel-based SERPA method. However, as with all library-based screening methods, oversampling is required to ensure that all members of the library are examined in SEREX: thus, if, for example, the cDNA library contains 104 unique clones, then at least 105 colonies would need to be screened for complete coverage, so even with the advent of colony picking and arraying robots, SEREX remains a relatively low-throughput method.

(vi) *Protein microarrays* are a versatile, miniaturized platform used to simultaneously characterize the biomolecular interactions of thousands of different proteins that are spotted in defined locations on a solid support; as such, protein microarrays represent a natural technological evolution from ELISA, SERPA, and SEREX. Protein microarrays, in principle, allow the quantitative analysis of binding of a wide variety of analytes, including Abs, proteins, DNA, RNA, small molecules, lipids, enzymes, as well as peptides. The three types of protein microarrays that are commonly used are analytical, functional, and reverse-phase microarrays. Analytical protein arrays, or antibody arrays, are ideal for quantification of different known proteins in a biological sample, monitoring protein expression levels and protein profiling in what amounts to miniaturized, highly multiplexed ELISAs. Functional protein microarrays can be subdivided into those based on recombinant proteins and native proteins and can be used for autoantibody and immune response profiling, biomolecular interaction profiling, and identification of enzyme substrates, among others. Reverse-phase protein arrays are composed of spots of different crude tissue homogenates or cell lysates and are suited for detection of known proteins in multiple tissues/cells based on blotting of the reverse-phase arrays with antigen-specific antibodies. In general, protein microarrays can be applied in diagnostic and therapeutic research, through new biomarker discovery for disease staging and monitoring, potential drug-target evaluation, and for identification of new drug targets. Of the different protein array types, functional protein arrays appear best suited to autoantigen discovery and autoantibody profiling and are discussed in more detail below.

(vii) *Recombinant protein production*: Different protein production systems can be employed to produce recombinant proteins in sufficient quantities for protein microarray fabrication. The key problem associated with recombinant protein production is identifying the best expression system for a particular protein. To date, there is no universally applicable protein expression system [2]. Each system has its advantages and disadvantages; therefore, the choice of expression system selection should be based on the properties of the recombinant protein as well as the scale of expression required. Although exploring multiple expression systems in parallel sounds enticing, factors such as protein solubility, yield, speed, and cost need to be taken into consideration as it involves substantial resources. Choosing the right system for protein expression can be particularly important in obtaining biologically active and functional recombinant proteins [3].

(viii) *Surface chemistry*: The microarray surface chemistry plays a critical role in determining protein microarray quality. Slide surfaces vary; aldehyde and epoxy-derivatized glass surfaces are used for random attachment through amines, whereas nitrocellulose, hydrogel, or metal surfaces are used for attachment of affinity-purified proteins. An ideal surface chemistry should resist nonspecific adsorption, while preserving the folded structure of the arrayed proteins.

AutoAbs as biomarkers have gained considerable attention in the clinical autoimmunity diagnostic field as candidate diagnostic and prognostic biomarkers because they are, in theory, detectable many years before clinical symptoms appear. However, identification and validation of autoAb biomarkers has historically been constrained by the available technological approaches and the high attrition rates during studies on larger cohorts.

In current practice, autoAbs are most often used to confirm diagnoses and classify systemic inflammatory and autoimmune diseases, but are also becoming increasingly important biomarkers that predict complications and/or comorbidities. In addition to autoAbs, other biomarkers included in OMICS-portfolios are also becoming key components of PPM-related care services. An important contribution of those biomarker-based multi-analyte (multimarker) panels is to help understand the pathogenic processes throughout the entire course of the disease. This understanding will be key to guide patient stratification for clinical trials and evidence-based interventions as opposed to the "trial-and-error" approaches that are prevalent today.

Advances in systems biology, network, and integrative medicine and translational research have generated many candidate biomarkers with potential clinical value. Before diagnosis, biomarkers could be used for screening and risk assessment. During diagnosis, markers can determine staging, grading, and selection of initial targeted therapy. During treatment, they can be used to monitor targeted therapy, select additional therapy, or monitor recurrent diseases.

Plasma proteomics is increasingly recognized as a fundamental approach to accelerate biomarker discovery and drug development, taking PPM resources to the next level: thousands of proteins can be analyzed in plasma to identify new biomarkers and to define the upgraded applications, key challenges, strategies, and discoveries in PPM-related proteomics.

A combination of genomic and phenomic (proteomic, metabolomics, cytomic, etc.) biomarkers is becoming of great significance to predict risks of the chronification. So, the concept of an actionable biomarker is based on the expectation that results of biomarker testing can be used to guide clinical management of disease.

Actionable biomarkers can inform clinical practice at many different stages of an autoimmune disease.

The most basic use of an actionable biomarker is in making a clinical diagnosis of symptomatic disease. For example, detection of autoAbs directed against specific disease-related biomarkers indicates the presence of autoimmune disorder. In systemic autoimmune diseases, diagnostic biomarkers are also central to clinical practice. The presence of rheumatoid factor (RF) and anticitrullinated protein Abs (ACPA) aid the diagnosis of rheumatoid arthritis (RA); and the presence and specificities of ANA facilitate diagnosis of systemic lupus erythematosus (SLE).

Although specific genetic mutations or polymorphisms are associated with certain autoinflammatory conditions, such as familial Mediterranean fever and Muckle–Wells syndrome, and minor subtypes of certain autoimmune diseases, SLE for example, most autoimmune diseases are polygenic, with individual gene polymorphisms conferring only a modest increase in disease risk.

3.1.1 Diagnosis of asymptomatic disease

Beyond their use in diagnosing symptomatic disease, actionable biomarkers can prove informative in the diagnosis of early, asymptomatic disease. Diagnosing a disease at the asymptomatic stage could enable early therapeutic intervention, with the ultimate goal of preventing the development of symptomatic disease or at least limiting the pathologic sequelae of the disease. Prophylactic treatment has the potential to reduce the incidence or severity of RA in this group of subclinical persons at risk, given that current treatments result in impressive clinical and radiological improvements when used to treat recent-onset, symptomatic RA. Likewise, specific autoAbs can be detected in the blood of individuals who will go on to develop symptomatic SLE, which might prove useful in guiding prophylactic treatment of this disease.

3.1.2 Assessment and prediction of disease activity

Biomarkers that aid assessment of disease activity are useful because they can reveal the presence or progression of disease despite the remission of symptoms. For example, such biomarkers are especially needed for osteoarthritis (OA), for which few, if any, disease-modifying therapies exist. Cartilage oligomeric matrix protein, a product of cartilage turnover, and metabolomic profiles are being explored as potential biomarkers for assessment of disease activity in OA. In addition to providing potential surrogate endpoints, biomarkers could also facilitate enrollment of patient subgroups more likely to response to a candidate therapeutic.

3.1.3 Dynamic biomarkers

Assessment of disease progression and response to therapy—another actionable use of biomarkers—is monitoring the progression of disease or response to therapy. Such "dynamic" biomarkers can facilitate prediction of the ultimate clinical outcome by reporting early changes in disease-associated biological processes. Results of dynamic biomarker profiling could prompt the clinician to initiate or intensify therapy in the setting of highly active disease or, conversely, to withdraw a specific treatment in the setting of an insufficient therapeutic response.

Beyond clinical practice, pharmacodynamic biomarkers that serve as surrogate endpoints, described by the FDA as biomarkers that can substitute for a clinical endpoint that reflects "how a patient feels, functions, or survives," can be useful in drug development. Pharmacodynamic biomarkers are used primarily to guide drug development, with clinical endpoints still forming the basis of regulatory approval.

3.1.4 Assessment of drug toxicity

The use of pharmacodynamic biomarkers in drug development can also enable early detection of drug-related toxicity. Examples of dynamic biomarkers of toxicity include traditional laboratory variables such as biochemical indicators of liver and kidney dysfunction, but also surrogates of cardiac toxicity, such as increased blood pressure, an increase in levels of serum lipids, and prolongation of the QT interval on an electrocardiogram. Going forward, it will be important to identify NGBs that can serve as early indicators of toxicity or other adverse events in clinical trials.

3.1.5 Predicting responsiveness to therapy

Just as useful as the pharmacodynamic biomarkers, which reflect the response to therapy, are biomarkers that predict responsiveness to therapy before therapy is initiated. Matching the right drug with each individual is important because a particular treatment will benefit only the subset of patients in whom the mechanism the drug targets is active. Thus, the ability to identify the individuals most likely to respond to a particular therapeutic would greatly benefit patients, preventing those who would not benefit from experiencing drug-related adverse events and incurring the costs of a treatment that is ineffective.

For many autoimmune diseases, a single biomarker can be informative on a population level. This inadequacy has shifted attention to the use of multiple biomarkers and, in parallel, to the development of technologies for the multiplex measurement of multiple variables. A panel of multiple biomarkers comprises different entities of the same type of variable, for example, a number of distinct circulating proteins or expressed genes representing a specific molecular pathway. Alternatively, the panel comprises a combination of disparate types of feature, such as a collection of radiographic, histologic, cellular, proteomic, and genetic variables. Combining known biomarkers in a *multiple biomarker approach* might in fact improve the prediction of the risk of the subclinical pathology or disease-related event. Moreover, the impact of adding the multimarker scores to the standard risk factors leads to the discovery of plasma biomarker candidates that have a larger potential to become the next-generation biomarker for usage in the clinic and can unlock the potential of the human proteome, uncovering actionable insights that are essential for biomarker discovery and targeted drug development. A number of limitations of *multimarker panels* should be acknowledged, including potential multiplexing and analytical challenges in screening, monitoring, and assaying multiple markers simultaneously, as well as the challenges of interpretation for the clinician due to different cut-offs for each of the separate markers. Nevertheless, it can be anticipated that scoring calculators and algorithms will increasingly use circulating biomarkers in combination with clinical variables to allow appropriate surveillance and fully informed counseling of patients suffering with chronic autoimmune disorders, their families and persons at risk, and other stakeholders in the process of the long-term care.

4. Specific autoimmunity conditions subareas

4.1 Biomarkers in clinical practice: Biomarkers as applicable to autoimmune myocarditis

Autoimmune myocarditis (AM) is one of the key unresolved issues in cardiology and requires new biomarker-based approaches using immune- and gene diagnostic techniques. Different types of antimyocardial autoAbs exist and thus it is crucial to understand the pathogenic significance of each.

Antimyocardial autoAbs constitute a rather heterogeneous group of Abs, including entire families of Abs against beta 1-adrenergic and muscarinic receptors, mitochondrial ATP/ADP-translocase, sarcolemma proteins, ANT-translocator, etc., but most often, against alpha- and beta-chains of myosin. AutoAbs secreted by autoreactive B-cells are directed against a wide array of different epitopes of the same myocardial (or mimicry-related) autoAg, which suggests the essential role of the primary antigenic stimulus for the induction of autoreactive B-cell clones. The detection of high-affinity antimyocardial IgG-class autoAbs in the sera indicates an important and direct role of myocardial autoAg in the induction of autoAb synthesis in AM. Such autoAbs are known to be a consequence of switching the synthesis of natural IgM-class autoAbs, which are detected at the preclinical stages of AM, to the synthesis of high-affinity IgG-class autoAbs containing a wide range of idiotypic determinants (including pathogenic properties) in their structure and characteristic for the clinical picture. Meanwhile, the appearance of natural IgM-isotype autoAbs in blood of clinically healthy donors is considered an early predictive sign of autoimmune disorders and as one of the key factors in the etiopathogenesis of autoimmune diseases.

At the preclinical phase of AM, autoreactive T-cells are in contact with the primary inducing epitope or pathogenic idiotype and capable of recognizing latent autoepitopes in the autoAg molecule since embryogenesis are usually not activated. The density of such autoepitopes on the surface of APCs usually is insufficient for induction of proliferation and activation of autoreactive effector T-cells to myocardial tissue.

A different situation occurs in persons who have had certain types of infections (chlamydiae, coxsackieviruses, CMV). Here, the pathogens might have epitopes cross-reacting with those of the heart, which leads to the formation of a huge number of processed peptides with immunogenic properties on the surface of professional antigen-presenting cells (APCs).

According to I. Takeda et al., many autoAbs are formed in patients with autoimmune diseases in continuation of apoptosis, and with the discovery of new, previously unknown Abs (including autoAbs with catalytic properties) the study of apoptosis mechanisms at various stages of autoimmune disease is directly related to the molecular and cellular effects of such autoAbs.

Catalytic autoAbs as a diagnostic tool for AM: Two classes of antimyocardial autoAbs are detected in these forms of autoimmune pathology, namely, autoAbs against alpha- and beta-chains of cardiomyosin. The presence of autoAbs to the alpha-chain alone satisfies the criteria of organ-specific autoimmune syndrome, illustrating the possibility not only of adequate diagnostics and monitoring but also of seroprediction of the disease course.

According to R. Mobini et al., anti-CM autoAbs disturb the structural integrity and physiological activity of cardiomyocytes, participating in the induction and progression of autoimmune inflammation. H.P. Schultheiss et al. demonstrated cross-reactivity of antimyocardial autoAbs both with mitochondrial membrane antigens in the classical version and with mimicry peptides of viral origin. They also reported, in blood of patients with AM, including those complicated by DCM development, the presence of autoAbs to intracellular proteins, e.g., creatine kinase, desmin, and heat shock proteins that cross-react with anti-KM autoAbs, which can aggravate ISS development and disease course in general.

Future therapeutic possibilities: As the pathophysiological significance of anticardiac autoAbs directed against various Ags has become increasingly clear, attention has shifted to methods to suppress their effects. One of the new approaches to the treatment of AM and its complications (in particular, DCM) can be immunosorption, providing suppression of Ab-mediated proteolytic activity against myocardium or removal of autoAbs themselves from the bloodstream and thus contributing to suppression of autoimmune inflammation activity, improving myocardial function, and overall condition of the patient.

4.2 Biomarkers in clinical practice: Biomarkers as applicable to multiple sclerosis

A large spectrum of myelin-associated autoAbs occur in MS patients. The crucial step in the pathogenesis of MS is demyelination mediated by autoAbs, which causes release of separate and pathogenically valuable myelin-associated epitopes into the bloodstream. Along with canonical Abs, some of the families proven to occur are Abs possessing catalytic activity (catAbs or abzymes).

Occurrence of autoAbs in MS: Myelin proteins synthesized by oligodendrocytes have the most pronounced immunogenic properties in the structure of the nervous system. More than 30 myelin proteins have been described, of which myelin basic protein (MBP), proteolipid protein, myelin-associated glycoprotein (MAG), myelin-oligodendrocytic glycoprotein (MOG), and a number of enzymes that form the protein framework of the nerve fiber myelin sheath are of most interest. All of the above proteins and glycoproteins are autoAgs, with MBP and MOG having the most pronounced immunogenic and encephalitogenic properties.

Three components of myelin, MBP, MAG, and MOG, play central roles in the induction of both T- and B-cell response at the stage of appearance of autoreactive CD4+ lymphocytes in myelin that crossed the blood–brain barrier.

MBP-hydrolyzing abzymes and their role in the destruction of myelin: Ab-mediated proteolysis of MBP produces a limited set of peptides, most of which come from peptide bond scission at one or more of five sites. The most immunogenic and encephalitogenic sites are found in three areas (numbering from the N-terminus): *43–170*, immunodominant area of the chain to include the sites defined; *81–103*, a sequence-specific immunogen; and *82–98*, a subsequence-specific highly immunogenic and highly encephalitogenic site.

By changing the sequence specificity of Ab-mediated proteolysis, one might reduce the frequency of cleavages in these areas and thus minimize demyelination.

Preearly diagnosis of MS using abzyme screening and monitoring: MBP-targeted Ab-protease activity has been detected in subclinical stages 1–2 years before clinical disease. Ab-protease activity correlates with the extent of demyelination and disability of patients. Sequence-specific Ab-proteases have proved to be greatly informative and thus valuable as biomarkers to monitor MS at both subclinical and clinical stages. Therefore, monitoring the activities and specificities of Ab-proteases could have high *subclinical* and *predictive* value.

Future therapeutic possibilities: The prospects for progress in developing novel approaches for the diagnosis and treatment of MS have been greatly encouraged by several observations, including:

1. the identification of a major immunogenic region in MBP, 82–98;
2. the demonstration that both B- and T-cell responses to MBP are focused on this region, particular for patients who express HLA-DR2;
3. clinical studies suggesting that patients treated with MBP peptide 82–98 can be made tolerant to the protein in association with delayed disease progression or reduced disease activity over time.

The translational potential of Ab-proteases and the knowledge is in the rational design of new diagnostic tools and new therapeutics based on principles of artificial biocatalysts and biodesign. In addition, the traditional goal of AB engineering is to combine different Ab domains to create individual Abs that have binding specificity, optimal half-life, and desirable effector functions. Abs can be engineered to produce proteins with higher affinity or smaller molecular variants that can retain or modify the functional properties of the original Abs.

4.3 Biomarkers in clinical practice: Biomarkers as applicable to type 1 diabetes

Type 1 diabetes (T1D) is a chronic disorder based on the autoimmune destruction of pancreatic β-cells within the islets of Langerhans (β-cell Is), leading to insufficient production of insulin and, as a result, to permanent life-threatening hyperglycemia.

Most of the serum biomarkers used in clinical practice have been well established for the diagnosis of T1D, showing normal and abnormal physiological or pathological processes and play an important role in clinical diagnosis, prognosis, and monitoring therapeutic responses. They include glucose level, hemoglobin Ale (HbA1c) (a product of nonenzymatic glycation), C-peptide, and autoantibodies (AAb). However, these molecules are identified in late-stage patients when more than 80–90% of the pancreatic insulin-producing β-cells have already been lost [4].

4.3.1 Genetics of T1D and genomic biomarkers

A large cohort of patients have human leukocyte antigen (HLA) alleles associated with T1D, localized on chromosome 6p21. The function of HLA molecules is to present peptides to T-lymphocytes. They constitute approximately 50% of the familial aggregation of T1D. There is a strong association with HLA DR and DQ, both of them are alpha-beta heterodimers and cell surface receptors that present antigens to T-lymphocytes. They are highly linked to each other and, to a lesser degree, to other HLA loci [5]. In the general population, about 40% of people have high-risk HLA alleles for T1D, but there are ethnic differences [6]. Major genetic determinants of T1D are polymorphisms of class II HLA genes encoding DQ, DR, DP, and alleles of the class I HLA-B gene. The combination of HLA alleles inherited from both parents is the key to the development of T1D [7].

To date, the HLA class II DRB1-DQA1-DQB1 genotypes are recognized as the strongest genetic risk factors for T1D. Ninety-five percent of individuals who develop diabetes have DR3-DQ2 or DR4-DQ4, while protective haplotype DRB1*15:01, DQA1*01:02 is present in only 20% of the general population and less than 3% of patients with T1D (Table 1).

For example, the most predisposing genotype to T1D is composed of haplotypes DRB1 *03:01-DQA1*05:01-DQB1*02:01 and DRB1*04:01/02/03/04/05/08-DQA1*03:01-DQB1*03:02/04 [8].

4.3.2 Non-HLA loci

After the HLA region, the insulin gene (INS) on chromosome 11p15 has the strongest association with T1D. It harbors three major insulin variable number tandem repeats. The insulin polymorphisms regulate the amount of insulin mRNA in the thymus and may influence the development of immune tolerance to insulin [9].

The risk of TD1 is explained by variation at HLA haplotypes and more than 50 non-HLA SNPs, including INS and PTNP22, cytotoxic

Table 1 DR and DQ haplotype associated with T1D risk.

DRB1	DQA1	DQB1
High risk		
04:01/02/03/04/05/08	03:01	03:02/04 (DQ8)
03:01	05:01	02:01 (DQ2)
Moderate risk		
08:01	0401	04:02
04:04	0301	03:02
01:01	0101	05:01
09:01	0301	03:03
Moderate protection		
04:03	03:01	03:02
07:01	02:01	02:01
11:01	05:01	03:01
Strong protected		
15:01	01:02	06:02 (DQ6)
14:01	01:01	05:03
07:01	02:01	03:03

T-lymphocyte-associated antigen (CTLA)-4 gene, interferon induced with helicase C domain 1 (IFIH1) genes, and Interleukin-2 receptor alpha-chain (IL2RA) [10,11]. PPTPN22, a gene encoding a lymphoid-specific phosphatase, influences the signaling of T-cell receptors. CTLA-4 is an immune-modulatory protein, expression of which in different forms is linked with susceptibility or resistance to T1D [12].

This high genetic heritability creates the potential for more powerful integration into genetic risk score used in a variety of clinical practice and research. The genetics of T1D is polygenic and complex. It is modulated by alternative splicing of islet autoantigen mRNAs (e.g., IA-2), by imprinting (e.g., INS), and through epigenetic modifications gene–environment interaction and gene–gene interaction [13–15].

The panel of prognostic biomarkers depending on ethnicity based on subclinical recognition of bioindicators long before the disease clinically manifests itself may be a pivotal approach to predict the disease development. The diagnosis and prognosis of T1D in clinical practice is focused on genotyping and the presence of multiple AAbs. T1D is a multifactorial disease including genetics, environmental triggering, and modifying factors. The current biomarkers to diagnose T1D still rely on the consequences of hyperglycemia.

4.3.3 Biomarkers connected with hyperglycemia

Diabetes may be diagnosed in patients with classic symptoms of hyperglycemia (polyuria, polydipsia, polyphagia, weight loss) or diabetic ketoacidosis. The criteria for diagnosis of diabetes by the American Diabetes Association (ADA) are the fasting plasma glucose (FPG) $\geq$126 mg/dL, or HbAlc $\geq$6.5%, or plasma glucose $\geq$200 mg/dL detected either in a 2-h plasma glucose of the oral glucose tolerance test (OGTT) or randomly. FPG and OGTT are more sensitive indicators in diagnosing diabetes than HbA1c as a reflection of chronic hyperglycemia. Despite the fact that HbA1c allows to measure average blood glucose for the past 2–3 months indirectly, it cannot be associated with the condition such as β-thalassemia or other hemoglobinopathies [16]. According to the World Health Organization, they are the most common genetically inherited single-gene disorders in the world. About 5.2% of the world population carry a significant variant caused by mutations and/or deletions in the α- or β-globin genes [17].

4.3.4 C-peptide

The process of insulin biosynthesis begins in the endoplasmic reticulum of β-cells by the formation of preproinsulin, which is broken down to proinsulin by microsomal enzymes. Proinsulin is transported to the Golgi complex, where it is split into the active insulin hormone and the equimolar amount of C-peptide. The level of stimulated serum C-peptide is considered a sensitive measurement of β-cell function and can be used to differentiate autoimmune diabetes from other diabetes subtypes as insulin, like other protein hormones, is rapidly destroyed after secretion (half-life time is 3–5 min) [18]. C-peptide measurement can be a biomarker to evaluate primary therapy outcomes, aimed at assessing the maintenance or improvement of endogenous insulin secretion. However, C-peptide measurement may not be as reliable when diabetes and obesity are combined. Patients with T1D and autoantibody-positive may be overweight, which can lead to insulin resistance, resulting in high levels of fasting insulin or C-peptide and loss of stimulated insulin secretion [6].

4.3.5 AutoAbs as biomarkers

T1D accounts for 5–10% of the total variety of diabetes types. One of the main criteria of T1D is the presence of specific autoantibodies, so Abbs serve as relatively good prognostic markers of the risk of eventual development of the disease. The presence of at least one type of Abbs occurs in 95% of patients with hyperglycemia and T1D [19]. Islets of Langerhans autoantibodies recognize insulin (IA), islet cell cytoplasmic, glutamic acid decarboxylase (GAD), protein phosphatase-like (IA-2), and zinc transporter 8 (ZNT8A), all are found in secretory granules within pancreatic β-cells [20,21]. Not all individuals with T1D are autoantibody-positive. At the same time, it is rare when the patient with progression of diabetes loses expression of all autoantibodies by the time of diagnosis [6]. There is no simple explanation of diabetes development because of heterogeneity of the disease. IAA and GADA are more frequently detected as the first AAbs in children, but they often disappear by the beginning of clinical manifestations (IAA in 25% and GADA in 10% cases, respectively). A small number of children have IA-2A or ZnT8A as the first Abb, which persist at the time of diagnosis [22,23].

The clinical symptoms of T1D represent the near end stage of β-cell destruction, with only 10–20% of β-cells still functioning at the time of diagnosis [24]. The asymptomatic period of their destruction may last from several months to decades [19]. Thus, there is a great demand for prognostic biomarkers to predict the development and/or progression of T1D to identify individuals at high risk for early prevention and intervention.

There is another opinion that AAbs are not the cause of T1D, but they reflect the degree of disease progression and can be used as a proof of immune activity against β-cell autoantigens in antibody-positive individuals. All five types of AAbs are used as the gold standard for diagnosis, classification, and prediction of T1D development.

The prevalence of T1D is increasing worldwide. In the US population, it is 0.3% (15–25/100,000), while in the Japanese population this indicator is only about 1/100,000. In Finland, the probability of getting T1D is 36–50 per 100,000 population. T1D is also diagnosed in adult age, with approximately 30% of patients developing symptoms after 18 years of age [25].

More than 85% of patients with T1D do not have first-degree relatives (FDRs) with T1D in their family history. However, if T1D was detected, it is highly recommended to screen FDRs of the patient or those having high-risk HLA genotypes for AAbs to assess the risk of its development. The probability of T1D development in an individual with any two of the five common Abbs is much higher compared to an individual positive for a single islet AAb after seroconversion. The rate of its progression will depend on many factors such as the type and number of AAbs, age, gender, genotype, level of physical fitness, etc. [19]. In children with a single AAb the probability of T1D development is about 10% within 10 years, while in children with more than two AAbs it is 11% every year, accounting for >70% within 10 years (Table 2) [19,27].

Table 2 Risk of T1D development in a person whose relatives have T1D (% in childhood when proband with T1D) [6,25,26].

% of the risk	Relative
	Offspring
3.2	Sibling (6% lifetime)
6	Dizygotic twin
1.3–4	Mother
6–9	Father
10	Both parents
50–70	Monozygotic twin

Despite obvious benefit of AAbs combined with genetic susceptibility in T1D prediction, the presence and levels of AAbs after seroconversion do not track disease progression or identify the best timing for intervention and therapy [28,29].

As T1D is an immune-mediated disease, most of T1D patients have more than one autoimmune disorder. One of the most common coexisting diseases of T1D is thyroid autoimmune disease and celiac disease (CD) [6]. Genetic factors in the etiology of T1D are confirmed by its occurrence in combination with genetic mutations, affecting key genes of immune functions [30].

For example, T1D is also part of rare monogenetic disorders: the autoimmune polyendocrine syndrome type 1 (APS-1), including hypoparathyroidism, mucocutaneus candidiasis, hepatitis, Addison's disease, oophoritis, lymphocytic hypophysitis, etc. Most patients have mutations (autosomal recessive type of inheritance) of the autoimmune regulator gene on chromosome 21. About 18% of patients with APS-1 develop T1D. Patients with the common diabetes-protective HLA allele DQB*10602 have some protection from T1D; however, it does not prevent from Addison's disease development (Table 3).

The X-linked polyendocrinopathy syndrome is associated with immune dysfunction, T1D, and severe malabsorption (IPEX) (X-linked recessive disease) and affects boys with mutations of the gene encoding Forkhead Box P3 (FOXP3). Foxp3 is a transcription factor that is essential for functioning and development of regulatory T-cells, which play an important role in maintaining self-tolerance. The effective therapy of this disorder is bone marrow transplantation with restoration of T-regulatory cells [31].

Table 3 Autoimmune diseases associated with T1D.
AutoAb

Disease	Type
Addison's disease	21-Hydroxylase, P450scc, 17α-hydroxylase
Celiac disease	Tissue transglutaminase (tG2A), endomysial (EmAIgA), gliadin
Pernicious anemia	Parietal cell
Thyroid diseases (Grave's and Hashimoto thyroiditis)	Peroxidase, thyroglobulin, TSH receptor
Hypogonadism	P450scc, 17α-hydroxylase
Hypoparathyroidism	Ca(2+)-sensing receptor (CaSR)
Hepatitis	Cytochromes P450:CYP21A2, CYP2C9, CYP2D6
Vitiligo, alopecia	Melanocytes tyrosinase (TYR), hair follicle tyrosine hydroxylase (TH)

Syndrome of STAT3 polyautoimmunopathy associated with mutations in STAT3 genes includes permanent autoimmune neonatal diabetes, thyroid disease, CD, and intrauterine growth retardation. It has been reported that activating mutation in STAT3 leads to inhibition of the transcription factor Isl-1 and decreased insulin expression and synthesis [32].

4.3.6 Potential new types of biomarkers

The abovementioned overview of classic diagnostic biomarkers of T1D is used in clinical practice. Focusing on new markers of β-cell destruction by high-throughput and sensitive "-omics" technologies for the study of proteins, nucleic acids, and metabolites is one of the priorities in discovery of potentially novel biomarker candidates. Fig. 2 illustrates the main omics technologies and platforms for discovery and validation of serum biomarkers in T1D (Fig. 13).

4.3.7 Proteomics analysis of serum is considered to be challenging

It is complicated by heterogeneity and individual variations, or different stages of the disease. As β-cells are about 0.002% of body mass, the development of specific serum biomarkers revealing the entire spectrum reflecting β-cell mass, their function, and stress or death as the result of an autoimmune attack has been challenging [33]. There are data in the

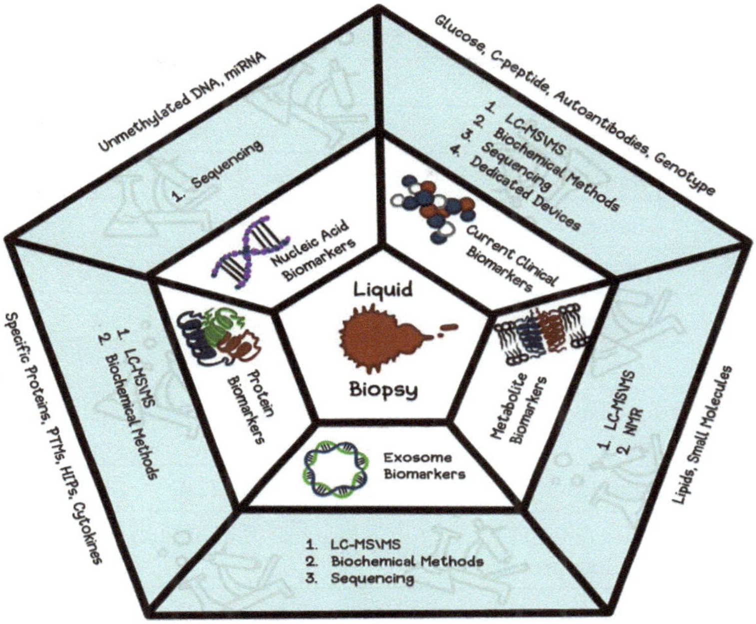

Fig. 13 Omics technologies used for discovery and validation of serum biomarkers in T1D.

literature on several promising candidates as biomarkers of T1D, but most of them are widely distributed in plasma and do not have specificity with respect to pancreatic islet cells. Unfortunately, most of these proteins showed opposite changes in terms of up/down expression regulation [3,34–37] (Table 4).

4.3.8 The ratio of proinsulin to C-peptide

The elevated ratio of proinsulin/C-peptide as the indicator of β-cell dysfunction precedes the development of T1D, especially in younger children. This ratio combined with other biomarkers can demonstrate the significance for predicting the onset of the disease in the asymptomatic period, being the reflection of alterations in the endoplasmic reticulum during processing and folding insulin protein [38,39].

4.3.9 Metabolomics biomarkers

A few amino acids and lipid metabolites are associated with T1D, making them the candidates of metabolomics biomarkers. Glucose is a simple and useful, but it is not a specific and unique biomarker used for the diagnosis

Table 4 Protein biomarker candidates.

Gene	Name of protein	Function
C3	Complement C3b	Innate immunity, complement activation
C4A	Complement C4-A	Innate immunity, complement activation
KNG1	Kininogen 1 isoform1	Innate immunity, DC activation
CLU	Clusterin	Cytoprotective capability
ADIPOQ	Adiponectin	Homeostasis of glucose
APOA4	Apolipoprotein A-IV	Adhesion of leukocyte
APOC4	Apolipoprotein C-IV	Viral infection
AZGP1	a-2-glycoprotein 1 (zinc)	Activity of lipid mobilization
BTD	Biotinidase	Activity of biotinidase
SERPINA6	Corticosteroid -binding protein	Correlate to insulin response or deficiency
SERPINF2	Alpha-2-antiplasmin	Serine protease inhibitor
TTR	Transthyretin	Involve in pancreatic islet cells stimulus-secretion coupling
LUM	Lumican	Extracellular matrix structural constituent

of diabetes. It is necessary to consider other metabolic indicators as the predictors of the disease, which are easier to identify by mass spectrometry and nuclear magnetic resonance spectroscopy. The amino acid methionine concentration in plasma (μmol/L) is involved in DNA methylation and can correlate with the time of AAbs appearance. Another investigation of lipid dysregulation in blood tests of high-risk patients has revealed that sphingolipids play an important role in inflammation and insulin signaling and can be used as a biomarker [3,40].

4.3.10 The new autoAgs and autoAbs

The autoimmune form of T1D represents a heterogeneous group, which can include patients with reactivity to multiple, novel, and rare types of antigens, e.g., islet cell surface antigen, tetraspanin-7 (a membrane glycoprotein), islet of Langerhans regenerating protein-la (REGIA), protein family members of commonly tested autoantigens GAD67 and IA-2β,

L-type voltage-gated calcium channels, pancreatic duodenal homeobox 1, aminoacyl-tRNA synthetase, golgin-160, amylase alpha-2A, etc. [3]. Circulating exosomes may also be an interesting source of biomarkers in T1D. Under insulitis condition β-cells release exosomes to APCs, activating them. These exosomes contain autoantigens, including proinsulin, GAD65, and IA-2 [41].

4.3.11 Posttranslational modifications

Despite the fact that potential benefit of posttranslational modification (PTM) for predicting T1D is still in its infancy, it is important to mention the importance of these developments. The significance of inflammation in the pathogenesis of T1D due to interaction of the immune system and β-cells is undeniable. Inflammation can lead to alternative splicing, misfolding of pancreatic β-cells proteins, and PTMs. Such excessive recognition of posttranslationally modified autoantigens by autoreactive T- and B-cells is not limited to a single disease; it is also observed in CD and RA. Additionally, due to their difference from native proteins, modified autoantigens may be recognized as foreign, which can overcome immune tolerance to β-cell antigens [42]. Since 2–5% of T1D patients can be negative to standard antibodies mentioned above, antibodies to neoantigens may serve as an update to the current T1D AAb panel for diagnosis and prediction of the disease [43]. Hybrid insulin peptides (HIPs) are a novel type of neoantigens. Present in β-cell secretory granules, HIPs are formed by covalent crosslinking of proinsulin fragments with other peptides. CD4 T-cells from a patient with T1D trigger a memory recognition response to HIPs detectable in mouse pancreas [3,44]. PTMs observed in T1D include oxidation, citrullination and deamidation of GAD65, phosphorylation of peripherin, deamidation of proinsulin peptide, etc. [45]. Glycation, phosphorylation, and lipidation also may play a role in immunogenicity of autoimmune diseases, including T1D [45–47].

4.3.12 Cytokines as T1D biomarkers

Cytokines are a powerful mediator of inflammation and microbial elimination, playing an important role in stimulation, regulation, and intercellular signaling of immune cells. They also may induce β-cell destruction. However, four cytokines, IL8, IL-IRa, MCP-1, and MIP-Iβ, have been discovered to be downregulated in the sera of T1D patients. At the same time, many cytokines, chemokines, and growth factors were elevated in antibody-positive individuals and were detected in patients with prediabetes

[48,49]. Unfortunately, cytokines are not organ-specific serum biomarkers of T1D and can change in response to injury and damage in other types of tissues and autoimmune disorders [50].

4.3.13 Nucleic acid-based predictive biomarkers

The most significant advance in prognostic biomarkers has been the discovery of small circulating DNA fragments. They can carry a variety of genetic and epigenetic alterations, related to progression and development of the disease. Unmethylated insulin and amylin DNA leak into the bloodstream and become detectable during the progression of β-cell death and the development of T1D. Therefore, measurement of the amount of unmethylated DNA of both of them in the peripheral circulation may be an important biomarker for early detection of β-cell death in T1D [51]. Amylin is a glucoregulatory hormone and an important regulator of energy metabolism, which is cosecreted with insulin from β-cells. Its expression correlates with that of unmethylated insulin DNA [52]. The ratio of unmethylated/ methylated insulin DNA detected by new methylation-specific real-time PCR and sequencing techniques enables differentiation of patients with T1D from normal individuals with 38% sensitivity and 95% specificity [53].

MicroRNAs (miRNAs) are noncoding RNAs, consisting of 21–23 nucleotides, that are considered regulators of gene expression, aiding mRNA stability. It has been suggested that miRNAs could be potential circulating biomarkers in the regulation of β-cell function and pathogenesis of T1D as increased serum levels of miR-21, miR-210, and miR-375 are found in subjects with T1D [54,55]. However, the significance of the miR-375 in providing insights into β-cell has been questioned [40].

4.3.14 Biomarkers of T1D complications

Moreover, T1D is often accompanied by severe complications, so early prediction and monitoring of progression, and prevention or reversal of the disease (especially at an early stage) are vital.

T1D entails many complications: diabetic nephropathy, diabetic cardiomyopathy, stroke, diabetic retinopathy, and diabetic neuropathy. The mechanism of glucose transport plays a role in the pathogenesis of these disorders. Under the influence of hyperglycemia, glucose transport is disrupted in the endothelium of the retinal capillaries, in the mesangial cells of the renal glomeruli, and in neurons. Glucose accumulates inside the cells. In addition to the direct destructive effect of glycemia, the cells are affected by hormones and mediators, for example, catecholamines, vasopressin, acetylcholine, endothelin, bradykinin, angiotensin II, thrombin, cytokines, lipoproteins, endotoxins, and other molecules. All of them, to one degree or another, can serve as a biomarker.

One of the main roles in vascular wall damage in T1D is played by advanced glycation end products (AGEs) and oxidation end products (OPs). AGEs are proteins, lipids, and nucleic acids that have undergone glycation. Glycation involves the formation of chemically reversible products of early glycosylation with proteins—Schiff bases and Amadori adducts (for example, glycated hemoglobin; HbA1c). Early adducts undergo slow and complex chemical changes followed by the formation of AGEs. These metabolites are chemically active. They affect intercellular interactions and can cross-link proteins of the extracellular matrix, depriving vessels of elasticity. In addition, by binding to receptors (RAGE) on the surface of endothelial cells, smooth muscle cells, and monocytes, they affect cellular metabolism and act on intracellular proteins.

The AGE/RAGE interaction has been proposed as a possible mechanism linking microangiopathy with diabetic nephropathy, retinopathy, and neuropathy. Also, the accumulation of AGE contributes to the release of proinflammatory cytokines and adhesion molecules, such as IL-6, TNF, vascular endothelial growth factors (VEGFs), E-selectin, etc. The totality of these pathophysiological events causes vascular endothelial dysfunction of vessels of different calibers. In addition, there is evidence of the interaction of AGE with intracellular Basic Fibroblast Growth Factor—FGF-β, as an important pathophysiological link in the development of diabetic cardiomyopathy. Thus, the increased AGE content is a reliable prognostic biomarker of cardiovascular complications in DM1, regardless of the presence or absence of other risk factors. N_ε-carboxymethyl lysine (N-ε-CML), the most abundant of circulating AGEs, has been extensively studied. N-ε-CML has been found elevated in the serum of diabetic patients.

Interestingly, after binding AGE to a specific receptor, it triggers a MARK-dependent cascade of reactions that leads to the accumulation of OPs such as superoxides, hydroxyl radicals, and peroxides. Reactive oxygen species, for example, the superoxide anion, react with NO radicals to form the peroxynitrite anion, a toxic oxidant, which leads to tissue damage. Data on the effect of NO on cells on pancreatic β-cells are contradictory. The researchers agree that the nature of the interaction depends on the concentration of the given molecule. The precursor of the brain natriuretic peptide NT-proBMP has proved to be a good candidate for prognostic biomarker of vascular complications, including in DM1.

Studies are underway on VEGF, its receptors (VEGFR1 and VRGFR2), as well as protein kinase C, as biomarkers and targets for targeted therapy. Hyperproduction of VEGF leads to increased permeability of retinal vessels, the development of macular edema, and inadequate angiogenesis. The DAG-PKC pathway is one of the most studied signaling pathways associated with diabetes mellitus.

In the subclinical form of diabetic cardiomyopathy, circulating miRNA-21 is a good diagnostic biomarker for a comprehensive assessment of information that also includes data on the duration of diabetes mellitus, lipid profile, and glycated hemoglobin levels. It is worth noting that overexpression of miR-21 improves mitochondrial function and reduces cardiomyocyte apoptosis.

Focusing on the variety of studies used in diabetology, we can say there is no single ideal predictive biomarker for T1D patients. It represents a multimarker panel of proteins, AAbs, PTMs, or nucleic acids from serum that will provide an accurate prediction or measurement of the particular disease progression. There is no precise interpretation for T1D prediction due to heterogeneity of the disease. Many factors contribute to T1D risks: the patient's genotype and age, numbers and Abbs phenotype, their appearance, epitopes, and affinity.

4.4 Biomarkers in clinical practice: Biomarkers as applicable to SLE

SLE is a chronic inflammatory disease characterized by diverse autoantibodies and varied clinical manifestations. The etiology of SLE is not completely clear, but it is known that genetic, environmental, and hormonal factors are important in the pathogenesis of the disease. There is much interest in biomarkers, because SLE manifestations and disease activity vary widely and this complicates both the diagnosis and the prediction of the further development of the disease.

Nowadays many biomarkers have been developed and validated, such as ANA, complement (C2, C3 and C4), anti-dsDNA Ab, SLE cells, etc. However, they have relatively low specificity or sensitivity. Thus, there is still a need for further investigation of the SLE disease pathogenesis, validated biomarkers, the discovery of new biomarkers, common diagnostic test panels, and NBBs in SLE.

Novel biomarkers have to be helpful in predicting the onset of SLE, early diagnosis and in differential diagnosis from clinically overlapped disorders, in correct evaluation and staging of organ involvement, and in predicting and monitoring response to therapies.

Biomarkers for SLE may be classified as epigenomic, genomic, and metabolomic. In a complex disorder such as SLE, a number of biomarkers have been proposed that fall into each of the above-described categories; many of them are well validated and still used in clinical practice. However, the search for and definition of novel biomarkers is of utmost importance, as many unmet needs still exist in the diagnosis and management of the disease.

4.4.1 Epigenomic biomarkers

miRNAs are a class of small noncoding molecules that function as regulators of gene expression at posttranscriptional level. miRNAs are implicated in the regulation of numerous pathways, including immune system response, as they control differentiation of the various subgroups of immune cells and their immunological functions. miRNAs were found to be differentially expressed in cellular subpopulations of the immune system, including T-cells (miR-21, miR-148, miR-125a), B-cells (miR-21, miR-125b), and dendritic cells (miR-146a, miR-126) as well as in specific organs affected by the disease such as the kidneys. Polymorphisms in the genes encoding miRNAs can also play an important role in susceptibility to the disease and/or in the development of specific clinical phenotypes. For example, the polymorphisms rs57095329 (A > G) and rs2431697 (T > C) in the MIR146A gene, which is involved in the pathogenesis of SLE through regulating the expression of TRAF6 and IRAK1, were associated with an increased risk of developing the disease. Indeed, the variant alleles seem to determine a lower expression of the miRNA itself. Thus, there was the supposition that miRNA may be a novel biomarker for SLE. A recent study found that miR-155, miR-499a, and miR-142 are involved in the pathogenesis of SLE and in its clinical phenotypes, both through genetic variability and their differential expression.

4.4.2 Genomic biomarkers

By combined long noncoding RNA–miRNA–mRNA network analysis and DNA methylation analysis, seven genes appear to be involved in the development of SLE, IFI44, IFI44L, IFIT1, IFIT3, PLSCR1, RSAD2, and OAS2. The results of quantitative reverse transcription PCR, immunohistochemistry, and DNA methylation analysis confirmed the potential of these genes as biomarkers.

4.4.3 Proteomic biomarkers

Antinuclear antibodies (ANAs, also known as antinuclear factor or ANF) are autoantibodies that bind to contents of the cell nucleus. ANA detected by indirect immunofluorescence (IIF) on HEp-2 cells has long been regarded as a pivotal immunological biomarker in serum for classifying a patient with SLE, as well as assessing eligibility for SLE. The ANA test is included in the ACR-1997, SLICC-2012, and EULAR/ACR-2019 criteria. If the patient is positive for ANA, further testing for Ag-specific ANAs, such as dsDNA, Sjögren's syndrome antigen A (SSA (Ro60)), Sjögren's syndrome antigen B, Sm, and ribonucleoprotein should be done. Despite the fact that antinuclear antibodies are detected not only in patients with SLE, it is highly characteristic of SLE and can be used as a biomarker for screening, classification, diagnosis, prognosis, and staging. ANA tests have high sensitivity, ranging from 90% to 95% in SLE patients but a relatively low specificity as they can occur in 5–20% of healthy controls, especially in older people. The sensitivity of ANA tests may be related to the early detection of SLE. Although ANA tests are essentially universal for SLE, a negative ANA test cannot rule out SLE diagnosis. Choi et al. reported that 6.2% of patients with SLE were ANA-negative. In fact, up to 30% of patients with SLE screened in clinical trials for new therapies are ANA-negative. The inconsistency of ANA results in SLE may be due to variability in IIF-ANA assays, which have discrepant antigen properties (e.g., the combining capacity between molecularly cloned proteins and HEp-2 cells), the effects of nuclear antigens by HEp-2 cells, variability in laboratory routines, or different threshold in positive ANA judgment. Therefore, a standard ANA test with a consistent fine specificity, detection method, and laboratory protocol should be established in the future.

Serum complement 3 (C3) and complement 4 (C4): Serum C3 and C4 are widely used to assess the presence of biologically active immune complexes and monitor disease activity. As one of the immunological criteria, low levels of both C3 and C4 are weighted higher than having low levels of either C3 or C4 alone in the EULAR/ACR-2019 classification criteria for SLE. However, measurement of C3 and C4 levels is not a sensitive tool to study SLE patients and does not accurately reflect ongoing complement activation in SLE sera, because of their limitations, such as partial compensation of the diminished levels of C3 and C4 by acute phase production of the two proteins. Moreover, the variability in C4 gene copy number in the population leads to a wide interval of "normal" C4 levels. Applying the analysis of C4 gene copy number to SLE patients, it has been reported that 3 out of 11 SLE patients with persistently low C4 levels in spite of low disease activity are carriers of only two copies of the C4 gene.

More information can be gained using functional assays that explore the whole complement cascade, such as the capacity of erythrocyte lysis (hemolytic potential) or the levels of solid phase or soluble membrane attack complex. On the other hand, it is widely accepted that measurement of complement split products represents a powerful tool to get a full picture of complement status in SLE. The rapid decay of these products has been the main obstacle to their detection, but novel approaches, including evaluation of cell-bound complement products, have been proposed to overcome it.

Detection by cytometry of C4d levels on erythrocytes and B-cells (cell-bound complement activation products, CB-CAPs) is a sensitive assay for the diagnosis and follow-up of SLE patients. Moreover, CB-CAP can predict evolution to classifiable SLE in patients in which the disease is suspected but ACR criteria are not yet fulfilled.

iC3b is the breakdown product of C3b, thus reflecting complement activation via the classical, lectin, and alternative pathway. Its half-life in serum (90 min) is long enough to allow detection but short enough to reflect ongoing complement activation. iC3b detection by lateral flow assay minimizes serum handling and in vitro activation of complement. Its evaluation, expressed as ratio of blood iC3b to serum C3 levels, has been proposed as a useful tool to analyze complement activation in sera. In fact, this ratio identifies active disease and is strongly associated with nephritis.

Antinucleosome antibodies (ANuA): The prevalence of ANuA in SLE varies from 50% to 100%, and ANuA can be combined with clinical findings and other laboratory tests for diagnosing SLE and drug-induced lupus. The presence of ANuA is related to glomerulonephritis and disease activity in SLE patients. Moreover, the sensitivity of the ANuA assay for SLE is 61%, and the specificity is 94%. The overall positive likelihood ratio of ANuA is 13.81, and the negative likelihood ratio is 0.38. The probability that a subject with positive ANuA has SLE is 41 times greater than a subject with negative ANuA.

Anti-dsDNA antibodies are biomarkers that are associated with SLE disease activity, and they can predict the development of LN. The level of anti-dsDNA antibodies can fluctuate over time because of their association with SLE disease activity. They can disappear during treatment and return during a flare, especially in active nephritis. As one of the most characteristic ANA types, anti-dsDNA antibodies have a high specificity (96%) but low diagnostic sensitivity (52–70%) for SLE because of their transient appearance.

The level of anti-dsDNA antibodies does not always correlate with active LN, and this makes anti-dsDNA an unreliable biomarker for assessing disease activity. Although anti-dsDNA antibodies can exacerbate LN by being deposited in the kidney or driving cytokine production, the levels of anti-DNA antibodies do not necessarily correlate with active disease. The reasons for anti-dsDNA antibody level variability may be the following: (1) antibody concentration varies with the SLE pathology over time, (2) antibodies may be deposited in renal tissue and not detectable in serum, (3) assays for anti-dsDNA antibodies are not sensitive and specific enough, (4) non-nephritic flares are not always accompanied by anti-dsDNA antibodies, and (5) not all subclasses of anti-dsDNA antibodies are pathogenic and associated with disease activity and renal disease.

Serum anti-Sm antibody: The antigens of the anti-Sm antibodies are the core units of the small nuclear ribonucleoproteins (snRNPs), termed A to G, and will bind to the U1, U2, U4, U5, and U6 snRNPs. Currently, anti-Sm antibodies serve as biomarkers in the classification of SLE. These antibodies are highly specific for SLE and are not found in other rheumatic diseases. However, despite the high specificity (99%), they have very low sensitivity (5–30%). Anti-Sm antibodies correlate with SLE disease activity and are also associated with LN, and therefore their presence in patients with SLE may be associated with the development of LN in the future. In addition, it has been shown that high titers of anti-Sm antibodies can be predictors of silent LN and may predict early adverse LN outcomes.

Anti-C1q antibodies: Hereditary deficiency of complement component 1q is closely associated with SLE. Anti-C1q antibodies reduce the level of C1q, which leads to inhibition of the clearance of immune complexes and apoptotic bodies or the deposition of immune complexes on the glomerular basement membrane, which in turn plays a pathogenic role in the development of LN. Elevated anti-C1q antibody titers with a sensitivity of 81–97% and a specificity of 71–95% predict renal flares in LN. Moreover, antibody titers correlate with active LN, and the absence of anti-C1q antibodies predicts an almost zero probability of developing lupus nephritis. Unfortunately, anti-C1q antibodies were not included in the SLE classification, as there is no standardized laboratory assay for them. In addition, anti-C1q antibodies can be found in other autoimmune diseases, including hypocomplementemic urticarial vasculitis (100%), Sjögren's syndrome (26%), RA (19%), and Stevens–Johnson syndrome (14%), and even in apparently healthy individuals (3–5%).

CXCL 10, Galectin-9: The level of the chemokine CXCL 10 is increased in the blood serum and affected tissues of SLE patients, where it is a recruiter of effector CXCR3+ and memory T-cells, NK cells, and plasma cells to inflammatory foci. Among interferon-regulated chemokines, CXCL10 has the highest correlation with disease activity and is the most accurate predictor of disease flares, but van den Hoogen et al. reported a higher correlation with disease activity of Galectin-9 compared with CXCL10 levels (*P* ¼ 0.003 vs *P* ¼ 0.21). At variance with CXCL10, serum levels of Galectin-9, a beta-galactoside-binding lectin, reflect organ damage and not only disease activity. Galectin-9 is detectable also in cerebrospinal fluid, suggesting a potential use of the lectin in the diagnosis of CNS involvement.

IL-1 family: Among IL-1 family cytokines and receptors, the soluble form of ST2/IL-1 receptor 4 (IL-1R4) has been recently proposed as new biomarker in SLE. ST2/IL-1R4, upon recruitment of the accessory chain IL-1R3, mediates IL-33 signaling; the soluble form of the receptor prevents the interaction of IL-33 with membrane receptor, thus behaving as a decoy receptor. Soluble IL (sIL)-1R4 levels are increased in active SLE patients, and highly correlated with disease activity index and with anti-dsDNA and anti-C1q antibody levels. Moreover, in patients with active nephritis, sIL-1R4 is correlated with urinary proteins. When the diagnostic value of sIL-1R4 is directly assessed by means of multivariate analysis, sIL-1R4 contributes to a similar extent as anti-dsDNA and IL-18BP to the identification of patients with active nephritis, which is the most relevant variable in discriminating active from inactive patients.

The family of IL-1, IL-18, IL-18BP, and sIL-1R2 are also used as SLE biomarkers, which correlate both with the activity of the disease and with organ involvement.

BAFF family: Type I and type II IFNs also regulate the expression and secretion of cytokines and receptors of the B-cell activating factor (BAFF) family, which play a fundamental role in B-cell development, maturation, and survival.

A study was made of BAFF levels in SLE, and it was reported that the increase in serum levels is more related to serological activity than to global disease activity. Higher levels of BAFF may characterize subgroups of patients. A study was recently conducted on three members of the BAFF family (BAFF, APRIL, and soluble BCMA) to evaluate their potential as biomarkers for diagnosing and monitoring patients with SLE.

Despite the underexpression of membrane BCMA on lupus B-cells, the soluble form of the receptor is detectable in higher amounts in SLE sera. Soluble BCMA may counteract BAFF and APRIL, negatively affecting their signaling. Thus, its increase, which probably represents the activation of a regulatory circuit, closely reflects the levels of BAFF and APRIL. The three cytokines are all present in higher amounts in sera from active patients and decrease in remission. They display a similar ability to distinguish control group from SLE patients, with APRIL and soluble BCMA showing a higher specificity and sensitivity than BAFF.

Metabolomic biomarkers: Thanks to the development of new technologies, in particular in the field of mass spectrometry, it is believed that the key role in diagnostics is gradually occupied by the metabolome. For example, in a recent study of fecal metabolome from 32 SLE patients and 29 healthy controls, it was found that a total of 23 most significant metabolites that were involved in the amino acids, lipids, purine, and vitamin metabolisms were altered in feces of SLE group, compared with HC group. The levels of proline, L-tyrosine, L-methionine, L-asparagine, DL-pipecolinic acid, glycyl-L-proline, xanthurenic acid, kynurenic acid, L-carnosine, monoacylglycerol (MG) 22:6, MG 16:5, lysophosphatidylethanolamine (lysoPE) 16:0, lysophosphatidylcholine (lysoPC) 22:5, phosphatidylglycerol (PG) 27:2, and 1,2-dioleoyl-rac-glycerol were increased in fecal samples of SLE, whereas the levels of adenosine, adenosine 5′-diphosphate (ADP), D-alaninyl-D-alanine (D-Ala-D-Ala), lauryl diethanolamide, sulfoquinovosyl diacylglyceride (SQDG) 26:5, thiamine pyrophosphate, trigonelline, and mucic acid were decreased in SLE. It was also found that L-valine, pyrimidine, erucamide, and L-leucine are excellent potential diagnostic biomarkers for SLE, and alterations of fecal metabolites are closely correlated with SLE. Another study related to proteomic profiling of kidney tissues in SLE showed that coronin-1A is a biomarker of LN, and the level of it in serum can distinguish LN patients from SLE patients without nephritis with 100% sensitivity and 100% specificity. Moreover, granzyme B is increased in the serum and kidneys of patients with SLE, and it is correlated with a poor prognosis of LN.

4.5 Biomarkers in clinical practice: Biomarkers as applicable to RA

Rheumatoid arthritis is a systemic autoimmune inflammatory disease characterized by synovial inflammation and predominantly joint damage.

ACR/EULAR (American College of Rheumatology/European League against Rheumatism) 2010 clinical criteria that are used for the diagnosis of RA included the levels of RF, ACPAs, CRP, and the erythrocyte sedimentation rate. There are two major subtypes of RA (seropositive and seronegative), according to the presence or absence of RF and ACPAs as laboratory indicators.

RA biomarkers that will be discussed below can be divided into several groups, antibody-derived RA biomarkers, nonantibodies RA biomarkers, genomic RA biomarkers, and epigenomic RA biomarkers.

Classic antibody-derived RA biomarkers: RF was the first autoantibody to be discovered in people with RA. RF is an autoantibody targeting the Cγ2–Cγ3 domain in the Fc region of immunoglobulin G. RFs can be of any immunoglobulin isotype (IgM, IgG, IgE). Most of the RFs measured clinically are IgM RFs. The complexes of RFs and IgG/PTM-modified proteins initiate complement activation and production mediators of inflammation. The sensitivity and specificity of RF is 69% and 85%, respectively.

Anticitrullinated protein antibodies are very important diagnostic biomarkers included with standard RA diagnostics criteria. ACPAs consist of a group of antibodies which react with a variety of citrullinated antigens. Citrulline is an amino acid formed during posttranslational arginine side chain deamination. Citrullination is catalyzed by the calcium-dependent enzyme peptidyl-arginine-deiminase (PADs), which include five subtypes (PAD 1,2,3,4,5). Factors that can initiate citrullinated processes may include smoking, periodontitis caused by chronic *Porphyromonas gingivalis* infection, and prolonged NETosis activation.

ACPA isotypes: IgG isotype is the specific biomarker for RA triggering macrophage activation through Fc gamma receptor engagement, and IgA activates neutrophils and is the antibody present in the mucous membranes.

Some patients are seronegative for RF and ACPP, which means that the diagnosis occurs much later, sometimes at an advanced clinical stage of the disease. Therefore, new biomarkers should be examined in clinical practice for early diagnostic in preclinical stages and highly effective target treatment.

New antibody-derived RA biomarkers: Antibodies against peptidyl-arginine-deiminase (anti-PAD)—one of interesting new RA biomarkers. As mentioned early, PADs are a group of calcium-dependent enzymes that catalyze citrullination. For high efficiency, PADs need very high calcium concentration in vivo which is physiologically unavailable. In vitro citrullination assays typically use 5–10 mM calcium to achieve maximal PAD4 activation, while extracellular free calcium concentrations are estimated to be 0.49–0.98 mM in synovial fluid (SF) and 1.1–1.3 mM in plasma. Anti-PADs increase the activity of these enzymes, reducing their calcium requirement into a normal range of calcium concentration (in vivo).

Antibodies against carbamylated proteins (anti-CarP)—also called antihomocitrullinated protein antibodies (AHPA)—are another new antibody-derived RA biomarker. Carbamylation is a chemical reaction that leads to nonenzymatic posttranslational modifications of proteins. The essence of the process is transformation of lysine in protein to homocitrulline catalyzed by cyanate. Cyanates are present as breakdown products of urea in the blood, and may come exogenously, from air or tobacco smoke.

Nonantibodies RA biomarkers: 14-3-3η protein (14-3-3η) is detectable in the SF and serum of RA patients, and is a regulator of proinflammatory factors associated with RA. In the course of research, it was found that the 14-3-3η protein has stimulating effect on matrix metalloproteinase (MMP). Given the key roles of MMPs played in degrading cartilage matrix, scientists research 14-3-3η protein as a diagnostic and prognostic biomarker. The sensitivity and specificity of 14-3-3η + ACPA combination and RF + ACPA combination for RA diagnosis was 71%/92% and 71%/84%, respectively, under the condition of basically equal sensitivity. 14-3-3η–ACPA pairing resulted in superior specificity compared with routine RF–ACPA pairing. Analysis of results from many studies supports the prognostic role of 14-3-3η. Elevated protein levels could predict the development of erosive arthritis and more rapid progression of the disease.

Genomic biomarkers: Correlation between RA and genetics began to be considered with the discovery of HLA-DRB1 in 1980. This gene is positioned on chromosome 6 (6p21.3). The HLA region within the MHC contributes to almost 50% of the genetic susceptibility for RA.

Further research has led to the detection of genetic biomarkers outside the MHC region, the most statistically significant of which are PTPN22 (protein tyrosine phosphatase nonreceptor type 22), TRAF1-C5 (TNF receptor-associated factor 1—complement component 5), and CD40 (cluster of differentiation 40).

Epigenomic biomarkers: Epigenetic mechanisms are key to understanding the pathogenesis of RA and make it possible to diagnose and predict the course of the disease early. The most attention focuses on three key epigenetic areas in RA, namely DNA methylation, histone modification, and expression and/or function of microRNAs.

4.5.1 DNA methylation

DNA methylation is a biological process by which methyl groups are added to the DNA molecule. Methylation can change the activity of a DNA segment without changing the sequence. When located in a gene promoter, DNA methylation typically acts to repress gene transcription. Thus, DNA hypo- or hypermethylation is associated with an increase or repression of gene transcription, respectively.

Significant hypermethylation of CpG islands in the promoters of the M20 peptidase gene containing domain-1 (PM20D1), SHROOM1, and homeobox protein engrailed-1 (EN1) was revealed in very early RASF compared to normal SF. The SHROOM1 gene is involved in the development of nervous tissue and the rearrangement of microtubules during cell division. Chondrocytes of the knee joints in patients with early RA and transient arthritis differ significantly in SHROOM1 methylation, which makes it a valuable biomarker for early diagnosis of the disease.

PTPN11 is upregulated in RASF and encodes the tyrosine phosphatase SHP2, which promotes invasion, migration, adhesion, spreading, and survival of RA FLS. Analysis of the PTPN11 enhancer in RASF revealed hypermethylation, which increases cell aggressiveness and their sensitivity to glucocorticoids. This finding elucidated the mechanism of action of PTPN11 in RASF, and also demonstrated that SHP2 is a potential therapeutic target in RA.

Histone modifications and biomarkers: Histone modification is one of the regulatory mechanisms that modulate the chromatin structure and thereby affect various DNA-templated processes, such as gene transcription, DNA replication, DNA recombination, and DNA repair in cells.

RA synovial fibroblasts overexpress the transcription factor T-box 5 (TBX5), on the promoter of which were found active histone marks, including H4K4me3. The overexpression of TBX5 affects expression of 790 genes including IL-8, chemokine C-X-C motif ligand 12 (CXCL12), and chemokine C-C motif ligand 20 (CCL20), leading to the development of RA.

High levels of the H3K4me3 trimethylation activating mark in the matrix metalloproteinase (MMP-1, MMP-3, MMP-9, and MMP-13) promoters were found in RASF. In addition, an increased level of H4 histone acetylation was also associated with MMP-1 activation.

MicroRNAs: MicroRNAs are small (on average 22 nucleotides) noncoding RNAs that are complementary in sequence to coding mRNAs. Almost 3% of the human genome encodes microRNAs, and microRNAs control the function of up to 30% of total human genes. The composition of a set of microRNAs shows promising results. The biggest value is attributed to miR-125a-5p. It has high sensitivity and specificity in aCCP-positive and -negative patients. Other microRNAs of possible clinical importance are decreased levels of miR-146a, miR-150, and miR-155.

4.6 Biomarkers in clinical practice: Biomarkers as applicable to systemic sclerosis

Systemic sclerosis (SS) is a rare chronic autoimmune disease characterized by irreversible tissue fibrosis, vasculopathy, and pronounced cellular and humoral immunity abnormalities. Although the pathogenesis of SS is not yet fully elucidated, there is a need to develop and validate new biomarkers for the prediction, early diagnosis, and monitoring of the disease.

To date, there are not many biomarkers that allow early and accurate diagnosis of SS. However, scientists are currently engaged in the search and development of new types of biomarkers. Nowadays the most frequently used biomarkers are autoantibodies, such as anticentromere (ACA), antitopoisomerase (ATA), and anti-RNA polymerase III (ARA) that can be detected in over 50% of SSc patients.

Epigenomic biomarkers: Epigenetic factors play a significant role in the development of SS. Combined heterozygous knockout of friend-leukemia integration factor 1 and Krüppel-like factor 5 in mice reflects the three SSc cardinal features: vascular changes, immune activation, and fibrosis. Nonselective inhibition of histone deacetylases has potent antifibrotic effects in vitro and in vivo by regulating several profibrotic pathways. Noncoding RNA–miRNAs with profibrotic (miR155) and antifibrotic (miR29a) functions have been described in SSc and other fibrosing disorders. Thus, targeting epigenetic mechanisms might be an attractive approach for the treatment of SS.

Genomic biomarkers: Analysis of skin samples from patients with dcSS revealed a four-gene biomarker panel consisting of THBS1, COMP, SIGLEC1, and IFI44; their expression correlated moderately well with the modified Rodnan skin score. Another study discovered a two-gene pharmacodynamic biomarker (THBS1, MS4A4A) which may prove useful in the longitudinal assessment of skin involvement.

4.6.1 Proteomic biomarkers

Antitopoisomerase: In patients with SS, autoAbs are formed directed against Topoisomerase 1 (ATA; anti-Scl70). The presence of ATA is associated with DcSS, which has a worse prognosis compared to LcSS due to early and more severe organ involvement. ATA shows a higher probability for developing micro- and macrovascular damage and is associated with interstitial lung disease and digital ulcers but not pulmonary arterial hypertension.

Anticentromere: ACA is one of the biomarkers of cutaneous SS and occurs in 45–50% of these patients. However, pulmonary arterial hypertension occurs in about 10% of patients positive for these antibodies. Because the centromere is not an immunodominant Ag, this increases the potential for β-cell clonal diversity, which in turn leads to the production of various Abs that coexist with ACA, such as anti-Ro52. The presence of ACA is associated with a lower inflammatory state, in line with the less severe phenotype observed in LcSSc patients. This is also reflected in the lower amounts of circulating Eotaxin3 and MCP1 and the higher levels of circulating IL-21.

Anti-RNA polymerase III is associated with an increased risk of SS renal crisis and a high risk of rapid skin thickening. In a recent study, it was shown that, as with ACA, ARA autoAbs are associated with higher levels of Eotaxin3 and MCP1. In addition, they found that there was less multi-specificity of other autoAbs. High levels of circulating IL-17A/F and IL-31 are seen in association with ARA, which could explain the rapid progressive fibrosis of the skin observed in these patients.

Lung-epithelial-derived surfactant protein (SP-D): Surfactant is a lipoprotein originally described for its essential role in reducing surface tension at the air–liquid interface of the lung and a key role component in lung immune defense. SP-D is a relevant diagnostic biomarker of SSc-associated interstitial lung disease, which has been proven by different studies.

Krebs von den Lungen-6 (KL-6) is a high molecular weight, mucin-like glycoprotein secreted by type II alveolar pneumocytes and bronchial epithelial cells in response to cellular damage and regeneration in patients with ILD. KL-6 is a mucin-associated glycoprotein, which may be a trigger for TGF-β signaling and fibrosis. KL-6 levels differentiated between fibrotic ILD, such as idiopathic pulmonary fibrosis and chronic hypersensitivity pneumonitis, and nonfibrotic lung disorders, including sarcoidosis and pulmonary alveolar proteinosis.

IL-21: Production of IL-21 is functionally linked to follicular T helper cells, which have been shown to play a pathological role in skin fibrosis in SSc. Production of IL-21 by these follicular T helper cells promotes B-cell activation and maturation.

IL-32: In the recent study was shown that in SSc-PAH patients, IL-32 sera levels were significantly higher when compared with SSc patients without PAH and patients affected by iPAH. The analysis of ROC curve showed that IL-32 sera levels above 11.12 pg/mL were able to predict patients with PAH (sensitivity = 90%, specificity = 100%).

Diagnostic profiles in SS: A study was conducted in which the capabilities of a separately determined biomarker were compared with pathophysiological panels. It has been shown that serum producer clustering, stratified by autoAb reactivity, can distinguish between different underlying pathophysiological profiles in SSc. In addition, combining markers into profiles can reveal subtle differences, offering new possibilities for classifying heterogeneous diseases such as SS.

Metabolomic biomarkers: Serum profiling of 110 metabolites from SS and healthy control patients was performed using targeted ultra-high-performance liquid chromatography coupled with tandem mass spectrometry (LC-MS/MS) analysis. Any changes in the levels of L-tyrosine, L-tryptophan, and 1-methyl-adenosine distinguished healthy controls from SSc and alterations in L-leucine, L-isoleucine, xanthosine, and AMP profiles differentiated between progressing and stable SS-ILD with L-tryptophan and L-leucine being the best performing discriminators in the respective groups. Increased serum levels of L-leucine and L-isoleucine in progressing compared with stable SSc-ILD patients were confirmed with an independent enzymatic assay with definition of a critical threshold.

5. Biomarkers in clinical practice: Biomarkers as applicable to personalized dietary protocols in autoimmune conditions

Autoimmune inflammation (Fig. 14) can become chronic due to loss of tolerance and/or when regulatory processes such as resolution do not occur. In a place where the inflammation becomes excessive, irreparable damage to the tissues of the body can occur and diseases develop. Such

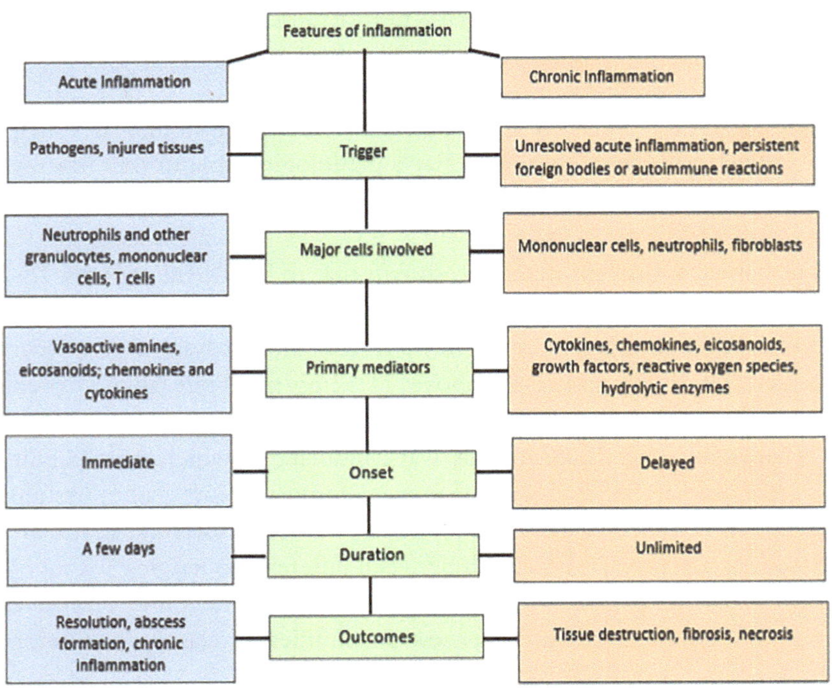

Fig. 14 Features of autoimmune inflammation.

diseases are characterized by markedly increased concentrations of autoimmune inflammatory markers and activated inflammatory cells at the site of tissue damage and in the circulatory system; this state of inflammation can be considered as a "high degree."

Many dietary factors can influence various aspects of autoimmune inflammation. Based on this, personalized nutrition and precision foodomics play a role in predisposition to conditions that have an inflammatory component and altered nutrition may be a useful tool for the prevention or treatment of such conditions. In this sense, micro- and macronutrients can be powerful dietary signals that influence the metabolic programming of cells and play an important role in homeostasis control. Also genes' role in human nutrition and the understanding that genetic predisposition can be an important factor contributing to the main causes of mortality associated with dietary priorities increases the demand for a new range of "-omic" sciences. The key is nutrigenomics, which involves the use of high-performance genomics tools in nutrition research.

Nutrigenomics contributes to a deeper understanding of the effect of nutrition on metabolic pathways and homeostasis: how this regulation is disrupted at an early stage of the disease and to what extent individual sensitizing genotypes contribute to the development of such diseases. Subtle changes in gene expression, even at the single cell level, can now be measured using quantitative methods such as real-time PCR and high-density microchip analysis, in monitoring nutrigenomic biomarkers to control the efficacy of autoimmune conditions-related personalized dietary protocols.

Nutrigenomics is able to extract useful biological information from the molecular and genetic collected data. The integration of *omics* technologies with the consolidated OMICS-portfolio provides a more mechanistic interpretation of the collected information, providing clinicians with a new generation of autoimmunity-related biomarkers.

The study of the role of nutrition on inflammation requires the identification of the main candidate biomarkers (Fig. 15), which are objectively measured and can be evaluated as indicators of normal biological processes, pathogenic processes, or pharmacological response to intervention. Most existing intervention studies focus on a single marker or a small number of markers of inflammation. The difficulty lies in the fact that it is not always possible to accurately determine the effect of specific nutrients on inflammation, as well as in the fact that the prepainful condition is characterized by small metabolic disorders that can slowly progress to the disease and there

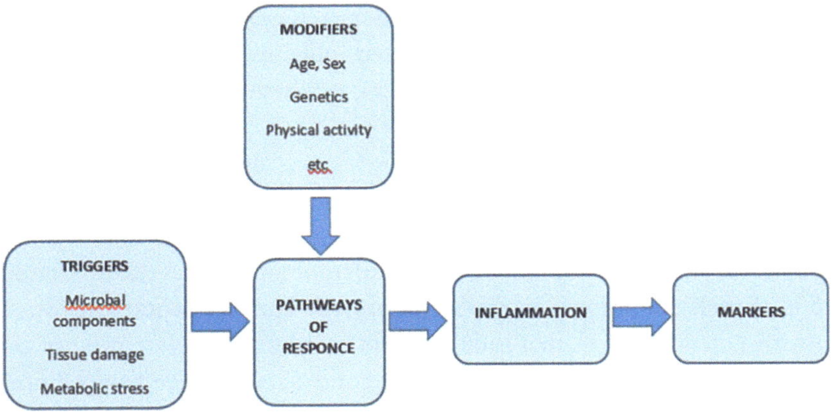

Fig. 15 The inflammation and the role of biomarkers to monitor the latter.

is no generally accepted "normal range" for most of these biomarkers, which at an early and reversible stage may have high prognostic significance. Therefore, it is likely that the combination of several biomarkers of inflammation and integrated indications, possibly based on kinetic analysis after certain tests, will be the most informative, composite biomarker of inflammation. It is also necessary to take into account that some inflammatory mediators leave the site of inflammation and enter the bloodstream, so inflammation in one place can cause changes in distal areas. Another problem is that the same markers are involved in both acute and chronic inflammatory processes, which means that it is impossible to use markers to distinguish between these two processes, so it worths to use a complex information apart from biomarkers.

Based on the data obtained and databases collected and mined, the antiinflammatory diet being developed does not adhere to strict rules regarding calories or portion sizes but suggests daily consumption of a variety of antiinflammatory products, without focusing on the consumption of one or two specific foods or nutrients. This provides a greater variety of protective food components, some of which can work synergistically to enhance immunity while allowing for the practitioners to monitor specific nutrition-related biomarkers.

Briefly, the antiinflammatory diet is a healthy eating plan that can help reduce chronic low levels of inflammation, which may otherwise increase the risk of various chronic diseases. Although research is limited, it may also help reduce markers of inflammation in people with an autoimmune type of inflammation.

The use of new data on existing biomarkers is a way to optimize existing treatment methods and, with the help of tools such as mathematical modeling to determine the best treatment methods, will lead to better and more cost-effective patient care. Guided by clusters of biomarkers, and in the future by a narrower and more accurate set of them, it will be possible to register the onset of various diseases at a deeper level and early stages and observe changes in the internal environment of the body. To date, studies are being conducted on the effects of food and diets on various biomarkers and a proposal is being made to create an information base containing information on the effects of various biomarkers on human health. The available data allow to predict the average results from a particular intervention and genotype, but it is impossible to predict the results for individual people. Ultimately, nutrigenomics will allow to develop effective strategies of dietary intervention to restore normal homeostasis and prevent diseases and will allow the use of individual counseling, favorably differing from the use of standard diets used for many years. The transfer of the research discipline of nutrition genomics into clinical practice is one of the most difficult tasks, so further randomized controlled trials are needed in the future to provide evidence and give scientific credibility to the concept of personalized nutrition.

6. Conclusion

Despite the current paucity of definitive biomarkers for autoimmune diseases, a deeper understanding of disease pathogenesis is starting to uncover putative mechanistic biomarkers, as well as identify mechanistic roles for certain known biomarkers. In the future, biomarkers might help to establish a molecular taxonomy of autoimmune diseases that are currently classified according to phenotype alone.

A growing area of biomarker research in autoimmunity conditions is the search for biomarkers that can predict successful drug-free remission. Stratifying diseases classified according to phenotype is not the only way that biomarkers can be used to forge a molecular taxonomy of disease: they can do so also by breaking down the boundaries of current classifications. That is, biomarkers can be used to uncover molecular similarities between diseases thought to be distinct.

As a field, autoimmunity has long been at the forefront of biomarker discovery; some of the biomarkers identified will prove to be descriptive, whereas others could prove to be mechanistic. Nevertheless, as our understanding of the molecular immunology of autoimmune disease progresses, we envision a future with biomarker-based molecular subtyping of disease that can guide clinical decision making.

References

1. https://pubmed.ncbi.nlm.nih.gov/21443578/.
2. https://www.sciencedirect.com/science/article/pii/S0303720720302446?via%3Dihub.
3. Yi L, Swensen AC, Qian WJ. Serum biomarkers for diagnosis and prediction of type 1 diabetes. *Transl Res*. 2018;201:13–25.
4. Eisenbarth GS, Flier JS, Cahill G. Type-I diabetes-mellitus—a chronic autoimmune-disease. *N Engl J Med*. 1986;314:1360–1368.
5. Cudworth AG, Woodrow JC. Letter: HL-A antigens and diabetes mellitus. *Lancet*. 1974;2:1153.
6. Meimed S, Polonsky KS, Reed Larsen P, et al. *Williams Textbook of Endocrinology*. 12th ed; 2011:1436–1441.
7. Thomson G, et al. Relative predispositional effects of HLA class II DRB1-DQB1 haplotypes genotypes on type 1 diabetes: a meta-analysis. *Tissue Antigens*. 2007;70:110.
8. Noble JA, Valdes AM. Genetics of the HLA region in the prediction of type 1 diabetes. *Curr Diab Rep*. 2011;11(6):533–542. https://doi.org/10.1007/s11892-011-0223-x.
9. Vafiadis P, Bennett ST, Todd JA, et al. Insulin expression in human thymus is modulated by INS VNTR alleles at the IDDM2 locus. *Nat Genet*. 1997;15:289–292.
10. Pociot F, Lernmark Å. Genetic risk factors for type 1 diabetes. *Lancet*. 2016;387: 2331–2339.
11. Akil AA-S, Yassin E, Al-Maraghi A, Al-Malki EAK, Fakhro KA. Diagnosis and treatment of type 1 diabetes at the dawn of the personalized medicine era. *J Transl Med*. 2021;9(1):137. https://doi.org/10.1186/s12967-021-02778-6.
12. Kaufmann KB, Büning H, Galy A, Schambach A, Grez M. Gene therapy on the move. *EMBO Mol Med*. 2013;5(11):1642–1661. https://doi.org/10.1002/emmm.201202287.
13. Diez J, Park Y, Zeller M, et al. Differential splicing of the IA-2 mRNA in pancreas and lymphoid organs as a permissive genetic mechanism for autoimmunity against the IA-2 type 1 diabetes autoantigen. *Diabetes*. 2001;50:895–900.
14. Pugliese A, Awdeh ZL, Alper CA, Jackson RA, Eisenbarth GS. The paternally inherited insulin gene B allele (1,428 FokI site) confers protection from insulin-dependent diabetes in families. *J Autoimmun*. 1994;7:687–694.
15. Bansal A, Pinney SE. DNA methylation and its role in the pathogenesis of diabetes. *Pediatr Diabetes*. 2017;18:167–177.
16. Klonoff DC. Hemoglobinopathies and hemoglobin A1c in diabetes mellitus. *J Diabetes Sci Technol*. 2020;14(1):3–7. https://doi.org/10.1177/1932296819841698.
17. Modell B, Darlison M. Global epidemiology of haemoglobin disorders and derived service indicators. *Bull World Health Organ*. 2008;86:480–487.
18. Leighton E, Sainsbury CAR, Jones GC. A practical review of C-peptide testing in diabetes. *Diabetes Ther*. 2017;8:475–487.
19. Ziegler AG, Rewers M, Simell O, et al. Seroconversion to multiple islet autoantibodies and risk of progression to diabetes in children. *JAMA*. 2013;309:2473–2479.
20. Lampasona V, Schlosser M, Mueller PW, et al. Diabetes antibody standardization program: first proficiency evaluation of assays for autoantibodies to zinc transporter 8. *Clin Chem*. 2011;57:1693–1702.
21. Lampasona V, Liberati D. Islet autoantibodies. *Curr Diab Rep*. 2016;16(6):53. https://doi.org/10.1007/s11892-016-0738-2.

22. Endesfelder D, Hagen M, Winkler C, et al. A novel approach for the analysis of longitudinal profiles reveals delayed progression to type 1 diabetes in a subgroup of multiple-islet-autoantibody-positive children. *Diabetologia*. 2016;59:2172–2180.
23. Ilonen J, Lempainen J, Hammais A, et al. Primary islet autoantibody at initial seroconversion and autoantibodies at diagnosis of type 1 diabetes as markers of disease heterogeneity. *Pediatr Diabetes*. 2018;19:284–292.
24. Knip M. Disease-associated autoimmunity and prevention of insulin-dependent diabetes mellitus. *Ann Med*. 1997;29:447–451.
25. Redondo MJ, Steck AK, Pugliese A. Genetics of type 1 diabetes. *Pediatr Diabetes*. 2018;19(3):346–353.
26. Mrena S, Virtanen SM, Laippala P, et al. Models for predicting type 1 diabetes in siblings of affected children. *Diabetes Care*. 2006;29:662–667.
27. Bonifacio E. Predicting type 1 diabetes using biomarkers. *Diabetes Care*. 2015;38: 989–996.
28. Jin Y, She JX. Novel biomarkers in type 1 diabetes. *Rev Diabet Stud*. 2012;9:224–235.
29. Ziegler AG, Nepom GT. Prediction and pathogenesis in type 1 diabetes. *Immunity*. 2010;32:468–478.
30. Yang Y, Chan L. Monogenic diabetes: what it teaches us on the common forms of type 1 and type 2 diabetes. *Endocr Rev*. 2016;37:190–222.
31. Ochs HD, Gambineri E, Torgerson TR. IPEX, FOXP3 and regulatory T-cells: a model for autoimmunity. *Immunol Res*. 2007;38:112–121.
32. Velayos T, Martinez R, Alonso M, et al. An activating mutation in STAT3 results in neonatal diabetes through reduced insulin synthesis. *Diabetes*. 2017;66:1022–1029.
33. Saisho Y, Butler AE, Manesso E, Elashoff D, Rizza RA, Butler PC. Beta-cell mass and turnover in human: effects of obesity and aging. *Diabetes Care*. 2013;36:111–117.
34. Metz TO, Qian WJ, Jacobs JM, et al. Application of proteomics in the discovery of candidate protein biomarkers in a diabetes autoantibody standardization program sample subset. *J Proteome Res*. 2008;7:698–707.
35. Zhi WB, Sharma A, Purohit S, et al. Discovery and validation of serum protein changes in type 1 diabetes patients using high throughput two dimensional liquid chromatography-mass spectrometry and immunoassays. *Mol Cell Proteomics*. 2011;10:M111.012203.
36. Zhang Q, Fillmore TL, Schepmoes AA, et al. Serum proteomics reveals systemic dysregulation of innate immunity in type 1 diabetes. *J Exp Med*. 2013;210:191–203.
37. von Toerne C, Laimighofer M, Achenbach P, et al. Peptide serum markers in islet autoantibody-positive children. *Diabetologia*. 2017;60:287–295.
38. Lee AH, Heidtman K, Hotamisligil GS, Glimcher LH. Dual and opposing roles of the unfolded protein response regulated by IRE1 alpha and XBP1 in proinsulin processing and insulin secretion. *Proc Natl Acad Sci USA*. 2011;108:8885–8890.
39. Sims EK, Chaudhry Z, Watkins R, et al. Elevations in the fasting serum proinsulin-to-C-peptide ratio precede the onset of type 1 diabetes. *Diabetes Care*. 2016;39:1519–1526.
40. Fox TE, Bewley MC, Unrath KA, et al. Circulating sphingolipid biomarkers in models of type 1 diabetes. *J Lipid Res*. 2011;52:509–517.
41. Cianciaruso C, Phelps EA, Pasquier M, et al. Primary human and rat beta-cells release the intracellular autoantigens GAD65, IA-2, and proinsulin in exosomes together with cytokine-induced enhancers of immunity. *Diabetes*. 2017;66:460–473.
42. Eizirik DL, Cnop M. ER stress in pancreatic beta cells: the thin red line between adaptation and failure. *Sci Signal*. 2010;3:pe7.
43. Massa O, Alessio M, Russo L, et al. Serological proteome analysis (SERPA) as a tool for the identification of new candidate autoantigens in type 1 diabetes. *J Proteomics*. 2013;82:263–273.
44. Delong T, Wiles TA, Baker RL, et al. Pathogenic CD4 T cells in type 1 diabetes recognize epitopes formed by peptide fusion. *Science*. 2016;351:711–714.

45. Freitas PAC, Ehlert LR, Camargo JL. Glycated albumin: a potential biomarker in diabetes. *Arch Endocrinol Metab*. 2017;61:296–304.
46. Bezu L, Sauvat A, Humeau J, et al. eIF2alpha phosphorylation is pathognomonic for immunogenic cell death. *Cell Death Differ*. 2018;25:1375–1393.
47. Ganesan L, Levental I. Pharmacological inhibition of protein lipidation. *J Membr Biol*. 2015;248:929–941.
48. Stechova K, Bohmova K, Vrabelova Z, et al. High T-helper-1 cytokines but low T-helper-3 cytokines, inflammatory cytokines and chemokines in children with high risk of developing type 1 diabetes. *Diabetes Metab Res Rev*. 2007;23:462–471.
49. Purohit S, Sharma A, Hopkins D, et al. Large-scale discovery and validation studies demonstrate significant reductions in circulating levels of IL8, IL-1Ra, MCP-1, and MIP-1 beta in patients with type 1 diabetes. *J Clin Endocrinol Metab*. 2015;100: E1179–E1187.
50. Mack CL. Serum cytokines as biomarkers of disease and clues to pathogenesis. *Hepatology*. 2007;46:6–8.
51. Akirav EM, Lebastchi J, Galvan EM, et al. Detection of beta cell death in diabetes using differentially methylated circulating DNA. *Proc Natl Acad Sci USA*. 2011;108: 19018–19023.
52. Ling W, Huang YM, Qiao YC, Zhang XX, Zhao HL. Human amylin: from pathology to physiology and pharmacology. *Curr Protein Pept Sci*. 2019;20(9):944–957. https://doi.org/10.2174/1389203720666190328111833.
53. Usmani-Brown S, Lebastchi J, Steck AK, Beam C, Herold KC, Ledizet M. Analysis of beta-cell death in type 1 diabetes by droplet digital PCR. *Endocrinology*. 2014; 155:3694–3698.
54. Seyhan AA, Lopez YON, Xie H, et al. Pancreas-enriched miRNAs are altered in the circulation of subjects with diabetes: a pilot cross-sectional study. *Sci Rep*. 2016;6:31479.
55. Osipova J, Fischer DC, Dangwal S, et al. Diabetes-associated microRNAs in pediatric patients with type 1 diabetes mellitus: a cross-sectional cohort study. *J Clin Endocrinol Metab*. 2014;99:E1661–E1665.

Index

Note: Page numbers followed by "*f*" indicate figures and "*t*" indicate tables.

A

Abs with catalytic activity (catAbs/abzymes)
autoAb-based biomarker discovery and validation
asymptomatic disease, diagnosis of, 241
disease activity, assessment and prediction of, 241–242
drug toxicity, assessment of, 242
dynamic biomarkers, 242
enzyme-linked immunosorbent assays (ELISAs), 236
mass spectrometry, 237
protein microarrays, 238
recombinant protein production, 239
responsiveness to therapy, 242–243
serological analysis of recombinant cDNA expression libraries (SEREX), 238
serologic proteome analysis (SERPA), 238
surface chemistry, 240
western blotting, 236
autoaggression, prevention of, 230
autoimmune diseases (ADs), 230
in clinical practice, biomarkers
biomarker sets, clinical validation of, 228, 229*f*
drug development, 228
targeted therapy optimization, guide for, 228, 229*f*
global NGBs market, 226
ideal biomarker, characteristics and features of, 226, 227*f*
interactome-based networks, 226
predictive biomarkers, utility of, 230, 231*f*
proteolytic Abs, 226
sequence-specific Ab-proteases, 226
Access To Treatment Committee (ATTC), 136–137
Actionable biomarkers, 241
Advanced glycation end products (AGEs), 257
AI algorithms, 114
Antinuclear antibodies (ANAs), 235, 259–260
Antisense oligonucleotide (ASO) technology, 134, 138–139
Artificial intelligence (AI) algorithms, precision medicine, 19
and big data, 60–61*f*, 61–63
decision making process, 58–59
deep learning algorithms, 63–64
ENCODE data, 63–64
"HapMap" project, 63–64
healthcare and medicine, 61–63
human genome and human Inhalative PK profiling, 63–64
OMICS research, sequencing technologies with, 63–64
"one size fits all" theory, 63
single-cell parameters, identification of, 58–59
target-based medicines, 58–59
Autoimmune polyendocrine syndrome type 1 (APS-1), 251

B

Biomarker research evolution, in autoimmunity conditions
Abs with catalytic activity (catAbs/abzymes)
drug development, 228
global NGBs market, 226
ideal biomarker, characteristics and features of, 226, 227*f*
interactome-based networks, 226
proteolytic Abs, 226
sequence-specific Ab-proteases, 226
biotarget, 225–226
classical biomarkers, 222
classification of, 223*f*
clinical autoimmunity conditions, personalized and precision healthcare services

Biomarker research evolution, in autoimmunity conditions (*Continued*)
antibody (Ab) biomarkers, 234
cellular biomarkers, 233
genomics-related biomarkers, 232
metabolomic biomarkers, 233
proteomics-related biomarkers, 235
single nucleotide polymorphisms and haplotype biomarkers, 233
in clinical decision making, 224
in clinical settings, 224
development process, 225, 225*f*
drug repurposing, 224
healthcare, 223–224
services, 222*f*
molecular biomarkers, 222
multibiomarker-based profiles, 221*f*
multimarker-based companion diagnostics, 224
NBB-related protein biomarkers, SLE, 225*f*
nutrigenomics, 224
personalized and precision medicine (PPM), 221–222, 222*f*
pharmacogenomics, 224
predictive biomarker, 223–224
prognostic biomarker, 223–224
qualitative biomarkers, 223
quantitative biomarkers, 223
specific autoimmunity conditions subareas, clinical practice
autoimmune inflammation, personalized dietary protocols in, 270–273
autoimmune myocarditis (AM), 243–245
multiple sclerosis, 245–246
rheumatoid arthritis (RA), 264–267
SLE, 258–264
systemic sclerosis (SS), 267–270
type 1 diabetes (T1D), 246–258
therapeutic intervention, 224
Body mass index (BMI), 106

C

Cancer, precision diagnostics in
clinical trial design, 51–52
NGS-based technologies
data analysis, 46–48
illumina "sequencing by synthesis" platform, 46, 47–48*f*
sequencing library generation, 44–46
ThermoFisher's Ion Torrent pH-sensing technology, 46, 47–48*f*
tumor sampling and extraction, 43–44
variant interpretation, 48–49
precision oncology, approaches to, 41
somatic oncology testing, 40–41
targeted panels, 42–43
therapy selection
checkpoint inhibitors and immunotherapies, 51
targeted therapies, 50
whole exome sequencing (WES), 42
whole genome sequencing (WGS), 41–42
Cell-bound complement activation products (CB-CAPs), 261
Checkpoint inhibitors, immunotherapies, 51
Chronic systemic inflammation (CSI), 129
Circular RNA (circRNAs), 233–234
CytoBead assay, 235, 236*f*

D

Deep phenotyping, precision medicine, 58–59
approaches for, 110–111*t*

E

Electronic health records (EHRs), 2, 106, 108–109
Endometrial stromal cells (ESC), 203
Epidermal growth factor receptor (EGFR)
anti-EGFR agents, 191
EGFR TKIs, acquired resistance of
first-generation, 193
second-generation, 193
third-generation, 193–194
HER family receptors, 191
lung cancer, precision medicine in, 194
target, in lung cancer treatment, 192
tyrosine kinases (TK) activity, 191
Epilepsy, precision medicine in
genetic epilepsy syndromes, precision medication approaches, 152–153*t*

ATP7A mutations, 155–156
FOLR1 gene, 156
GAMT/AGAT/SLC6A8 (creatinine deficiency syndromes), 156–158
GRIN2A, B and D genes, 158–159
KCNA2, 159–160
KCNQ2 gene, 160
KCNT1 gene, 159
LDH7A1/PNPO/PROSC, 151–155
MOSC1/MOSC2/GPHN, 161
POLG1 gene, 161–162
PRRT2 mutation, 162
SCN8A, 165
SCN1A mutation, 163–164
SCN2A mutation, 164–165
SLC19A3, 167
SLC35A2 gene, 167
SLC2A1 mutations, 166
solute carrier family 6 member 1 (SLC6A1) gene, 166–167
syntaxin binding protein-1 (STXBP1), 167–168
transient receptor potential melastatin 6 (TRPM6) gene, 169–170
tripeptidyl peptidase 1 (TPP1) gene, 168–169
tuberous sclerosis complex 1 (TSC1), 170–171
tuberous sclerosis complex 2 (TSC2), 170–171
genetic testing tools
microarrays, 148–149
multiomic data, in research, 150–151
targeted gene panel testing, 149
whole-exome sequencing (WES), 150
wholegenome sequencing (WGS), 150
neurostimulation, 148
seizure remission, 148

F

Fasting plasma glucose (FPG), 249
Findable, accessible, intelligent, and reproducible (FAIR) approach, 112–114
First-degree relatives (FDRs), 250
Forkhead Box P3 (FOXP3), 251

G

Gas chromatography–mass spectrometry (GC–MS), 106

H

Health Information Technology for Economic and Clinical Health (HITECH) Act, 2
Hybrid insulin peptides (HIPs), 255

I

Illumina "sequencing by synthesis" platform, 46, 47–48*f*
Inhibitor of κB (IκB), 203
Inhibitors of NF-κB kinase (IKK), 203

L

Laboratory information management systems (LIMS), 108–109
Liquid chromatography–mass spectrometry (LC–MS), 106
Long non-coding RNAs (lncRNAs), 199

M

Machine learning (ML) algorithms, 114
single-cell parameters, identification of, 58–59
Median progression-free survival (mPFS), 193
Metabolomics, 106
approaches, 107*t*
Micro-rare patients, 129
Mobile "health and well-being" app (m-health app), 28
Multi-omics strategies, precision medicine, 104–106
Myeloid differentiation factor 88 (MyD88), 203–204

N

Nano-rare patients
broadening treatment opportunities, 141–142
challenges, 144–145
cost-effectiveness, 132–133
drug discovery technologies
predictability, 133

Nano-rare patients (*Continued*)
scalability, 133–134
supportive regulatory environment, 135
versatility, 133
facilitating earlier diagnosis, 142–143
genotype, phenotype and clinical investigator, 131
mutation focused drug discovery, 131
n-Lorem, 135
application, 136–137
challenges, 144
clinical tests, 138–139
demand for treatment, 140
donors and collaborators, 140–141, 141*f*
newborns, genomic screening of, 143
therapeutics, 141–142
nonprofit treatment, industrialized approach
access to treatment, 136–137
aggregate performance analysis, 139
appropriately characterized patients, access to, 135–136
baseline (natural history) characterization and ASO treatment effects evaluation, 138–139
demand for treatment, 140
network of collaborators, creating, 140–141
optimal ASO, 139–140
personalized ASO treatment, 135
quality systems, 136
stakeholders, sharing experience with, 139
treatment goals and assessment methods, 137–138
parsing patient populations, 129–130
personalized experimental treatments, 132
rapidity, 132
reducing isolation, 143
National Human Genome Research Institute (NHGRI), 4
Network-based biomarkers (NBBs), 224
Next-generation sequencing (NGS) technologies, 4, 16, 105–106, 108–109, 109*f*
NGBs, 226, 242
NGS-based technologies
cancer, precision diagnostics in
data analysis, 46–48
illumina "sequencing by synthesis" platform, 46, 47–48*f*
sequencing library generation, 44–46
ThermoFisher's Ion Torrent pH-sensing technology, 46, 47–48*f*
tumor sampling and extraction, 43–44
variant interpretation, 48–49
Noncoding RNAs and EGFR, in precision medicine
CBM complex, 204
inhibitor of κB (IκB), 203
LncRNAs and EGFR-TKIs, 199
miRNAs, in EGFR -targeted therapy, 197–199, 198*t*
myeloid differentiation factor 88 (MyD88), 203–204
ncRNAs and NF-κB signal, 204–205
NF-κB signaling
in cancer, 200
in chronic inflammation to cancer, 205–206
p50 and p65, 200–201
TGFβ-activated kinase-1 (TAK1), 204
tRNA -derived RNA fragments and EGFR, 200
tumor necrosis factor-α (TNF-α), 201–202
Nuclear factor-kappa B (NF-κB) signaling and activation
canonical NF-κB pathway, 196
EGFR, 196–197
non-canonical NF-κB pathway, 196
noncoding RNAs, 197
inflammation-related signaling pathways, 195
non-coding RNA, 195
transcription factor, 195
Nutrigenomics, 271

O

OMICS, 222, 225–226, 236, 240, 271
Oral glucose tolerance test (OGTT), 249
Orphan Drug Act, 128–129
Oxidation end products (OPs), 257

P

Peptidyl-arginine-deiminase (PADs), 265
Personal Health Records (PHRs), 2
p-Health care
  advanced data integration, 16–25, 17*f*
    decision level integration, 23–25
    feature level integration, 21–23
    raw data level integration, 20
  clinical informatics, 5–6*t*, 6–9
  data integration, 3*f*
  EHR systems, 2
  genomic and genomic medicine, 5–6*t*, 9–16, 12*t*
    databases and tools for, 14–15*t*
  infant cardiac surgery, by integrating proteomic data with genomic variations, 25–27
  privacy and security, 27–29
Plasma proteomics, 240
Precision medicine, 111–115
  benefits of, 102–103
  big data management challenges, 112
  cross-disciplinary collaborative research environment, 112
  deep phenotyping, 107–110
    approaches for, 110–111*t*
  end-stage diseases, 102–103
  gene-based collapsing analysis, 114–115
  healthcare data analytics, 103–104
  integrative clinical and multi-omics data analysis, 102–103
  integrative omics data challenges, 112
  multi-database management framework, 113*f*
  multi-omics strategies, 104–106
  streamline patient treatment, 103–104
  user-friendly graphical interfaces, 114–115
Precision oncology, 41

Q

Qualitative biomarkers, 223
Quantitative biomarkers, 223

R

Receptor of tyrosine kinase (RTK), 191
Rel homology domain (RHD), 195
Rheumatoid arthritis (RA) biomarkers
  ACPA isotypes, 265
  Antibodies against carbamylated proteins (anti-CarP), 265
  anticitrullinated protein antibodies, 265
  Anti-PADs, 265
  classic antibody-derived RA biomarkers, 265
  DNA methylation, 266–267
  epigenomic biomarkers, 266
  genomic biomarkers, 266
  nonantibodies RA biomarkers, 266

S

Serum biomarkers, 252
  metabolomics biomarkers, 253–254
  Omics technologies, 253*f*
  proinsulin to C-peptide, ratio of, 253
  proteomics analysis of, 252–253, 254*t*
Somatic oncology testing, 40–41
Systemic lupus erythematosus (SLE), 225*f*
  epigenomic biomarkers, 259
  genomic biomarkers, 259
  proteomic biomarkers, 259–264
    anti-C1q antibodies, 262
    anti-dsDNA antibodies, 261–262
    antinuclear antibodies (ANAs), 259–260
    antinucleosome antibodies (ANuA), 261
    B-cell activating factor (BAFF) family, 263
    CXCL 10, 262–263
    Galectin-9, 262–263
    IL-1 family, 263
    metabolomic biomarkers, 264
    serum anti-Sm antibody, 262
    serum complement 3 (C3) and complement 4 (C4), 260
Systemic sclerosis (SS)
  epigenomic biomarkers, 268
  genomic biomarkers, 268
  proteomic biomarkers
    anticentromere, 268–269
    anti-RNA polymerase III, 269
    diagnostic profiles in, 269
    IL-21, 269
    IL-32, 269

Systemic sclerosis (SS) (*Continued*)
Krebs von den Lungen-6 (KL-6), 269
lung-epithelial-derived surfactant protein (SP-D), 269
metabolomic biomarkers, 269–270

T

TGFβ-activated kinase-1 (TAK1), 204
ThermoFisher's Ion Torrent pH-sensing technology, 46, 47–48*f*
Transferrin receptors (TFR), 200
TRNA-derived RNA fragments (tRFs), 200
Tumor necrosis factor-α (TNF-α), 201–202
Type 1 diabetes (T1D), 246–258
autoAbs, as biomarkers, 249–252
β-cell destruction, 252
complications, 256–258
C-peptide, 249
cytokines, as T1D biomarkers, 255–256
and genomic biomarkers, 247
with hyperglycemia, 249
new autoAgs and autoAbs, 254–255
non-HLA loci, 247–248
nucleic acid-based predictive biomarkers, 256
posttranslational modifications, 255
serum biomarkers, 252
proteomics analysis of, 252–253

U

Ultra-rare patients, 129
Undiagnosed Disease Network (UDN), 131

V

Valid biomarker, 228–230

W

Whole exome sequencing (WES), 42
Whole genome sequencing (WGS), 41–42

FSC
www.fsc.org
MIX
Papier | Fördert
gute Waldnutzung
FSC® C083411

Kolibri 360 GmbH
Ferdinand-Jühlke-Straße 7
99095 Erfurt, Deutschland
produktsicherheit@kolibri360.de